POLARIZED LIGHT MICROSCOPY

by

WALTER C. McCRONE
Director
McCrone Research Institute, Inc.
Chicago, Illinois

LUCY B. McCRONE
Microscopist
Walter C. McCrone Associates, Inc.
Chicago, Illinois

JOHN GUSTAV DELLY
Microscopist
Walter C. McCrone Associates, Inc.
Chicago, Illinois

230 Collingwood, P.O. Box 1425, Ann Arbor, Michigan 48106

Library of Congress Catalog Card No. 78-61047
ISBN 0-250-40262-9

Manufactured in the United States of America

PREFACE

This volume is in reality the eighth edition of a course manual first prepared by Walter C. McCrone and Marvin Salzenstein in 1959. It has been printed, revised, re-edited and reprinted every other year since then; John Delly contributed the sections on the Michel-Lévy chart, fluorescence and parts of the chapter on Photomicrography and Lucy McCrone among other contributions is responsible for the chapter on "Crystal Optics". Until now it has been used only in the courses taught by the McCrone Research Institute throughout North America and Europe. Increasing requests from microscopists who have happened to see students' copies have convinced us to authorize this first hardcover edition for general distribution.

Walter C. McCrone

TABLE OF CONTENTS

POLARIZED LIGHT MICROSCOPY

I. INTRODUCTION

There are many fields of application for the microscope: bacteriology, cytology, pathology, metallography etc., and microscopists in these fields are called, respectively, bacteriologists, cytologists, pathologists, metallographers etc. There is, however, a field of interest for the microscopist not covered above which is, by far, the broadest application of all — this is the field of chemical microscopy and microscopists versed in this broad field are called, of course, chemical microscopists.

The chemical microscopist must be broadly trained. He must be, first of all, a good chemist and especially a good physical chemist. He must be a crystallographer since most of the things he looks at, whether animal, vegetable or mineral, are crystalline. He must have a deep-seated curiosity about the normally unseen. The dyed-in-the-wool microscopist will continually collect fibers, powders, scales, surfaces, seeds, bugs, suspensions, emulsions, chemicals etc. for a closer look with the microscope.

At the same time a microscopist who knows only the microscope and how things look thereunder will wither on the vine in the industrial research laboratory. He must understand the industrial research modus operandi. His clients, his fellow workers, know very little about his tools and techniques and generally leave him very much alone. He must seek out the problem, study it from a "research director" point of view, visualize his part and sell it. He must know what and where to sample to obtain significant results. He must know how to translate his results into answers meaningful to his client. He must spend hours of "unnecessary" time taking pictures, writing reports and explaining his results.

He may spend less than 5 percent of his time on any given problem looking through the microscope. If he cannot visualize his colleague's problem and his part in it no one else will. He will find that he has to do far more than chemical microscopy to complete his job. If he can do this he will soon need an assistant. If he cannot, the use of the microscope in his company will not fluorish nor will he.

If a co-worker comes to you and requests a particle size determination, you as a microscopist know that he may want such a measurement, but you know too that he may not need such a measurement. Actually he has a material, unknown to you, that may be a paint pigment, abrasive, cleaner or one of thousands of different materials. His problem may be that the particles do not suspend well in a liquid medium, may settle too quickly, may have strange rheological properties. The material may lack expected covering power as a pigment or expected polishing action as an abrasive. Your co-worker is not generally aware of all you can see with the microscope and thinks he can interpret the results of a particle size determination — the only operation he is confident you can perform.

At this early stage you have to be able to learn more about the sample without offending your source of information. Ordinarily he thinks you need nothing except the sample and the request. You must be able to find out more about his problem, remembering that he doesn't want any help from you other than the tabulated particle size data. There are, at this stage, various approaches depending on the two individuals involved. It is important for the microscopist not to brag of all he can do on this problem if he is only given a chance. Maybe he can (and maybe he can't). In any case it is easier to sell results you already have than results you hope to obtain.

Assume now that you know the problem and, in most cases, a particle size determination will be a help; however there are many other observations that may be helpful in the overall problem. For example:

1. Is the sample representative of the bulk of material?

2. Are the particles aggregated and is the size of the aggregates more important than the size of the individual particles?

3. Are the particles easily dispersed and do they remain dispersed?

4. Is the sample pure?

5. What is the predominant shape of the particles?

6. Are they a ground or a precipitated material?

These same questions can be answered with little extra work at the time the particle size is measured. At the same time photomicrographs should also be taken to illustrate any of the above points.

Now when your client returns you are in a position to discuss the answers with him in his own language. It is appropriate for you to try to impress him with all you know about his material, especially if he was not very cooperative with information when he gave you the sample. Comments such as these are teasers and often elicit further information and discussion:

1. These crystals of yours are very unusual; they have curved faces and look like crescent moons.

2. I notice that your particles were precipitated in such a way that they all grew to the same size and shape.

3. It looks as if you've got twinned crystals (or an unstable polymorph, or a very pure product etc.).

At this point your client sees the possibility of a real contribution on your part and he suddenly becomes effusively cooperative. This is the dangerous stage because you now have to produce. The important point now is to speak his language. Instead of a statement such as this, "Your sample contains single crystals flattened parallel to the 010 plane and showing the 101 and 110 prisms," you should, instead, say something like this: "Your sample has thin platelike particles that should give that new paint formulation a very high covering power". In this way, you also invite questions as to ways to increase covering power or work to correlate crystal size and shape with covering power. This approach, of course, increases the chances of his returning with additional problems.

There is almost no problem in chemistry that cannot, to at least a small extent, be helped by a good microscopist and often the problem can be completely solved by microscopy.

II. OPTICS

An understanding of elementary optics is essential to the proper use of the microscope. The microscopist will find that he can handle unusual problems in illumination and photomicrography much more effectively if he understands the underlying physical optics.

A. REFLECTION

1. Specular and Diffuse

A good place to begin is with reflection at a surface or interface. Specular (or regular) reflection results when a beam of light leaves a surface at an angle equal to the incident angle. This type of reflection occurs with highly polished smooth surfaces. It is stated more precisely as the law of reflection: the angle of reflection, r, equals the angle of incidence, i (Figure 1). Diffuse (or scattered) reflection results when a beam of light strikes an irregular surface and different portions of the incident light are reflected at different angles. Light reflected from a piece of white paper or a ground glass exemplifies diffuse reflection.

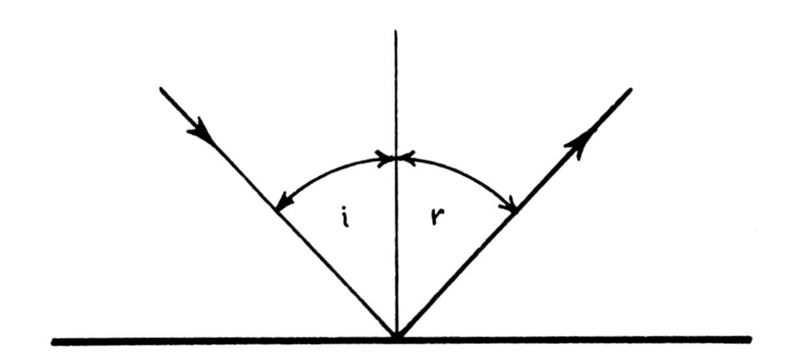

Figure 1. Specular reflection

Strictly speaking, of course, all reflected light, even diffuse, obeys the above law of reflection. Diffuse, reflected light is made up of many rays, specularly reflected each at a different angle from each tiny element of surface, and appears diffuse because the reflecting elements are very numerous and very small. The terms diffuse and specular, referring to reflection, describe not so much a difference in the nature of the reflection as a difference in the type of surface. A polished surface gives specular reflection, a rough surface gives diffuse reflection.

It is also important to note and remember that specularly reflected light tends to be strongly polarized in the plane of the reflecting surface. This is true because rays with vibration directions closest to the plane of the reflecting surface are most strongly reflected. This effect is strongest when the tangent of the angle of incidence is equal to the refractive index of the reflecting surface. This particular angle of incidence is called the Brewster angle. Polished metal has no Brewster angle, and reflected light from a highly polished clean metal surface is not well polarized.

2. Image Formation

a. Plane mirrors

Considering reflection by mirrors, we find (Figure 2) that a virtual image appears to be behind a plane mirror at a distance equal to the object-mirror distance. It is also reversed right to left but has the same size as the object. The word virtual means that the image appears to be in a given plane but that a ground glass screen or a photographic film placed in that plane would show no image. The converse of a virtual image is a real image, one that would be registered on a ground glass or a photographic film placed in that plane.

b. Concave mirrors

Mirrors may also be either convex or concave. The surface of the mirror usually represents a portion of the surface of a sphere. The center of curvature is the center of that sphere. The focus lies halfway between the center of curvature and the mirror surface (Figure 3).

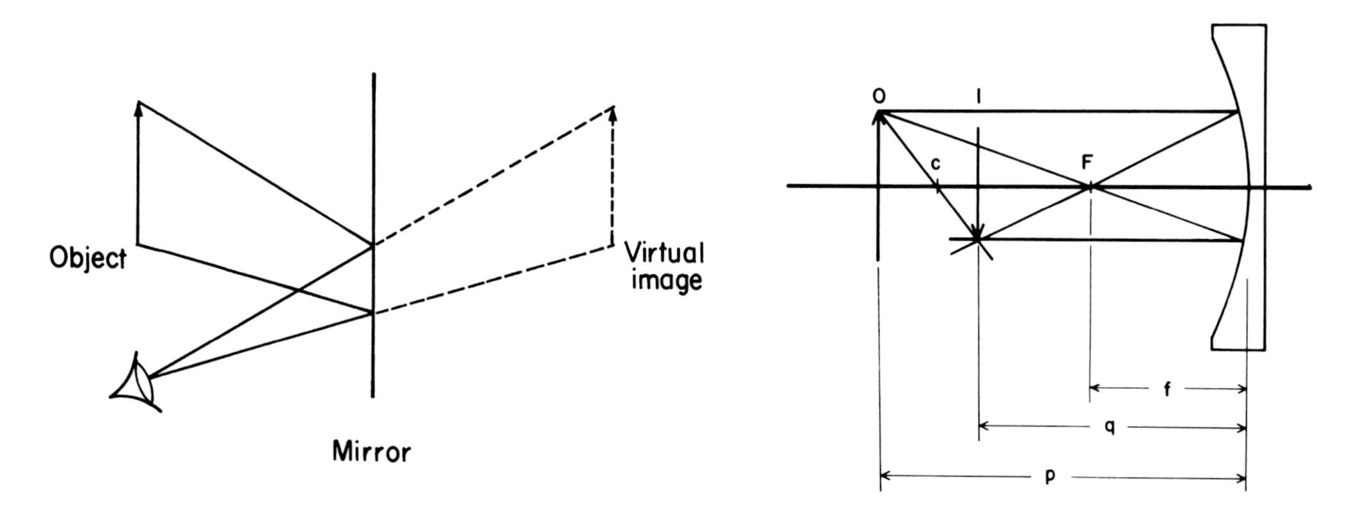

Figure 2. Image formation by a plane mirror

Figure 3. Image formation by a concave mirror

A concave mirror forms an image according to the following two premises:

1) A ray of light traveling parallel to the mirror axis must pass through the focus after reflection.

2) A ray of light which passes through the center of curvature must return along the same path.

A corollary of the first premise is:

3) A ray of light which passes through the focus is reflected parallel to the mirror axis.

The image of an object formed by reflection from a concave mirror can be found by construction based on the above premises (Figure 3) or by calculation from the familiar lens formula:

$$\frac{1}{p} + \frac{1}{q} = \frac{1}{f} \qquad (1)$$

where, p is the distance from the object to the mirror

q is the distance from the image to the mirror

f is the focal length.

The concave side of the substage mirror exhibits this kind of reflection.

c. Convex mirrors

A convex mirror forms no real image in the sense that a concave mirror does. A convex mirror of a 20-cm focal length will form a virtual image, 16.7 cm behind the mirror, of an object placed 100 cm in front of it. Construction of such an image position, though trivial, can be accomplished by following premises analogous to those stated above for a concave mirror (Figure 4).

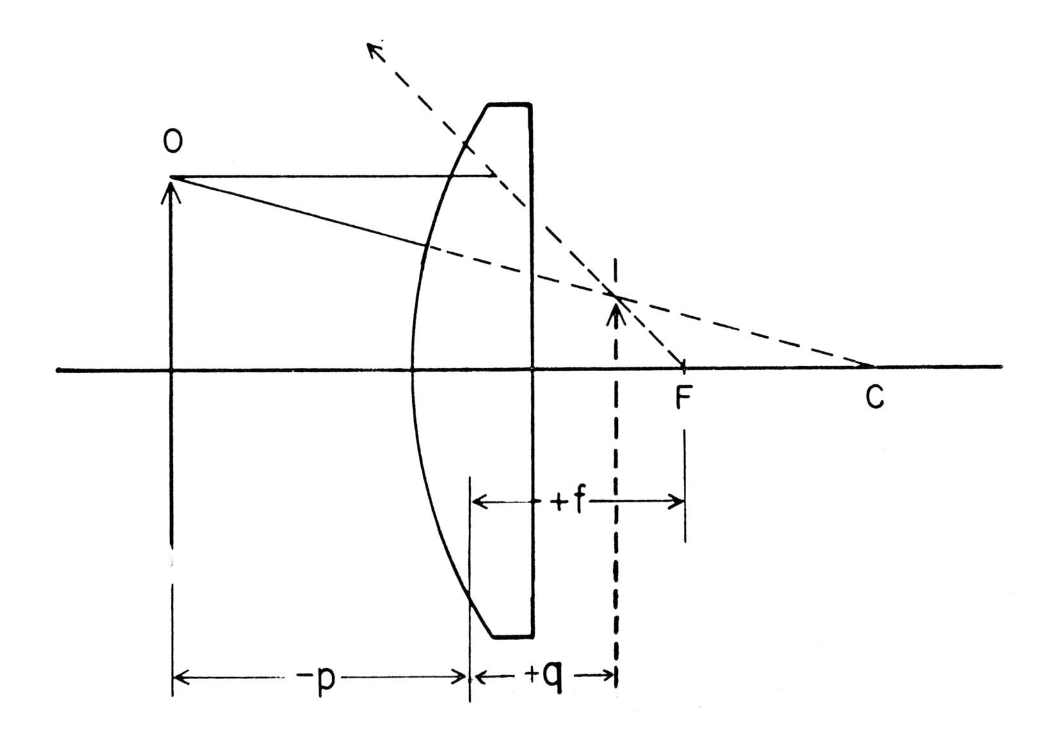

Figure 4. Image formation by a convex mirror

The same formula (1) applies if one uses different signs to show that the image and the object are on opposite sides of the mirror. Thus, from (1),

$$\frac{1}{-100} + \frac{1}{q} = \frac{1}{20}$$

$$q = 16.7$$

3. Spherical Aberration

No spherical surface can form a perfect image. The most serious imperfection, spherical aberration, occurs in spherical mirrors of large aperture (Figure 5). Light rays from a single point in the object reflected from the outer zone of a spherical mirror are not focused at the same point as the more central rays. This type of aberration is reduced by using only a portion of the aperture or by using aspheric surfaces.

B. REFRACTION

Turning now to lenses rather than mirrors, we must consider refraction, rather than reflection. Refraction refers to the change of direction and/or velocity of light as it passes from one

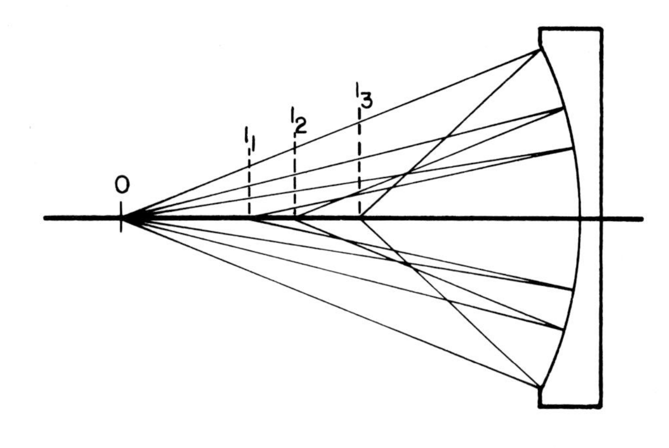

Figure 5. Spherical aberration by a concave mirror

medium to another. A change in velocity without a corresponding change in direction occurs only when the incident ray travels normal to the interface between media of different refractive index.

1. Refractive Index

The ratio of the velocity of light in air (or, more correctly, in a vacuum) to the velocity in the medium is called the refractive index. Table I lists some typical refractive index values measured with monochromatic light (sodium D line). The refractive index generally increases with the atomic number of the constituent atoms. Iodides, for example, have higher indices than chlorides. A high density for a compound also results in a higher refractive index. Conjugated aromatic hydrocarbons have higher densities and higher refractive indices than straight-chain aliphatic hydrocarbons. Substitution in either type of molecule by bromine, iodine or even nitro groups can cause large increases in refractive index.

Table I. Refractive indices of common materials measured with sodium light at 25° C

Vacuum	1.0000000	Flint glass	1.627
Air	1.0002918	Rock salt	1.5443
CO_2	1.0004498	Diamond	2.417
Water	1.3330	Lead sulfide	3.912

2. Temperature Coefficient of Refractive Index

The refractive indices of all substances vary with temperature. The degree of variation depends on the composition and the state of aggregation; that is, on whether the substance is solid or liquid. The temperature coefficient of refractive index, dn/dT, for liquids is usually more than 100 times as large as for solid substances. Most solids show coefficients of the order of 0.000001-0.00001 whereas most liquids show coefficients of 0.0003-0.0009. In general, the higher the refractive index, the greater the coefficient. Organic liquids show widely varying coefficients depending on their composition. Straight-chain paraffin hydrocarbons will have coefficients near the minimum, 0.0003, while aromatic molecules will be at least 0.0004, and much higher if they are conjugated or if the molecule has nitro, bromine or iodine substituents.

As an example, an immersion liquid with a refractive index of 1.6600 for the sodium D line, and at 25° C may have a temperature coefficient of 0.00042. That is, the refractive index of the liquid will decrease by 0.00042 refractive index units for every 1° C increase in temperature. In general, temperatures higher than 25° C will result in lower refractive indices, and temperatures lower than 25° C will result in higher refractive indices. At 20° C the above liquid shows refractive index for sodium light of 1.6600 + (5 x 0.00042), or 1.6621.

3. Dispersion of Refractive Index

Dispersion is another important property of transparent materials. This is the variation of refractive index with color (or wavelength) of light. When white light passes through a glass prism, the light rays are refracted by different amounts and separated into the colors of the spectrum. This spreading of light into its component colors is called dispersion and results from the change of refractive index with wavelength; as the wavelength increases, refractive index decreases.

Because of dispersion, the particular wavelength used in determining a refractive index must be designated. Light from a sodium lamp has a strong, closely spaced doublet with an average wavelength of 589.3 nm. Called the D line, it is commonly used as a reference wavelength. Table II illustrates the change of refractive index with wavelength for a few common substances.

Table II. Dispersion of refractive indices of several common materials

	Refractive index		
	F line (blue) 486.1 nm	D line (yellow) 589.3 nm	C line (red) 656.3 nm
Crown glass	1.5240	1.5172	1.5145
Flint glass	1.6391	1.6270	1.6221
Water	1.3372	1.3330	1.3312
Cargille medium 1.520	1.5303	1.5200	1.5160
Carbon disulfide	1.6523	1.6276	1.6182

A convenient measure of dispersion of any material is defined as:

$$\nu = \text{dispersion} = \frac{n_D - 1}{n_F - n_C} \tag{2}$$

where n is the refractive index of the material at the particular wavelength designated by the letter D, F and C which are spectral lines of sodium and hydrogen. It should be noted that as dispersion increases the value of ν decreases. The same composition factors that cause high refractive index and a high temperature coefficient of index cause high dispersion. Within this general framework are some chemical compositions that show unusually high dispersion. Among inorganic solids these include chromates and dichromates, cyanides, halide complexes and vanadates. Organic compounds with unusually high dispersion include any having a cinnamyl group.

4. Refraction at Normal Incidence

Light striking a surface at normal incidence changes in velocity only; there is no change in direction. Refraction causes an object immersed in a medium of higher refractive index than that of air to appear closer to the surface than it actually is (Figure 6). This effect may be used to determine the refractive index of a liquid with the microscope. A flat vial with a scratch on the bottom (inside) is placed on the stage. The microscope is focused on the scratch and the fine adjustment micrometer reading is noted. A small amount of the unknown liquid is added, the scratch is again brought into focus and the new micrometer reading is taken. Finally, the microscope is refocused until the liquid surface is in sharp focus. The micrometer reading is taken again and, with this information, the refractive index may be calculated from the simplified equation:

$$\text{refractive index} = \frac{\text{actual depth}}{\text{apparent depth}} \qquad (3)$$

Notice that this equation also tells us that if we are trying to determine the depth or thickness of a given transparent crystal by focusing on the lower and upper surfaces of that crystal, taking the difference in fine adjustment drum reading, and multiplying by the μ m/div. value of the drum, we will not have the correct thickness. We will have the apparent thickness only; not until we multiply the apparent thickness by the refractive index of the specimen will we obtain the actual thickness.

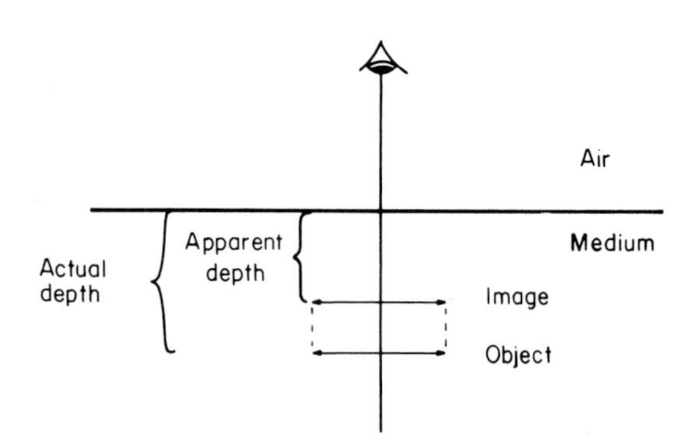

Figure 6. Shortening of optical path in an optically dense medium

5. Refraction at Other Than Normal Incidence

Light striking a surface at any angle other than normal incidence changes direction as well as velocity (Figure 7). A light ray traveling from a region of lesser to greater index is bent toward a normal to the interface; a ray traveling from a region of greater to lesser index is bent away from the normal. In 1621, Snell discovered that the sine of the angle of incidence, i., divided by the sine of the angle of refraction, r, equals the refractive index of the refracting material (assuming one medium to be air):

$$n_r = \frac{\sin i}{\sin r} \qquad (4)$$

This is now known as <u>Snell's Law</u>. Recognizing that the change in angle depends on the refractive indices of the two substances forming the interface, we can write the expression in a more general form:

$$\frac{n_r}{n_i} = \frac{\sin i}{\sin r} \qquad (5)$$

where n_i and n_r are the refractive indices of the incident and refracting media, respectively.

6. Critical Angle of Reflection

In passing from a medium of high refractive index to one of low index, the incident angle may become so large that no light is refracted into the less dense medium (Figure 8). This takes place when the angle of refraction becomes equal to 90° and equation (5) becomes:

$$\sin i = \frac{n\ (\text{low})}{n\ (\text{high})} \qquad (6)$$

This angle, i. is then called the critical angle, C. If the medium n_1 is air, the relationship becomes:

$$\sin C = \frac{1}{n\ (\text{high})} \qquad (7)$$

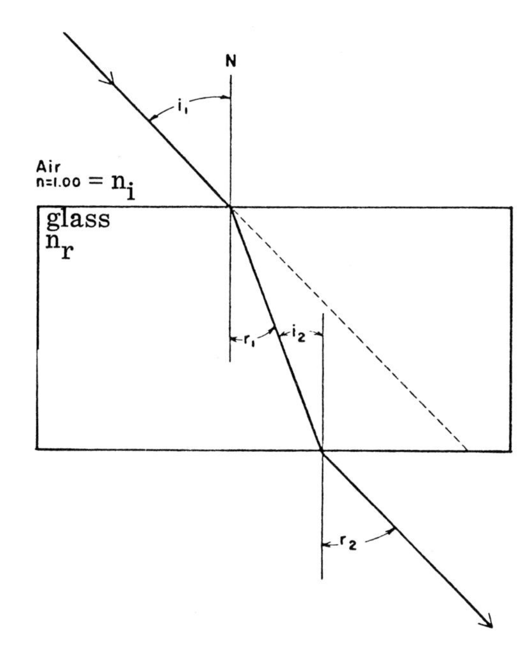

Figure 7. Refraction of light

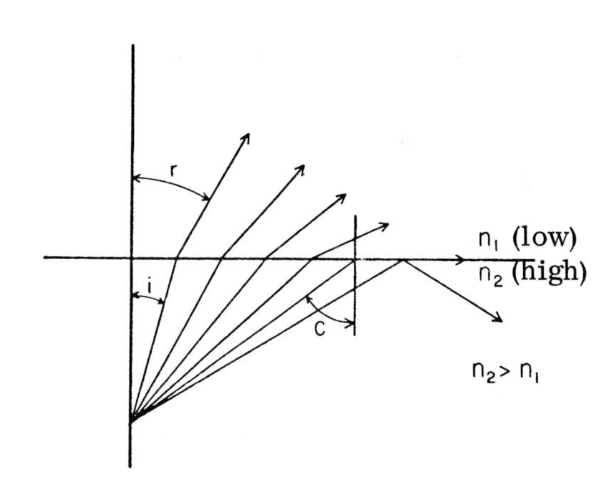

Figure 8. Reflection at the critical angle, C

The total reflection that results when the critical angle is exceeded is often taken advantage of in microscope illumination systems with right-angle prisms. In many microscopes a simple right-angle prism is often used in place of a first-surface mirror to obtain total reflection. Such prisms must be used as shown in Figure 9 to ensure total reflection.

partial

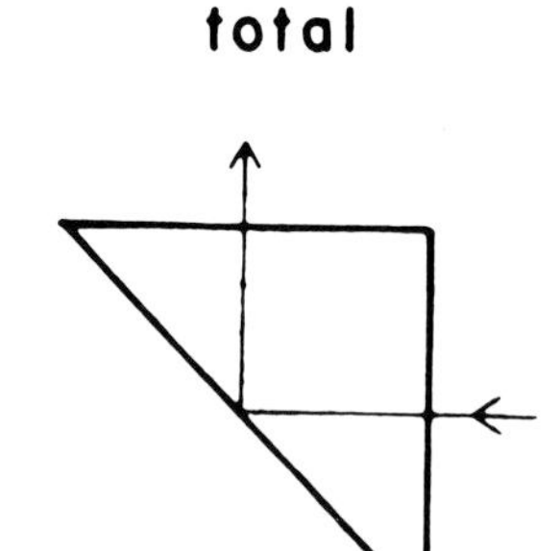

Figure 9. Proper use of prism to obtain total reflection

C. IMAGE FORMATION BY REFRACTION

1. Lenses

There are two classes of lenses, convergent and divergent, called also <u>convex</u> and <u>concave</u>, respectively. The focal point of a convergent lens is defined as the point at which a bundle of light rays parallel to the lens axis appears to converge after passing through the lens. The focal length of the lens is the distance from the lens to the focal point (Figure 10). The divergent or concave lens is shown in Figure 11.

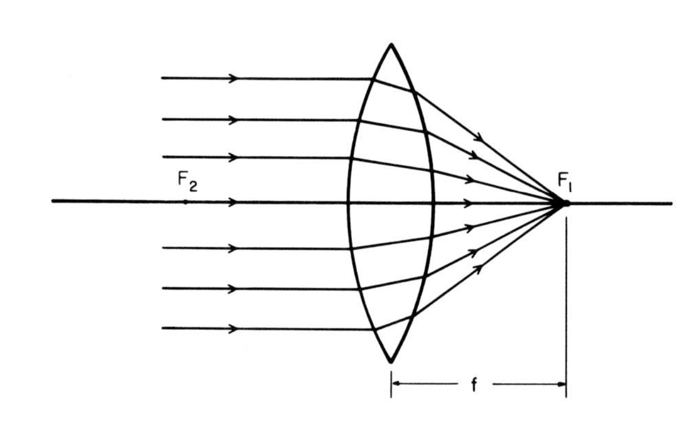

Figure 10. Convergence of light at the focal point, F_1, by a convex lens

2. Simple Magnifier

The earliest microscopes, modern hand magnifiers and the eye-lens of modern compound microscopes consist of a single lens or single lens system. Because of its use as an essential component of the microscope it is desirable that we fully understand the properties of this simple lens system.

Object size is determined by the angle formed at the eye by the extreme rays of the object. Bringing the object closer to the eye increases that angle (called the <u>visual angle</u>). This also increases the apparent size. However, a limit of accommodation of the eye is reached, at which the eye can no longer focus easily. At this distance, about 10 in. or 25 cm, the magnification of an object ob-

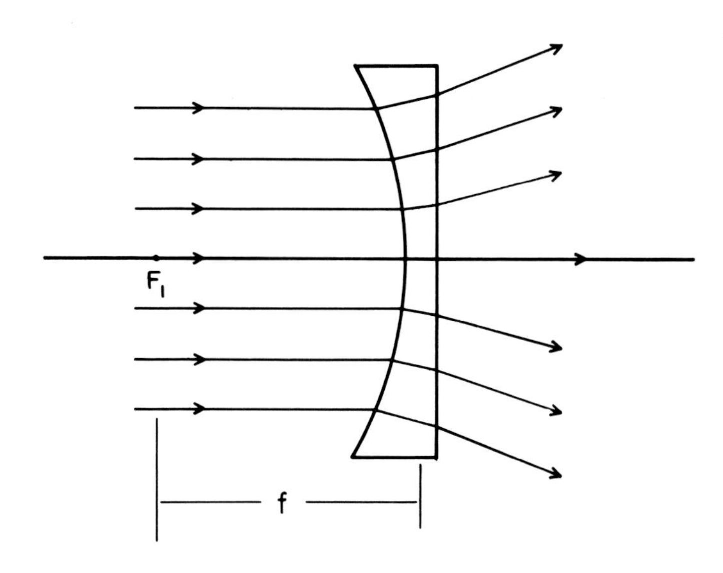

Figure 11. Divergence of light by a concave lens

served by the unaided eye is said to be unity. The eye can, of course, be focused at shorter distances but not usually in relaxed condition.

A positive, or convergent, lens can be used to help the eye focus on an object closer than 10 in. (Figure 12). The visual angle of the object is increased (as is its apparent size) while the image of the object appears to be 10 in. from the eye, where it is best accommodated.

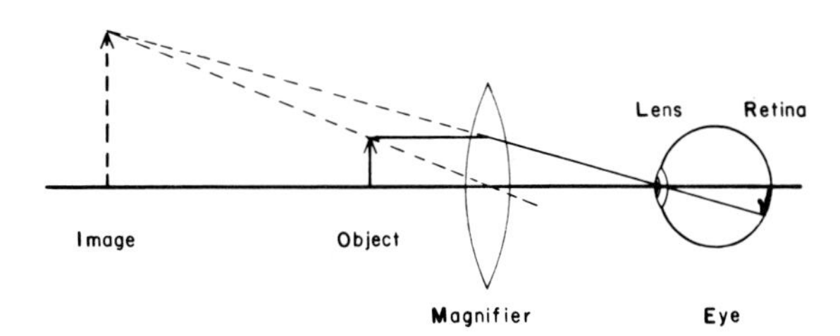

Figure 12. Simple magnifier

The magnification, M, of a simple magnifying glass is:

$$M = \frac{25}{f} + 1 \qquad (8)$$

where, f is the focal length of the lens in centimeters
25 is the normal reading distance in centimeters.

Theoretically, magnification can be increased with shorter focal length lenses. However, such lenses require placing the eye very close to the lens surface. They also show considerable image distortion. The practical limit for a simple magnifying glass is about 20X although Leeuwenhoek and, somewhat more recently, Svihla[1] have made and used single lenses magnifying up to about 300X.

---

1. Svihla, G., Microscope 15, 289-300 (1967).

3. Projected Image Formation

The formation of a projected image by a single lens or a cemented lens system (Figure 13) follows rules analogous to those already given for mirrors.

a. Light traveling parallel to the lens axis will be refracted to pass through the focus of the lens.

b. Light traveling through the geometrical center of the lens will be unrefracted.

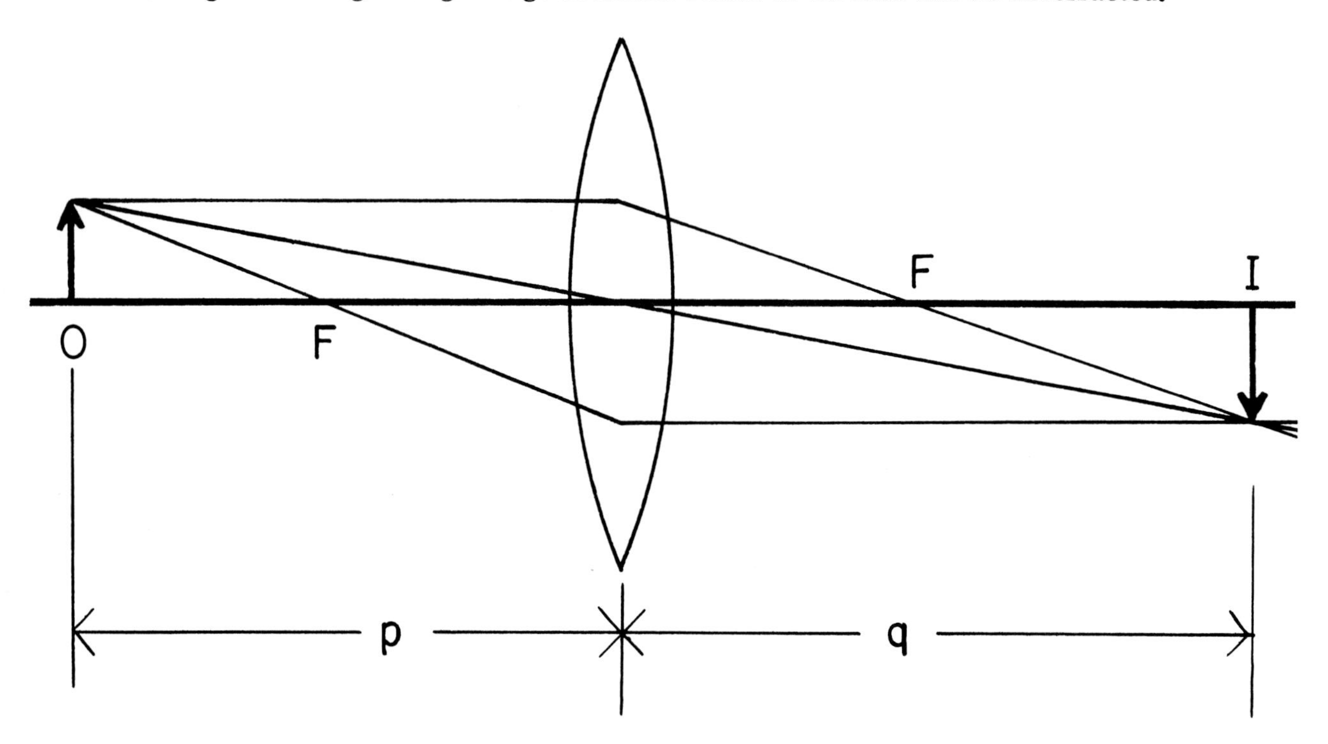

Figure 13. Image formation by a convex lens

The image position can be determined by drawing construction lines as directed by the above rules or by using the classical lens formula (1). Note that an object at infinity will be imaged at f and vice versa. The focal length of a lens can be measured by imaging a distant light source on a sheet of paper and measuring the distance between lens and paper.

The magnification, M, of an image of an object produced by a lens is given by the relationship:

$$M = \frac{\text{image size}}{\text{object size}} = \frac{\text{image distance}}{\text{object distance}} = \frac{q}{p} \qquad (9)$$

4. Lens Aberrations

Lenses have several types of aberration which, unless corrected, cause loss of detail in the image. Spherical aberration (Figure 14) is especially apparent in lenses having spherical surfaces. Light paths near the center of the lens focus at different points compared to light paths near the periphery. This can be reduced by diaphraming the outer zones of the lens or by grinding special aspherical surfaces in the lens system.

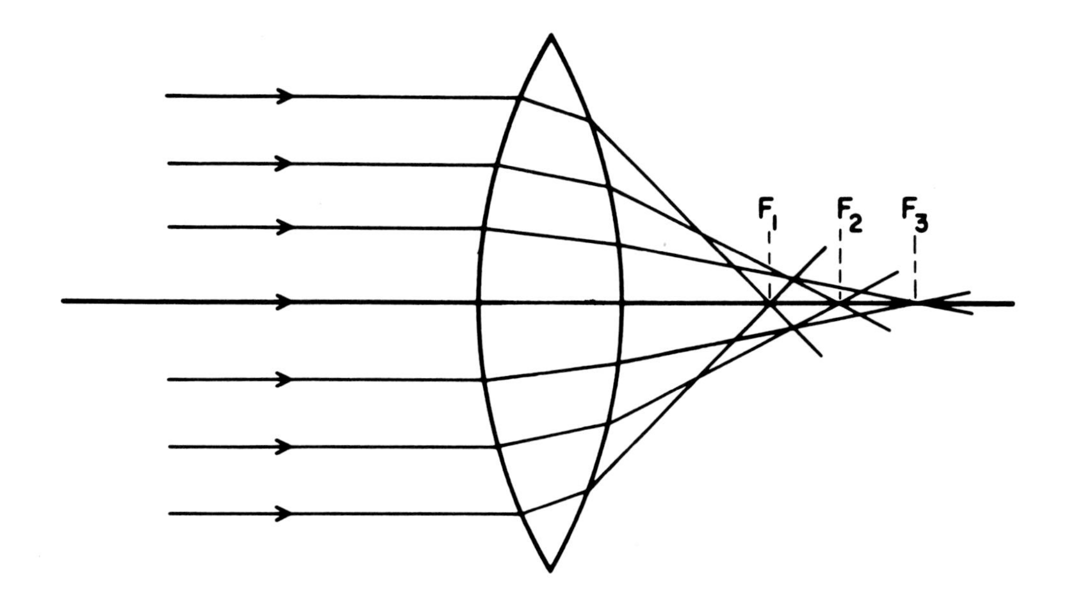

Figure 14. Spherical aberration by a convex lens

Chromatic aberration (Figure 15) is caused by refractive index variation with wavelength (dispersion). Thus, a lens receiving white light from an object will form a blue image closest to the lens, a red one farthest away. Achromatic lenses, employed to minimize this effect, are combinations of two or more lens elements of materials having different dispersive powers, for example, crown and flint glasses (Table II).

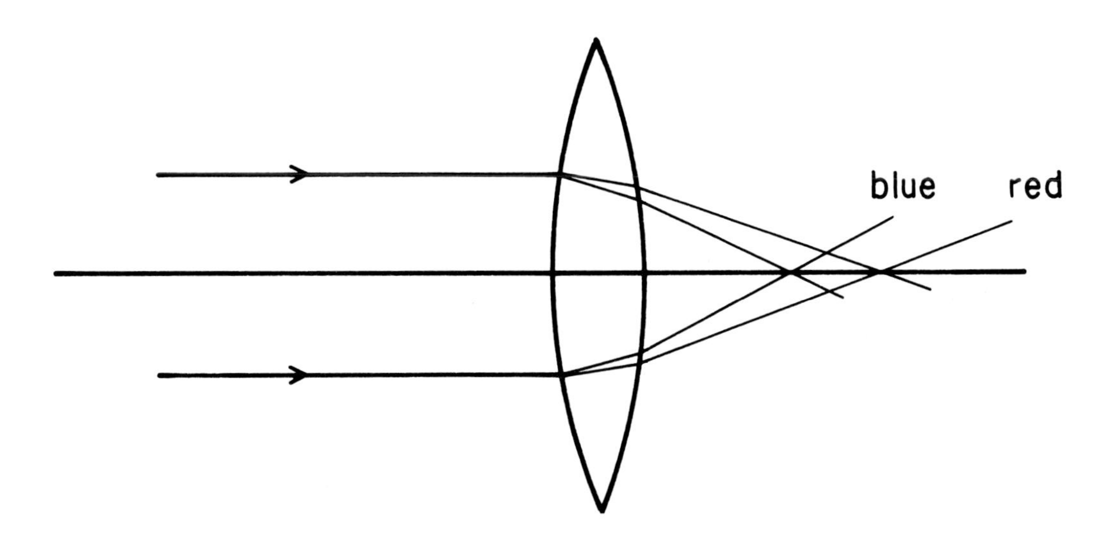

Figure 15. Chromatic aberration

Figure 16 shows two general methods for the minimization of chromatic aberration. In Figure 16a the two lenses must be of the same material and separated by one-half the sum of their focal lengths. In Figure 16b the center diverging lens must have higher dispersion than the two identical outer lenses, e.g., one flint and two crown lenses. The use of monochromatic light, an obvious way to eliminate chromatic aberration, often leads to a loss of conformation and detail when viewing colored objects.

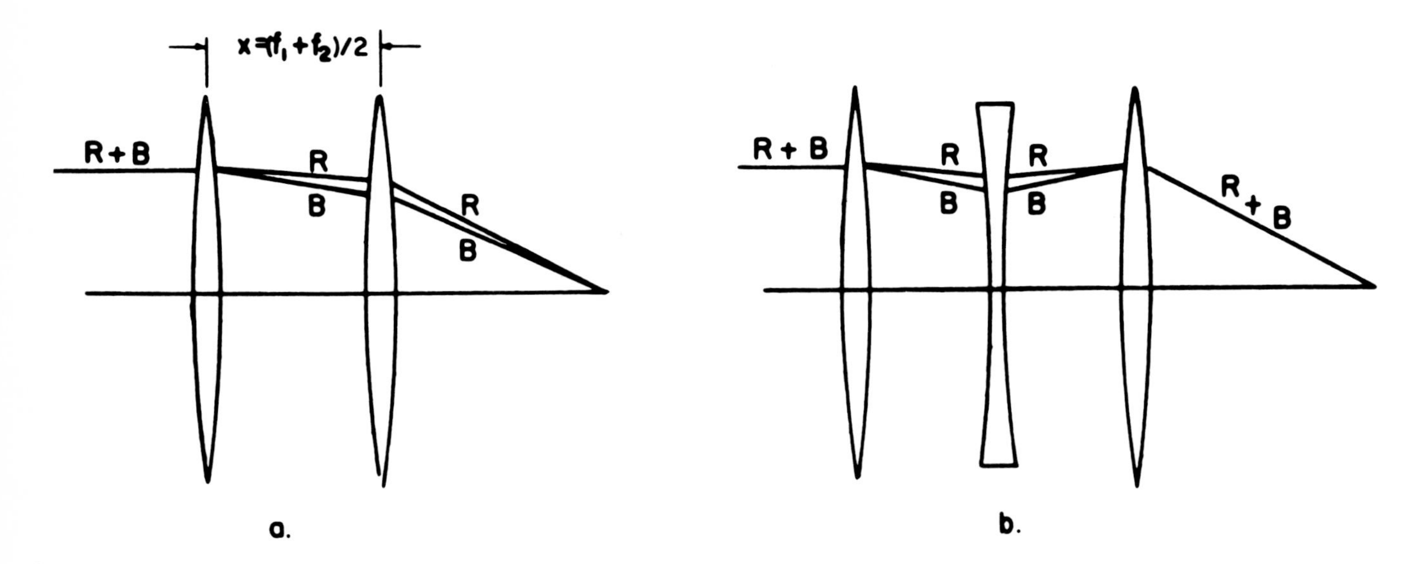

Figure 16. Achromatizing lens systems

Other important lens aberrations are astigmatism, coma and field curvature. Field curvature is a natural result of using lenses with curved surfaces. The image plane produced by such lenses will be curved (Figure 17). This kind of image occurs in microscopy unless plano (flat-field) objectives are used. The image across the entire field of view cannot all be in good focus at any one focus setting. In Figure 17, either the center of the image is in focus and the edges blurred, F_2, or the edges are in focus and the center is blurred, F_1, where F_1 and F_2 represent two different focus settings. This kind of aberration is particularly noticeable and troublesome in photomicrography, although most manufacturers now produce excellent microscope objectives and lens systems corrected to give very flat fields.

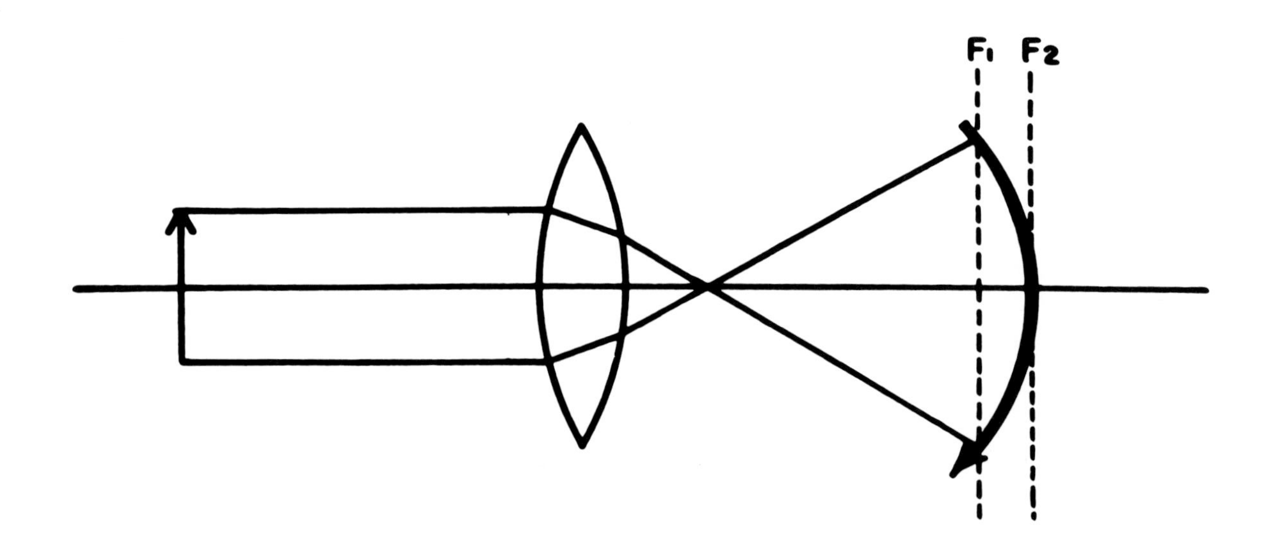

Figure 17. Field curvature

III. COMPOUND MICROSCOPE

The integral parts of the compound microscope are a condenser, objective and ocular. The condenser collects the light from the illuminator and projects it into the plane of the specimen. The objective projects an illuminated image of the specimen into the intermediate image plane. Finally, the ocular acts like a hand magnifier to add a second magnification stage for the human eye. The ocular may also function as a projection system in photomicrography and microprojection (Figure 18).

A. OBJECTIVES

Objectives are the most important optical component of the compound microscope. They form the intermediate image of the object, which is subsequently examined with the ocular. Some important characteristics of objectives are given in Table III and Figure 19. These and other important characteristics of objectives will be discussed separately.

1. Achromatic, Fluorite and Apochromatic Objectives

Objectives are divided into types according to their degrees of correction: achromats, fluorites (or semiapochromats), apochromats and plan- (these may be either plan-achromats or plan-apochromats). In achromats, chromatic aberrations are generally corrected for two colors, and spherical aberrations are corrected for one color. They have good working distance, moderate numerical apertures (NA - a measure of the resolving power) and are relatively easy to manufacture. They are very serviceable objectives, adequate for most applications and probably constitute 90% of all objectives in use. A special variety of achromats is made for polarized light work termed "strain-free". They are carefully made, from the slow cooling of the glass to the final assembly of the lens components, so as to avoid causing strains in the glass which would become visible when used in polarized light microscopy.

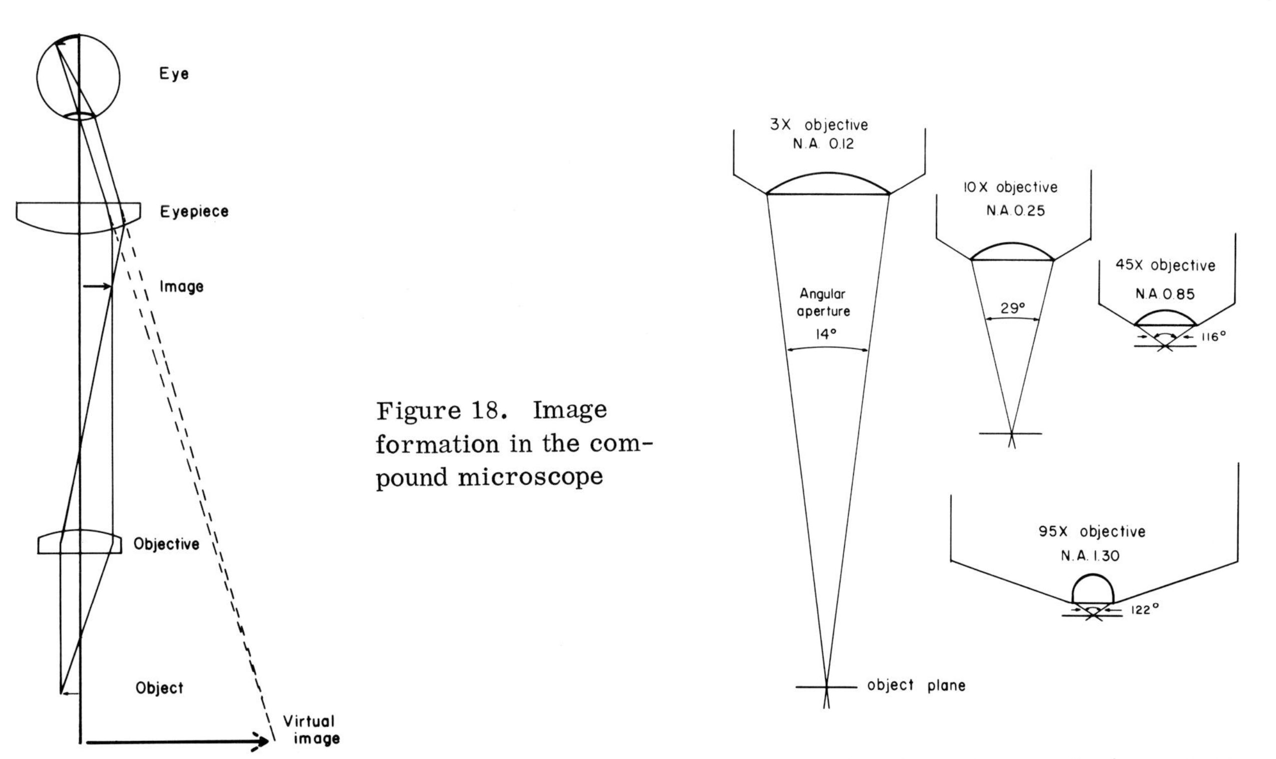

Figure 18. Image formation in the compound microscope

Figure 19. Angular and numerical aperture of objectives

Table III. Nominal characteristics of usual achromatic microscope objectives

Nominal focal length mm	Nominal magnification	Numerical aperture	Angular aperture degrees	Working distance mm	Depth of field μm	Diameter of field mm*	Resolving power, white light, μm
56	3X	0.08	9	40	50	8.5	3.8
32	5	0.10	11	25	16	5	3.0
16	10	0.25	29	7	8	2	1.22
8	20	0.50	60	1.3	2	1	0.61
4	40	0.66	83	0.7	1	0.8	0.47
4	40	0.85	116	0.5	1	0.6	0.36
1.8	90	1.30	122	0.2	0.4	0.4	0.20

* Measured approximately with a 10X eyepiece.

Fluorite objectives, or semiapochromatic objectives as they are sometimes called, are more highly corrected than achromats. Chromatic and spherical aberrations are both corrected for two colors. This higher correction is achieved through the use of fluorite, which has a refractive index-dispersion relationship not obtainable with glass. At a given magnification, fluorites have a somewhat higher NA than achromats. Because natural, clear fluorite of optical quality is difficult to find, it is expensive. Recently, synthetic fluorite has been used to make microscope objectives. Fluorites are an excellent compromise between the more universal achromats and the more highly corrected apochromats.

Apochromats are even more highly corrected objectives. Through the use of fluorite and additional glass elements, chromatic aberrations are corrected for three colors, and spherical aberrations for two. The NA is higher (for the same magnification) than for achromats and images are sharp and crisp. Unfortunately, field curvature, or the inability of an image to be in focus all the way across the field of view at one focus setting, is increased and the working distance (Figure 19) from the front lens to the top of the coverslip, is decreased. It is essential to use flat-field oculars whenever apochromats are used. These may be called compensating oculars or have, as part of their name, the syllable "plan", e.g., periplane, hyperplane or planoscopic.

2. Flatfield Objectives

Besides the three types of objectives classified according to chromatic and spherical corrections, there are plano or flatfield objectives. These flatten the image, correcting for field curvature through the use of additional lenses; these lenses are usually within the objective itself but in at least one case (Bausch and Lomb), the lenses are in the body tube. Bausch and Lomb flatfield objectives therefore cannot be used on any other manufacturer's stand. Plano or flatfield objectives are important in photomicrography where it is desirable to have all of the image in focus across the picture area. They are also of particular use in scanning thin films, such as blood smears.

Flatfield achromats, fluorites and apochromats (with the prefix — Plan) are available and, though expensive, give sharp, clear and flat images with excellent chromatic and spherical corrections. They are essential for the highest quality work.

3. Reflection Objectives

Another system of objectives employs concave and convex mirrors. Reflection objectives, because they have no refracting elements, do not suffer from chromatic aberrations as do ordinary refraction objectives. Based entirely on reflection, reflecting objectives are extremely useful in the infrared and ultraviolet regions of the spectrum. They also have a much greater working distance than the refracting objectives because they are easier to make with large diameters.

Still other long working distance objectives combine refraction and reflection. Such objectives in the 20-40X range typically have working distances of 12-16 mm. Unfortunately, some numerical aperture is sacrificed for the longer working distance. An ordinary 40X dry objective will have a numerical aperture range of 0.65 to 0.95. The reflection-refraction, long-working distance 40X objective has an NA no greater than 0.57. The reflection ray paths in this kind of objective yield an erect image as in a low-power stereomicroscope.

The diameter of the field of view depends greatly on the ocular — its diameter, and the diameter of the ocular diaphram — but is usually approximately half the reciprocal of the objective numerical aperture expressed in millimeters.

4. Numerical Aperture

The magnification of a lens depends upon its focal length and that of an objective depends upon the focal length of the entire system (Table III). It is achieved through curvature of the lens — the higher the curvature, the shorter the focal length. Lenses of different focal length therefore have different angular acceptance of diffracted light waves from the specimen (Figure 19). Lenses of shorter focal length (higher magnification) have the greatest angular apertures; that is, the largest acceptance angle of image-forming rays. The numerical aperture, NA, is related to this angular aperture, AA, as follows:

$$NA = n \sin\frac{AA}{2} \tag{10}$$

where, n is the refractive index of the space between the coverslip and the objective front lens. The highest theoretical numerical aperture for a dry objective, where the refractive index of air is 1.00, is, therefore, 1.00. In actual practice, a numerical aperture of 0.95 is the highest available in a dry objective.

Numerical aperture is related to resolving power (see section E below), the higher the NA, the greater the resolving power, and therefore high numerical aperture is desirable. To increase NA, we can increase the angular aperture and/or increase the refractive index of the space between specimen and objective. Increasing angular aperture means decreasing the focal length of the objective. This results in a front lens which can be so small that the lens itself becomes extremely difficult to mount. Besides, the working distance, which also decreases, becomes impractically small. Extreme examples have been made including one objective from the "Golden Age" of microscopy which had an initial magnification of 750X. Focal lengths of around 1.8 mm are the most generally useful maximum and these are all so-called immersion objectives (see below).

Given a practical maximum focal length and, therefore angular aperture, the best chance of increasing the numerical aperture is by increasing the refractive index of the medium in the space between the coverslip and the objective. The objective could be designed for any liquid since all liquids have refractive indices higher than 1.00. However, there are good reasons for confining the liquids to a specific few. One reason is that the cement used to mount the objective front lens may not be compatible with the liquid chosen. A much more important reason is that for optical reasons it is usually desirable to have the liquid match the front lens in both refractive index and dispersion.

Objectives intended for immersing in a liquid are called immersion objectives. There are water immersion objectives that have a theoretical maximum NA of 1.33, but practically are made about 1.10-1.25. Oil immersion objectives, which are the most common, have a theoretical maximum NA of 1.51; practically they are usually 1.25 to 1.32, although they may be as low as 1.00 or as high as 1.40. Quartz objectives made for use with short ultraviolet light use glycerine as an immersion medium. Extremely high NA can be achieved with special objectives which use monobromonaphthalene or methylene iodide as the immersion liquid. A given immersion objective must, however, be used with a liquid having specific refractive index-dispersion properties.

The angular aperture of the illumination cone of light from the condenser must be equal to or greater than the angular aperture of the objective for highest image resolution. This means that the condenser, an extremely important optical component, must itself have a high numerical aperture. Again, theoretical and practical limitations apply to the space between the top of the condenser lens and the bottom of the microscope slide. In more practical terms, this means that if a condenser is used dry with an oil immersion NA 1.40 objective, the final NA will not be greater than 1.0. To realize the benefits of objectives with NA's greater than 1.00, the condenser _must_ be oiled to the bottom of the slide. Oil is frequently omitted from the condenser-slide space largely through ignorance and/or because the procedure is considered messy and time consuming. Also, it may not be necessary or desirable to use the full numerical aperture. It should be understood, however, that unless an oil immersion condenser is oiled to the bottom of a slide when using high NA objectives, the resulting NA will not be greater than 1.0. If the observed microscopical preparation possesses little inherent contrast, _i.e._, transparent, colorless particles with refractive indices close to the mountant, then the substage condenser iris may, of necessity, have to be closed somewhat in order to see the specimen. Partial closing of the substage iris reduces the condenser NA often below 1.00. If the iris must be closed at least to a point at which the resulting condenser NA is 1.00, then immersion oil between the condenser top lens and the microscope slide serves no useful function. For best results, change the specimen preparation technique so that sufficient inherent contrast results and the condenser (properly oiled) can be used with an NA > 1.00, _i.e._, with the substage iris open completely or almost completely.

5. Magnification

The magnification engraved on an objective signifies the initial magnification of the specimen at the intermediate image plane before subsequent ocular magnification. The total magnification of the microscope is obtained by multiplying the objective magnification by the ocular magnification. Thus, a 10X objective and 10X ocular yield a 100X microscope magnification. In some of the larger microscopes where the mechanical tubelength is much longer than the standard 160 mm or 170 mm, lens systems are introduced to optically shorten the long tube. Additional magnification is one of the necessary consequences of using these lens systems. This additional magnifying component, called the _tube factor_, must be taken into account when computing total microscope magnification. The tube factor for these long bodytube microscopes is commonly 1.25. Thus, a 20X objective and 10X ocular when used on a microscope with tube factor of 1.25 produces a final image magnification of 20 x 10 x 1.25, or 250X. The magnifications engraved on objectives are usually only nominal. A "100X" objective, for example, may actually have a magnifying power of 98.8 to 100.6X; a "40X" objective may actually be 39.2-40.7X. In cases where _exact_ magnification is required, the final image size should be determined by viewing the image of a stage micrometer scale on your microscope with a positive (Ramsden) ocular (see Chapter VII.B.5). If the actual dimensions of the ocular graticule are known, the magnification of the stage scale by the objective can be measured directly.

6. Objective Mounts

The objective support may be either an objective clutch changer or a rotating nosepiece. The objective clutch changer ("quick-change" holder) permits mounting of only one objective at a time on the microscope. It has a centering arrangement, so that each objective need be centered only once with respect to stage rotation. Changing objectives is more awkward with this system than with the rotating nosepiece.

The rotating nosepiece allows mounting of two to six objectives on the microscope at one time. With such systems, either the objectives or the stage may be centerable. Several manufacturers provide centerable objective mounts so that each objective on the nosepiece need be centered only once to the fixed rotating stage. Objectives are better protected from dust by the rotating nosepiece since they are always covered. This, as well as the inconvenience of the so-called "quick-change" objective holder, makes it worthwhile to have the microscope fitted with a rotating nosepiece.

7. Coverslip Thickness

An important consideration in using different objectives is coverslip thickness. Just how important is the coverslip in the formation of sharp microscopical images? Spinell and Loveland[2] answer this question very completely in a paper entitled "Optics of the Object Space in Microscopy." For the average skilled microscopist, coverslip thickness should be close to the recommended thickness of 0.17 mm (continental and oriental microscopes) or 0.18 mm (U.S. and British microscopes). The allowable variation before detectable image deterioration occurs depends on the numerical aperture of the objective: it is $\pm 8\ \mu m$ for a 0.95 NA dry objective; $\pm 15\ \mu m$ for 0.85 NA; $\pm 45\ \mu m$ for 0.65; for lower NA objectives it makes little difference whether a coverslip is used or not. It is also best to use a coverslip of correct thickness for immersion objectives.

Coverslips vary greatly in thickness and some manufacturers' coverslips are better than those of others. In one test reported by Loveland, a group of Corning 1-1/2 coverslips averaged 184 μm in thickness and about half of the slips were suitable for use with the 0.95 NA dry objective; nearly 95% were suitable with a 0.85 NA objective (assuming that the objectives are corrected for use with 0.18 mm coverslips). It is best to buy No. 1-1/2 coverslips and to micrometer them when doing critical work.

B. OCULARS

The ocular, or eyepiece, is necessary in the second step of magnification. It functions as a simple magnifier viewing the image formed by the objective (Figure 18).

Oculars may be classified in several different ways. One convenient way is positive (Ramsden) or negative (Huyghenian); the latter have the diaphram between the upper and lower lens elements (Figure 20). Oculars may also be classified as compensating, high-eyepoint, projection, wide-field etc. The Huyghenian (or Huyghens) ocular is designed to be used with achromats while the compensating type is used with achromats of high NA, fluorites and apochromatic objectives. Oculars specially corrected to give a flat field are employed in photomicrography or projection and can be used with most objectives. It is usually best to follow the manufacturer's recommendations for the proper combination of objective and ocular.

The usual magnifications available in oculars run from about 4X up to 25 or 30X. The 4X is generally too low for anything but microprojection while 25 and 30X oculars have both slightly poorer imagery than that of intermediate powers and a very low eyepoint. The most useful oculars are in the 10 to 20X range.

---

2. Spinell, B. M., and R. P. Loveland, J. Roy. Micros. Soc. 79, 59-80 (1960).

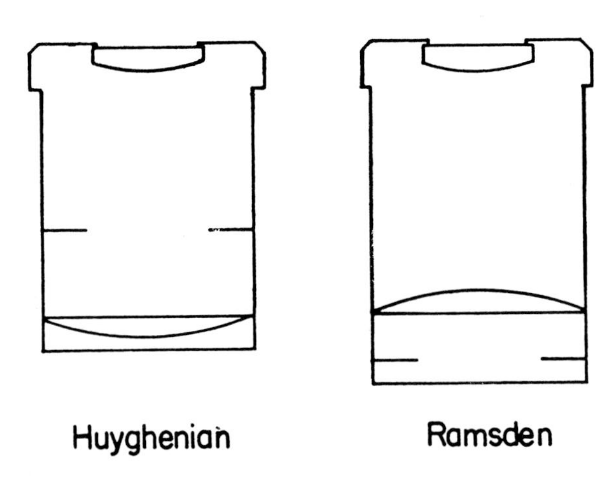

Figure 20. Sections through a positive (Ramsden), right, and a negative (Huyghenian), left, ocular

A convenient working rule to assist in the proper choice of oculars states that the maximum useful magnification (MUM) for the microscope is 1000 times the numerical aperture (NA) of the objective. Magnification in excess of MUM gives little or no additional resolving power and results in what is termed empty magnification. Table IV shows the results of such combinations and a comparison with the 1000 x NA rule. The underlined figure shows the magnification nearest MUM and the ocular required with each objective to achieve MUM. From this table it is apparent that only higher-power oculars can give full use of the resolving power of the objectives. It is obvious that a 10X, or even a 15X, ocular gives insufficient magnification for the eye to see detail actually resolved by the objective.

The factor of 1000 in Table IV is a reasonable working rule. The actual value depends on the quality of the optics (your eye and the microscope), how well your microscope illumination is set up and how well mounted your preparation is. The maximum useful magnification might be 300 NA

Table IV. Microscope magnification calculated for various objective-ocular combinations

Objective			Total magnification eyepiece					
Focal length	Magnification	NA	5	10	15	20	25	MUM* (1000 NA)
56 mm	3X	0.08	15X	30X	45X	60X	75X	80
32	5	0.10	25	50	75	100	125	100
16	10	0.25	50	100	150	200	250	250
8	20	0.50	100	200	300	400	500	500
4	40	0.85	200	400	600	800	1000	850
1.8	90	1.30	450	900	1350	1800	2250	1300

* MUM = maximum useful magnification.

if you have a good well-adjusted microscope and illumination system, a contrasty preparation and good eyes. For photomicrography, 100 NA or less might be justified because photographic film can be enlarged several times to bring out additional detail.

C. SUBSTAGE CONDENSER

The substage comprises, as the name suggests, the components beneath the stage. These components consist of the condenser, the substage (aperture) iris diaphram, the polarizer and the mirror.

The condenser provides a converging cone of light that illuminates the specimen. The ordinary condenser is a two-lens Abbe type (Figure 21) though there is also a three-lens Abbe condenser. Probably 90% of all brightfield condensers are of the two-lens type.

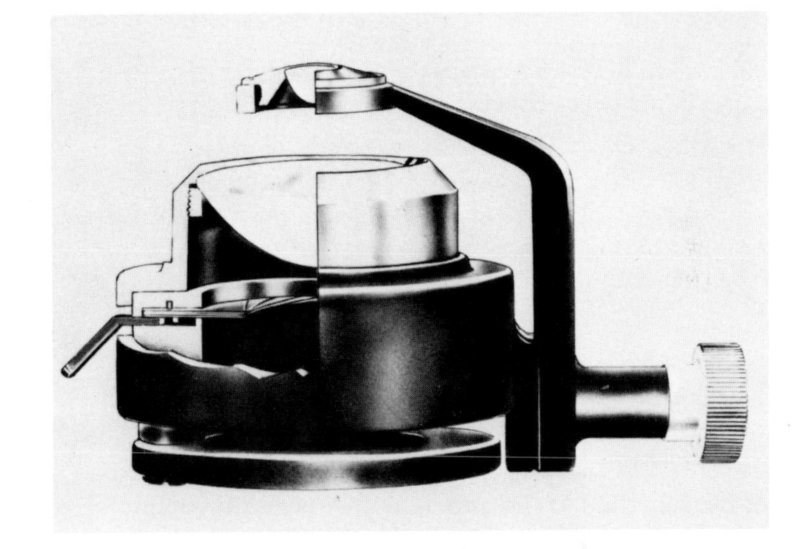

Figure 21. Two-lens Abbe condenser (drawing courtesy of Wild Heerbrugg)

1. Numerical Aperture

The condenser NA should be equal to or greater than the highest objective NA — usually about 1.25-1.32 for the 100X objective. The effective NA and therefore the resolving power of the system, is, at best, no better than one-half the sum of the objective and condenser aperture. If the condenser NA is less than 1.00 (some, for example, are 0.95), it is not intended for immersion. It may be an excellent dry condenser but the resolving power of an NA 1.32 objective can never be realized with this condenser and its effective NA is only 0.95 if the substage diaphram is wide open (which is seldom the case).

If the condenser is marked with an NA greater than 1.0, say 1.3, and if the top lens of the condenser is not oiled to the bottom of the slide, the effective NA will be that of the air space, 1.00. In summary, to get full resolving power from objectives of NA greater than 1.0, one must use a condenser with an NA matching or exceeding the objective. If the objective has an NA greater than 1.00, one must oil the condenser to the bottom of the slide. Some condensers are provided with interchangeable top lenses to change the NA. Others may be screwed or unscrewed into several different combinations to provide different NA's. Still others have a lever-operated, flip-out top lens, so that there are at least two maximum NA's.

2. Corrections

An ordinary two-lens condenser will form an image of the field diaphram in the field of view that will be somewhat fuzzy and surrounded with color fringes. This is due to chromatic aberration in such condensers and the fact that the rays are not all focused in the same plane. Condensers that focus light in one plane are termed aplanatic. There may be three, four or even five lenses in such a condenser. In achromatic-aplanatic condensers, chromatic aberrations are also corrected. Such condensers may have five, six, seven or more elements and are essential for the highest form of critical microscopy (Figure 22). A two- or three-lens condenser is perfectly adequate, however, for a student microscope, for one used only occasionally or for noncritical routine procedures.

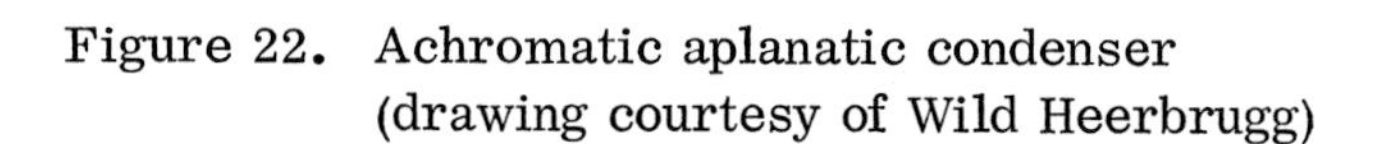

Figure 22. Achromatic aplanatic condenser (drawing courtesy of Wild Heerbrugg)

If there is a choice of condenser mount, one should definitely consider the centering mount because this is the only type that permits perfect alignment of the condenser on the microscope optical axis.

Special purpose condensers are usually required for darkfield, interference-transmission, interference-contrast, phase contrast, long-working-distance and strain-free condensers for use with polarized light. Quartz condensers and reflection condensers are made for use with ultraviolet light.

A mirror is provided as part of the substage when a built-in light source is not included. The mirror is about 50 mm in diameter and is usually two-sided, one plane and the other concave. The mirrors are second-surface mirrors which will form at least three images of the field diaphram. The plane side of the mirror is always used when a condenser is in position. The concave side of the mirror is used for very low NA objectives when a condenser is not used at all. First-surface mirrors are available; these may be polished stainless steel or evaporated aluminum protected with a thin silicon monoxide layer.

Polarizing microscopes have a graduated, rotating polar in the substage position. If polars are purchased for qualitative polarized light work, they should be selected for their color. Those nearest to absolute gray and not green or brown serve best where color photomicrography is important.

D. BODYTUBE

The microscope bodytube supports the objective at the bottom (over the object) and the ocular at the top. Some metallographs and other inverted microscopes have the positions of objective and ocular reversed. The tubelength is maintained at 160 mm except in Leitz instruments, which have a 170-mm tubelength, and metallographs, in general, which have a 215-mm tubelength.

The bodytube may simply be a cylinder that supports the viewing head on the top end and the nosepiece on the lower end, or it may be a complex assembly of lenses, adjustment knobs and slots. In polarizing microscopes, the bodytube contains the Bertrand lens which may be centerable and/or focusable and which may be inserted or removed from the optical axis by means of a knob at the side of the bodytube. The bodytube also contains the analyzer which can be taken out of the light path and which may be graduated and rotating. Compensators are also introduced into the bodytube. Finally, a lens just below the compensator renders the light parallel to avoid tubelength changes on introducing the compensator. A second lens restores the light beam to its original path. All of these additions make the polarizing microscope more expensive than other microscopes but the additional information they provide more than compensates for the increased cost.

Magnification changers may be incorporated in the bodytube with zoom-type lenses or with lenses providing stepped increases in magnification. Such systems include the Bausch and Lomb Dynazoom, Leitz Variotube and Zeiss Optovar.

E. STANDS

It is important to emphasize the difference between biological or medical microscopes on the one hand and chemical, polarizing or petrographic microscopes on the other. We must also emphasize that the latter should be used for the study of small particles.

The biological microscope is basically a stand on which a condenser, objective and ocular, all of varying degrees of sophistication, are mounted. The polarizing microscope has, in addition, two polarizing elements, a rotating stage and a Bertrand lens. All of these extras are useful in studying crystalline substances and especially in characterizing and identifying small particles. The Bertrand lens is also very useful for observing the back focal plane of the objective in order to determine or improve the quality of illumination.

Any microscopist planning to identify small particles must use the polarizing microscope. Instruments are available for any sum from $300 to over $20,000. A good microscopist can undoubtedly do excellent work with the lowest priced instruments. In fact, the beginning microscopist will have less difficulty doing good work with the lower priced instruments. To do a good job with the more sophisticated polarizing microscopes requires care, study and a basic understanding of microscope optics and illumination.

The more sophisticated instruments can, in trained hands, give better and faster results. Table V shows very roughly the cost of microscopes incorporating successive refinements.

Figures 23 and 24 show two different versions of a polarizing microscope. The Olympus shown in Figure 23 is used by the McCrone Research Institute for many of its intensive courses in microscopy. The Zeiss Ultraphot III shown in Figure 24 is clearly a superior microscope but not for most beginners.

F. STEREOBINOCULAR MICROSCOPE

The widefield, low-power stereobinocular microscope (Figure 25) gives an erect, three-dimensional image which makes it ideal for manipulating samples. Basically it is a stand with two separate compound microscopes, both focused on the same subject but from different angles (12° to 16° apart). Because it is impossible for two objectives of higher than 6-8X to be focused simultaneously on the same object when they are aligned on axes only about 14° apart, the stereomicroscope is limited to relatively low magnification.

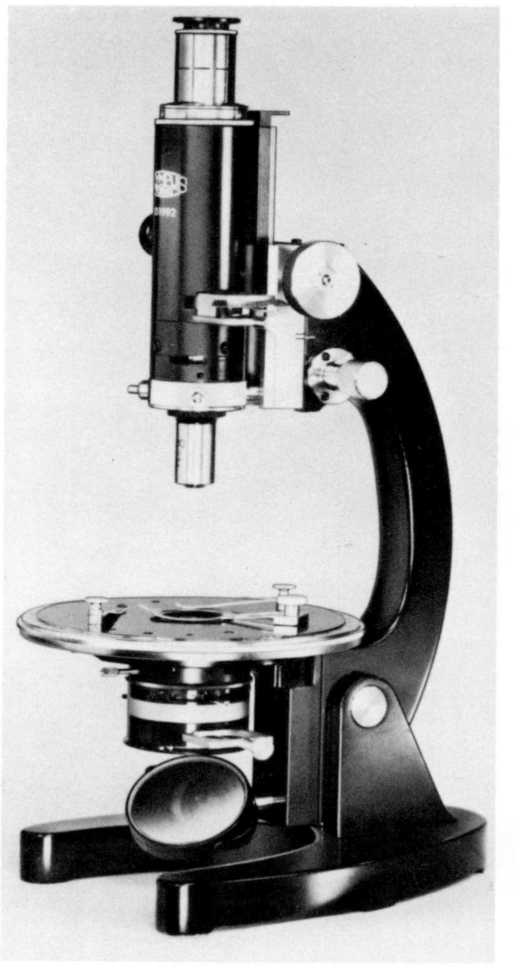

Figure 23. A simple, but wholly adequate, polarizing microscope (courtesy Olympus, New York)

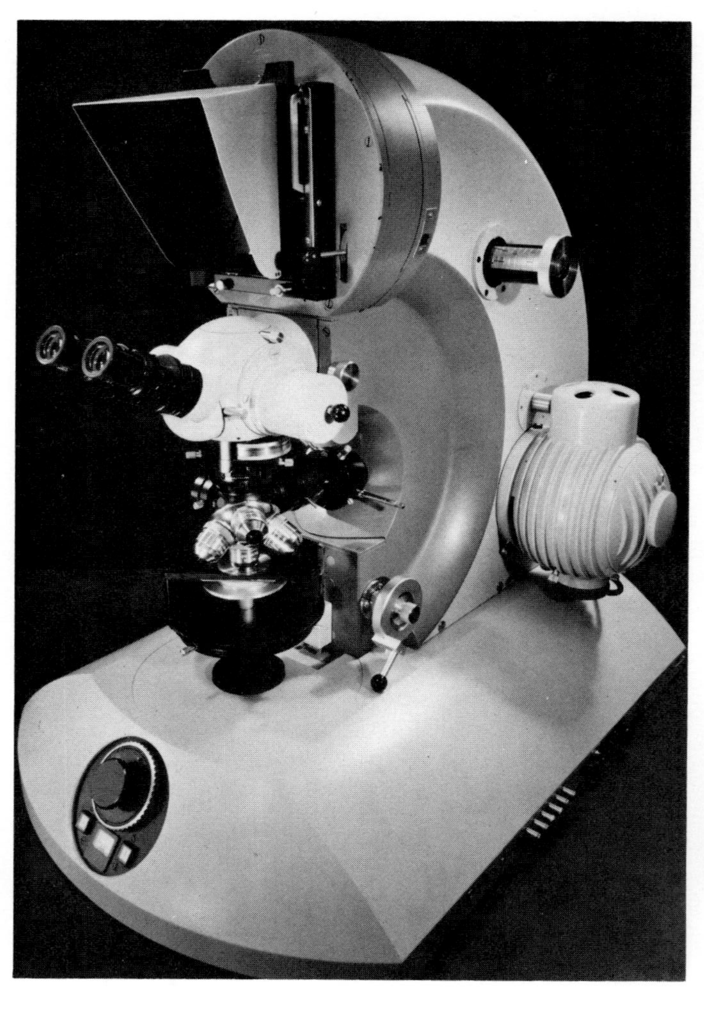

Figure 24. The highly sophisticated Ultraphot III of Carl Zeiss (courtesy Carl Zeiss, New York)

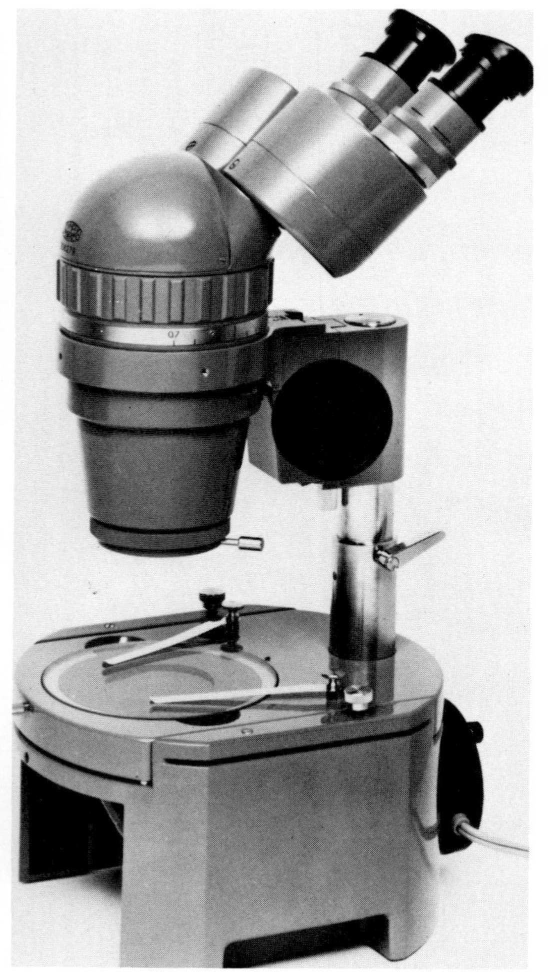
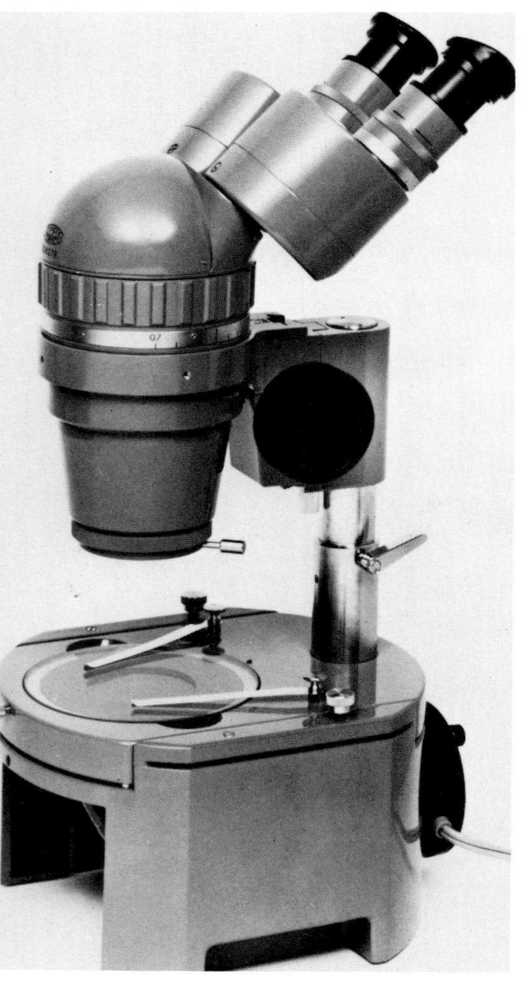

Figure 25. Transmitted light stereo-binocular microscope, reflected light capability is available (courtesy Olympus, New York)

Table V. Price and degree of refinement of polarizing microscopes

Approximate price, $	Degree of refinement
1,000	three centerable objectives up to 4 mm, 0.65 NA
1,500	4 mm, 0.85 NA; oil immersion objective
2,400	1.30 NA centerable condenser, focusable Bertrand lens
2,000	simultaneously rotating polars
2,000	apochromats, including oil immersion*
8,000	larger stand, flexibility of adding attachments
8,000	top lighting as built-in accessory
5,000	trinocular bodytube and automatic camera
2,000	phase attachments*
5,000	interference*
6,000	fluorescence attachments*
4,000	flatfield apochromats*
50,000	full universal with all accessories

* on a simple, say, $1,000 stand.

The objectives consist of cemented doublets or triplets with numerical apertures of about 0.14. This makes the maximum useful magnification of the stereobinocular about 140X. With supplementary magnification lenses and 33X oculars, it is possible to achieve magnifications of, perhaps, 300X, but image quality deteriorates. Samples requiring this degree of magnification are best examined with the single objective, binocular microscope.

The stereobinocular microscope is unsurpassed for low-power examinations, and for the manipulation of particles. Some models can be fitted for polarized light examinations in which case a special base with a light source or mirror is used to facilitate transmitted illumination. The base is equipped with a polar and rotating stage. An analyzer and compensator can be attached to the objective or bodytube . Both transmitted and incident illumination may be used with the stereobinocular microscope.

These microscopes can be purchased at various prices (Table VI) which reflect the degrees of refinement.

Table VI. Price and degree of refinement for stereomicroscopes

Approximate price, $	Accumulated degrees of refinement
300	fixed low objective magnification, incident lighting stage
800	variable, stepwise, objective magnification
1,200	variable, continuous, objective magnification
1,500	stand for transmitted light
2,000	stand for large object viewing
3,500	rotating stage and polarized transmitted light
4,000	trinocular head for photomicrography

IV. RESOLVING POWER AND ILLUMINATION

A. RESOLVING POWER

1. Diffraction

A knowledge of diffraction will help us understand much better how our microscope actually works. In Chapter IV.A.3 we see that a microscopic object will be resolved only if one, at least, of the diffracted rays enters the objective aperture along with the direct ray. We can see at once (Figure 27) why oblique rays from the condenser are necessary for resolution of the finest detail.

We should return now to diffraction itself. In geometrical optics, it is assumed that light travels in straight lines but this is not always true. A beam passing through a slit towards a screen creates a bright band wider than the slit with alternate bright and dark bands appearing on either side of the central bright band. The band intensity decreases as a function of the distance from the center. This phenomenon, diffraction, limits image reproduction. For example, the image of a pin point of light produced by a lens is not a pin-point but is a somewhat larger patch of light surrounded by dark and bright rings. The diameter, D, of this diffraction disc (to the first dark ring) is:

$$D = \frac{1.22\ \lambda}{\sin \theta} \qquad (11)$$

where, λ is the wavelength of the light

θ is one-half the lens angular aperture.

For a small diffraction disc to be maintained at a given wavelength, the lens aperture should be as large as possible. It should be noted that a shorter wavelength produces a smaller diameter diffraction disc.

2. Resolution

If two small objects are to be distinguished in an image, their diffraction discs must not overlap more than one half their diameters. The ability to distinguish such image points is called <u>resolving power</u> and is expressed as one half the diameter of the diffraction disc. The theoretical limit of resolving two discrete points, a distance X apart, is

$$X = \frac{0.61\lambda}{NA} \qquad (12)$$

where, λ is the wavelength of the light

NA is the numerical aperture of the objective.

Substituting a wavelength of 450 nm and a numerical aperture of 1.3, the practical limits for visible light wavelength and oil immersion objectives, we find that two points about 200 nm (or 0.2 µm) apart can be seen as two separate points. Further increase in resolving power can be achieved for the light microscope only with light of shorter wavelength. Ultraviolet light near 200 nm lowers the limit to about 0.1 µm, the lower limit for the light microscope. The resolution of detail actually observed in a microscopical image also depends on the illuminating conditions, specimen contrast, condition of the optics and quality of the human eye.

Under the very best illumination and contrast conditions, some unaided human eyes can resolve two points only about 60 µm apart. Under less optimum conditions, 120 µm might better be used and under average conditions, 300 µm may be the lower limit. Considering, then, the lower limit of resolution of the oil immersion objective with white light (0.2 µm), one can calculate the total magnification necessary under these three sets of conditions by dividing 0.2 µm into 60, 120 and 300 µm.

This gives 300, 600 and 1500X. This signifies that the ocular should furnish enough magnification to bring the total magnification up to the level required under the conditions at hand. Table VII shows the ocular magnification required by each of the common objectives under these three sets of conditions.

Table VII. Total and eyepiece magnification necessary to resolve detail shown by objective

Objective			Total (and eyepiece) magnification required		
NA	RP[a]	Magnification	200 NA[b]	400 NA[c]	1000 NA[d]
0.08	3.8	2.5X	16(7)	32(14)	80(35)
0.10	3.0	5	20(4)	40(8)	100(20)
0.25	1.22	10	50(5)	100(10)	250(25)
0.50	0.61	20	100(5)	200(20)	500(25)
0.66	0.47	43	132(3)	264(6)	660(15)
0.85	0.36	45	170(4)	340(8)	850(20)
1.30	0.20	90	260(3)	520(6)	1300(15)

[a] RP = resolving power
[b] excellent preparation, excellent illumination, excellent eye
[c] good preparation, good illumination, good eye
[d] average preparation, average illumination, average eye

The above reasoning has led to the generally helpful rule of thumb that useful magnification will not exceed 1000 times the numerical aperture of the objective. Although somewhat higher magnification may be used in specific cases, usually no additional detail will be seen.

It is curious, considering the figures in the table, that most, if not all, microscope manufacturers routinely furnish a 10X ocular as the highest power. A 10X ocular is useful but critical work requires a 15-25X ocular; 5-10X oculars are best used for scanning although it is nowadays possible to purchase widefield oculars of high power which are very suitable for scanning.

There are many equations given for resolving power[3]. Employing the usual one,

$$X = \frac{0.61\lambda}{NA} \qquad (13)$$

we can graph the resolution (μm) as a function of the numerical aperture (Figure 26) and wavelength.

3. Abbe's Theory

One of the most cogent theories of resolution is that of Ernst Abbe (1840-1905). He suggested that microscopic objects act as diffraction gratings (Figure 27) and that the angle of diffraction, therefore, increases with fineness of detail. He proposed that a given microscope objective would re-

3. Zieler, H. W., Microscope 17, 249-70 (1969).

solve a particular detail if at least two of the three transmitted rays (one direct ray and two diffracted rays) entered the objective. In Figure 27 the detail shown would be resolved in a and c but not in b. This theory, borne out by simple experiment, shows how to improve resolution. Since shorter wavelengths will give a smaller diffraction angle, they offer a better chance of resolving fine detail. Also, since only two of the transmitted rays are needed, oblique light and a high NA condenser will aid in resolving fine detail.

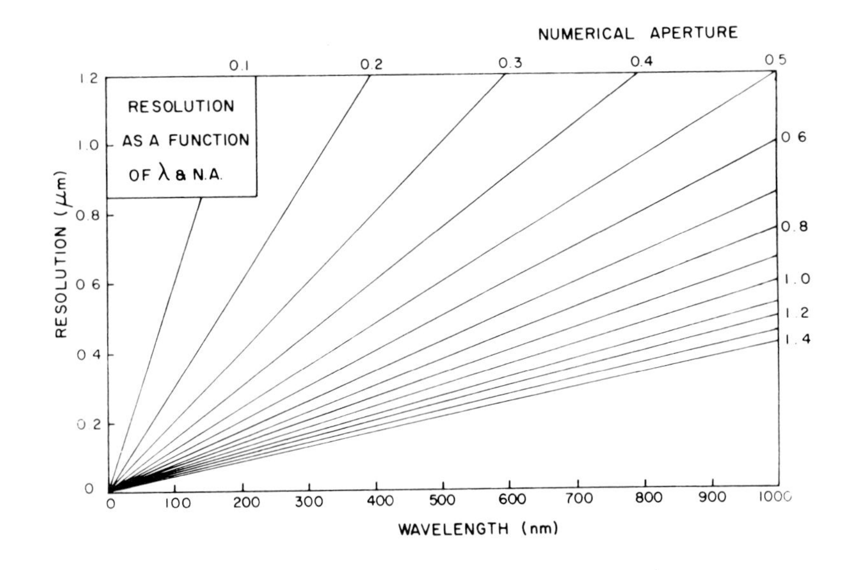

Figure 26. Resolution as a function of numerical aperture and wavelength

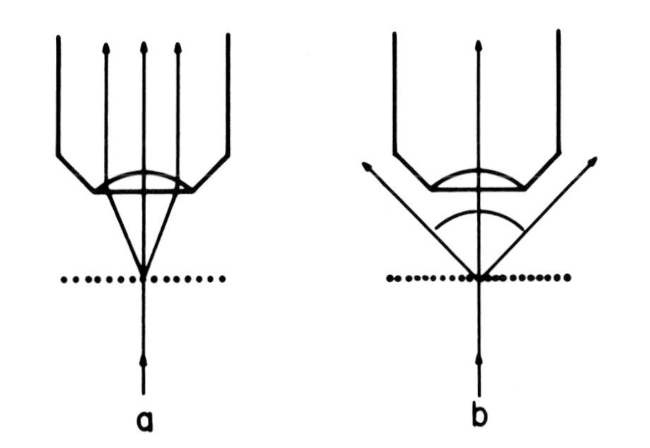

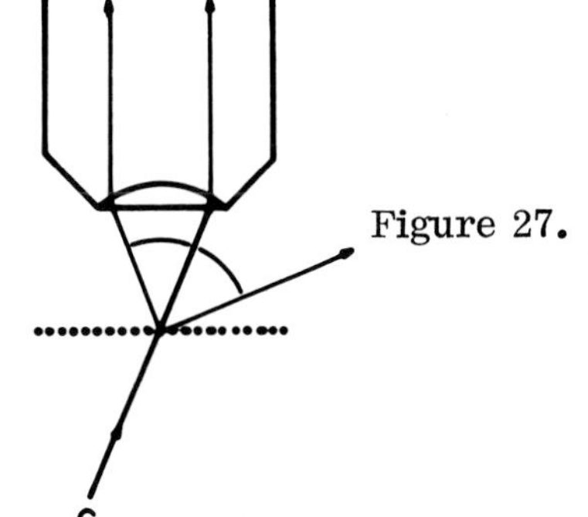

Figure 27. Schematic representation of Abbe's theory of resolution

4. Improving Resolving Power

The following list summarizes the practical approaches to higher resolution with the light microscope:

a. The specimen should have sufficient inherent contrast to permit use of a full substage aperture. It should be illuminated by either Nelsonian or Köhler illumination. The objective back focal plane should be fully illuminated and the field diaphram should be open just sufficiently to fully illuminate the field of view.

b. The condenser should be well corrected, with a numerical aperture as high as that of the objective.

c. An apochromatic oil immersion objective should be used with a compensating ocular. Proper immersion oil with the refractive index and dispersion required for the objective should be used.

d. Immersion oil should be placed between condenser and slide and between coverslip and objective. The preparation itself should be surrounded by a liquid with a refractive index not less than 1.515.

e. The illumination should be reasonably monochromatic and as short in wavelength as possible. An interference filter transmitting a wavelength of about 480-500 nm solves this problem. Ultraviolet light should be used to decrease the wavelength still further.

B. ILLUMINATION

Good resolving power and optimum specimen contrast are prerequisites for good microscopy. Though the optics (ocular, objectives and substage condenser) may be suitable, proper illumination is of paramount importance. The requirement for a good illumination system is uniform intensity over the entire field of view with independent control of light intensity, size of the illuminated field of view and angular aperture of the illuminating cone. Light intensity should be controlled for visual work by neutral density filters or by a variable voltage transformer on the light source. A field diaphram on the lamp housing usually controls the size of the illuminated field of view. The angular aperture of the illuminating cone is controlled with the substage iris.

The microscopist soon realizes that he must usually compromise to obtain optimum illumination. For good contrast, the substage iris must usually be closed down slightly. This, however, cuts down the condenser aperture and decreases resolving power. It is necessary to operate with the substage iris open as far as possible, consistent with image contrast, to have good resolution. Obviously, any other means of achieving good contrast will permit the substage iris to be opened farther with consequently improved resolution. Mounting the specimen in a medium of very different refractive index is one obvious way to improve inherent contrast.

Three types of illumination (Table VIII) are generally used:

Nelsonian illumination (after Edward Nelson, 1851-1938) is used when high intensity is necessary for oil immersion, darkfield, fluorescence, low birefringence or photomicrographic studies. Since the lamp filament is imaged in the plane of the specimen, a ribbon filament or arc lamp is required. The lamp must be focusable and the filament position must be adjustable in all directions.

Köhler illumination (after August Köhler, 1866-1948) is also useful for intense illumination, and may be obtained with any lamp not fitted with ground glass. The illuminator must, however, be focusable, have an adjustable field diaphram (iris) and the lamp filament position must be adjustable in all directions.

Diffuse illumination, so called because a ground glass is interposed between the lamp filament and the microscope condenser, gives illumination of high quality although of lower intensity because of the ground glass. As much as 70% of the initial illumination may be lost with a ground glass. The illuminator and lamp filament need no adjustments although an adjustable iris on the illuminator is helpful so that it may be used as a field diaphram.

Table VIII. Comparison of Nelsonian, Köhler and diffuse illumination

	Nelsonian	Köhler	Diffuse
Lamp filament	ribbon filament	any type	any type
Lamp condensing lens	desirable	required	not required
Lamp iris	useful	required	useful
Ground glass at lamp	none	none	present
Image of light source	in object plane	at substage iris	none
Image of field iris	near object plane	in object plane	near object plane
Image of substage iris	back focal plane of objective	back focal plane of objective	back focal plane of objective

All three types of illumination require that the microscope substage condenser be focused on the specimen to give a full illuminating cone and to permit the substage iris to be focused in the objective back focal plane. The differences in these three types of illumination lie in the adjustment of the lamp condensing lens. With diffuse illumination there is no lamp condenser, hence no adjustment. The lamp should be placed close to the microscope so that the entire field of view is always illuminated. If the surface structure of the ground glass becomes apparent in the field of view, the lamp can be moved slightly towards or away from the microscope, alternatively the substage condenser can be lowered just sufficiently to defocus the ground glass image.

With Nelson illumination the lamp condenser focuses the lamp filament at infinity; with Köhler illumination the lamp filament is focused in the plane of the substage condenser iris (also coincident with the lower focal plane of the substage condenser). The function of the substage condenser iris in controlling the angular aperture of the illuminating cone is precisely alike for all three types of illumination and the quality of the illumination can be checked quickly by inspecting the objective back focal plane (insert the Bertrand lens or remove the ocular).

1. Nelsonian Illumination

With Nelsonian illumination (Figure 28) the lamp condenser is focused to give parallel rays; focusing the lamp filament on a far wall is sufficient. Aimed, then, at the substage mirror, the substage condenser focuses the lamp filament in the object plane. The substage condenser iris will now be found imaged in the back focal plane of the objective and it serves as a control over convergence of the illumination. Although the substage iris also affects the light intensity over the field of view, it should most decidedly not be used for this purpose (see section A above). Illumination intensity may be varied with neutral density filters or, unless color photomicrography is anticipated, with variable voltage on the lamp filament.

Nelsonian illumination is seldom used because it requires a special lamp filament and because it shows little advantage over well-adjusted Köhler illumination.

2. Köhler Illumination

To arrange the microscope and illuminator for Köhler illumination, it is well to proceed methodically through the following steps:

a. Remove the diffusers and filters from the lamp.

b. Tilt the lamp until the beam is centered on the microscope mirror. Open the lamp diaphram (also called field diaphram, field iris or radiant field stop).

c. By moving the lamp condenser, focus a sharp image of the filament on the plane of the microscope substage iris. The filament image should be large enough to fill, even though unevenly, the microscope substage condenser opening. If it does not, move the lamp away from the microscope to enlarge the filament image and refocus.

d. Place a specimen on the microscope stage and focus sharply with a 16 mm (10X) objective. Open the substage diaphram completely. If the light is too bright, temporarily place a neutral density filter in the lamp.

e. Close field iris somewhat and adjust the mirror to center it in the field of view.

f. Move the specimen so that a clear area is under observation. Place the Bertrand lens in the optical path, or remove the ocular and insert an auxiliary telescope (sold as a phase contrast accessory) in its place, or remove the ocular and observe the back focal plane of the objective directly. Now observe the lamp filament through the microscope.

g. If the filament does not appear to be centered, swing the lamp housing in a horizontal arc centered at the field diaphram. The purpose is to maintain the field diaphram on the lamp in its centered position. If a vertical movement of the filament is required, loosen the bulb base and slide it up and down. If the base is fixed, tilt the lamp housing in a vertical arc around the field diaphram (again endeavoring to keep the lamp diaphram centered). If you have mastered this, you have accomplished the most difficult step. (Better microscope lamps simplify this step with adjustments to move the bulb independently of the lamp housing.)

h. Put the specimen in place, replace the ocular and the desired objective and refocus.

i. Open or close the field diaphram until it falls just outside the field.

j. Observe the preparation and adjust the contrast by opening or closing the substage iris. It must be as wide open as possible.

k. Observe the back focal plane of the objective, preferably with the Bertrans lens or the auxiliary telescope, and note the position of the substage iris. If it is not open at least two thirds of the diameter of the back focal plane, the preparation has too little inherent contrast or you are a bad judge of good illumination. It is instructive to vary the opening of the substage iris and observe the image and the objective back focal plane critically during this manipulation.

l. If the illumination is too strong, insert an appropriate neutral density filter between the illuminator and the condenser. Do not use the condenser iris or the lamp field diaphram to control illumination intensity.

3. Diffuse Illumination

Both Nelsonian and Köhler illumination require expensive illuminators with adjustable focus, lamp iris and adjustable lamp mounts. Diffuse illumination requires an illuminator although an expensive illuminator may be used if its expensive features are negated by inserting a ground glass diffuser or by using a frosted bulb. Admittedly an iris diaphram on the lamp would be helpful though it is not necessary. The steps required to set up proper diffuse illumination are:

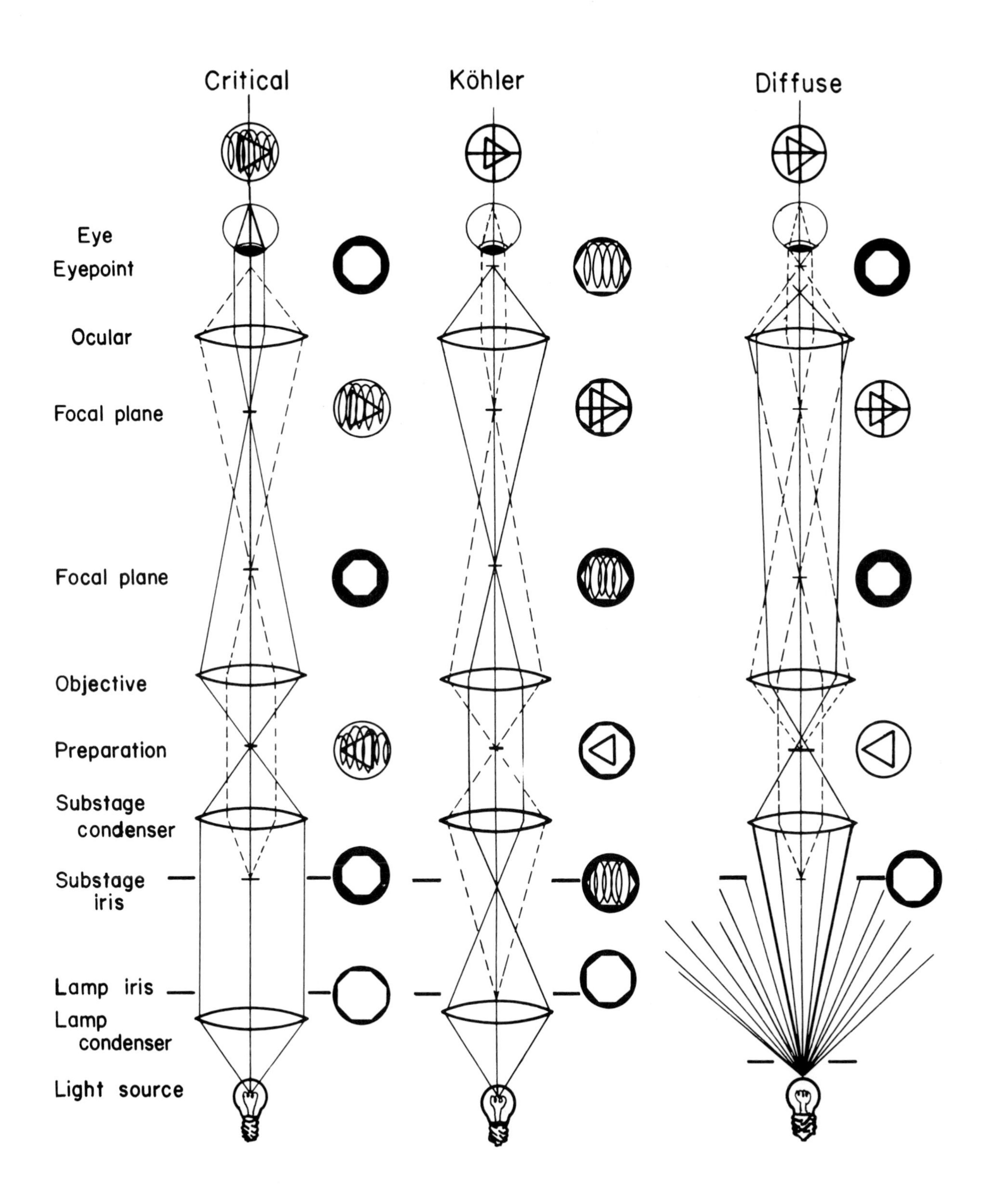

Figure 28. Microscope illumination

a. Be sure the illuminator has a frosted bulb or a ground glass diffuser. It should be possible to direct it towards the substage mirror, either very close to it or even in place of it.

b. Tilt the mirror, if in use, to illuminate the specimen.

c. Focus on any preparation using the 16-mm objective.

d. If necessary, remove the top lens of the condenser and, by racking the condenser up or, more often, down, bring into focus (in the same plane as the specimen) a finger, pencil or other object placed in the same general region as the ground glass diffuser on the lamp. The glass surface itself can then be focused in the plane of the specimen.

e. Ideally the ground glass surface will just fill the field of view when centered by the substage mirror; if it does not, move the lamp closer to or farther from the microscope (the position might well be marked for each objective used) or cut paper diaphrams of fixed aperture (one for each objective). A lamp iris would be useful here.

f. Lower the condenser just enough to defocus the ground glass surface and render the field of illumination even.

g. Observe the objective back focal plane and open the substage condenser iris about 75%. Make final adjustments of the substage iris while observing the preparation; the iris should be open as far as possible, still giving good contrast.

h. The illumination intensity should be adjusted either with neutral density filters or by changing the lamp voltage.

Many large, modern research microscopes with built-in illumination have a diffusion disc permanently located somewhere in the system. It may even be found as one side of a lens. These diffusion plates are not true ground glasses. They are usually etched, and are sometimes referred to as "orange peel". The question which arises when using these stands is what kind of illumination to set up. The answer is a modified Köhler. The departure from Köhler is necessary because the filament will not be visible in the back focal plane. In this event, the lamp condenser is focused or the entire light source is moved in or out of its mount while watching for evenness of illumination in the back focal plane.

Proper illumination is most important. It is easy to judge a microscopist's ability by glancing at his field of view and the objective back focal plane. The back focal plane should always be as nearly fully illuminated as possible, no matter what kind of illumination is used.

4. Incident Light Illumination

The particle analyst has a real need for reflected light examination (top lighting). Without it, one cannot completely describe an opaque particle. In fact, without top light he can only say "an opaque particle was observed" — not very satisfactory if the particle must be identified.

Many large, settled dust particles can be examined with the stereobinocular microscope, equipped for top lighting. Direct examination of particles on membrane filters or impactor slides is easy with such an instrument. When particles are smaller, however, better resolving power and higher magnification are needed. Also, the problem becomes one of observing any given particle preparation with transmitted polarized light, perhaps simultaneously, or at least in rapid sequence, with the same microscope.

Strangely enough, many transparent substances appear opaque when viewed in transmitted light with a polarizing microscope. Opacity can be due to total absorption (Table IX) or to a combination of total reflection and refraction (Table X). Highly refractive solids, such as rutile, diamond, silicon carbide, realgar and lead oxide (PbO), reflect and refract so much light in Aroclor®, and the lower index media, that they may seem opaque, especially when very small. The extent to which such substances appear opaque depends on the refractive index difference of the solid particle and its surroundings. Even very low-index solids (Table X) may appear opaque in Aroclor®. The particles will reflect light to the same degree as high-index particles.

The particles listed in Table X are often thought to be opaque when they are very small, especially when many such particles are held together in a matrix. A dried white paint particle or a white sidewall tire particle is a good example (Figures 29-30). Such particles are dense dispersions of colorless, transparent, but highly refractive particles of titanium dioxide, zinc oxide, basic lead carbonate or a similar substance in a transparent, colorless, organic binder. These particles, unless very thin, are "opaque" — no light is transmitted through the particle into the microscope objective. Even air bubbles, or perhaps especially air bubbles, can appear opaque by this mechanisn. Ethyl cellulose particles with and without a top light illustrate this. These polymer particles are very porous, *i.e.*, filled with air bubbles. The refractive index difference between air and ethyl cellulose is great, and, even though the index between ethyl cellulose and Aroclor® is not too great, the particles seem opaque (Figures 31-32). In the smallest particles of polymer, individual "opaque" air bubbles can be observed. Glassy flyash particles often contain enough small air bubbles to render the colorless glass sphere "opaque". Some substances are, of course, truly opaque.

Two particles, which appear opaque, are shown in Figures 29-32, first with transmitted light only, then with a top light added. The need for top light is thus graphically illustrated.

Table IX. Totally absorbing particles

anthracite	carbon black
magnetite, Fe_3O_4	asphalt
metals and alloys	coke
wolframite, (Fe, Mn) WO_4	oil soot
franklinite, (Fe, Mn, Zn) $(FeO_2)_2$	graphite
spent catalyst	stibnite, Sb_2S_3
copper oxide, CuO	potassium permanganate, $KMnO_4$
lead suboxide, Pb_2O	pyrolusite, MnO_2
molybdenum tetraiodide, MoI_4	mercurous oxide, Hg_2O
nickel iodide, NiI_2	niobium nitride, NbN
platinum oxides, PtO and PtO_2	niobium oxides, NbO and NbO_2
platinum hydroxide, $Pt(OH)_2$	osmium oxides, OsO and OsO_2
platinum iodide, PtI_2	strontium carbide, SrC_2

Table X. High and low refractive index particles that often appear opaque when < 1 μm

High refractive index	Low refractive index
copper oxide, Cu_2O	sodium fluoride, NaF
thallium iodide, TlI	potassium silicofluoride, K_2SiF_6
zinc sulfide, ZnS	potassium fluoride, KF
chromium oxide, Cr_2O_3	ammonium silicofluoride, $(NH_4)_2SiF_6$
cadmium sulfide, CdS	lithium fluoride, LiF
$CaFe_2O_4$	rubidium fluoride, RbF
mercuric sulfide, HgS	sodium silicofluoride, Na_2SiF_6
$AgAsS_2$	potassium fluoborate, KBF_4
arsenic tri-iodide, AsI_3	magnesium silicofluoride, $MgSiF_6 \cdot 6H_2O$
lead oxide, PbO	potassium fluoride, $KF \cdot 2H_2O$
silicon carbide, SiC	cesium fluoborate, $CsBF_4$
aluminum carbide, Al_4C_3	manganese silicofluoride, $MnSiF_6 \cdot 6H_2O$
$Pb_3Cr_2O_9$	cobalt silicofluoride, $CoSiF_6 \cdot 6H_2O$
mercuric iodide, HgI_2	iron silicofluoride, $FeSiF_6 \cdot 6H_2O$
antimony iodide, SbI_3	zinc silicofluoride, $ZnSiF_6 \cdot 6H_2O$
mercuric sulfide, HgS	$CoF_2 \cdot 5HF \cdot 6H_2O$
copper oxide, CuO	sodium cyanate, NaCNO
iron oxide, Fe_2O_3	NH_4HF_2
selenium, Se	nickel silicofluoride, $NiSiF_6 \cdot 6H_2O$
diamond	$NiF_2 \cdot 5HF \cdot 6H_2O$
Titanium dioxide, TiO_2	sodium sulfate, $Na_2SO_4 \cdot 10H_2O$

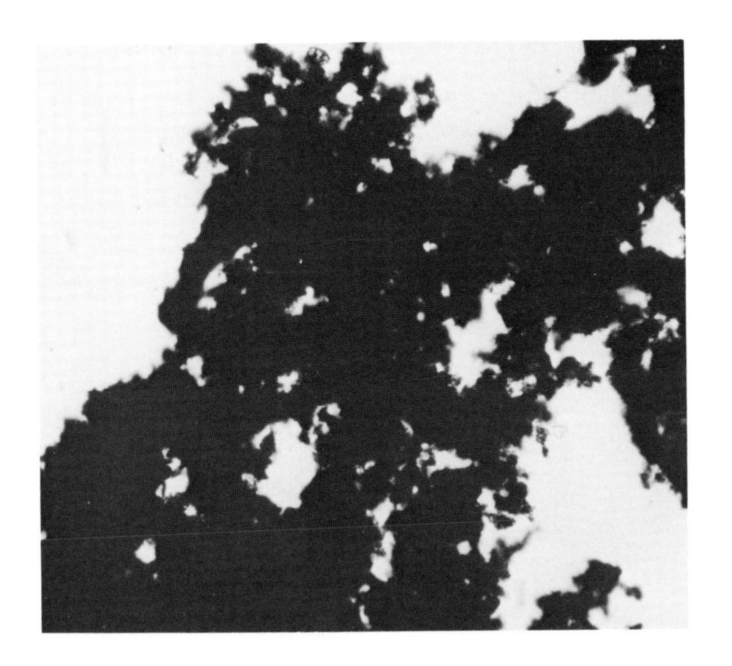

Figure 29. Particles from buffing white sidewall automobile tires, transmitted light only

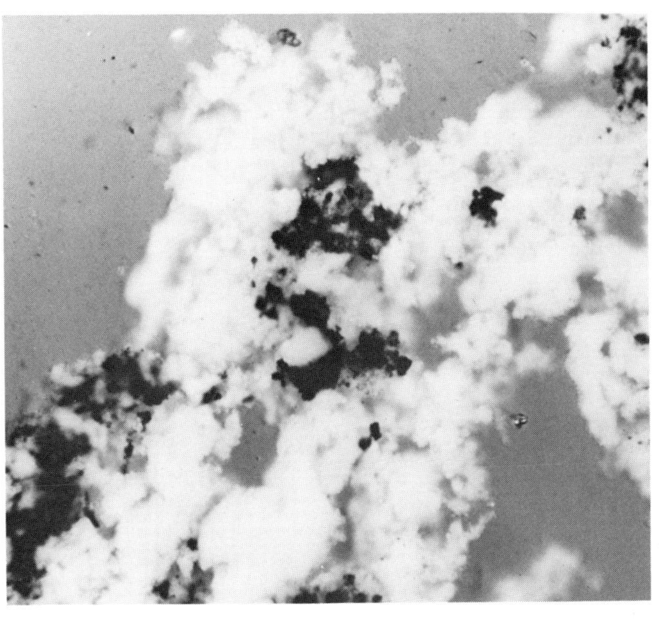

Figure 30. Same as Figure 29 but with top light as well as transmitted light

Figure 31. Ethyl cellulose, transmitted light only

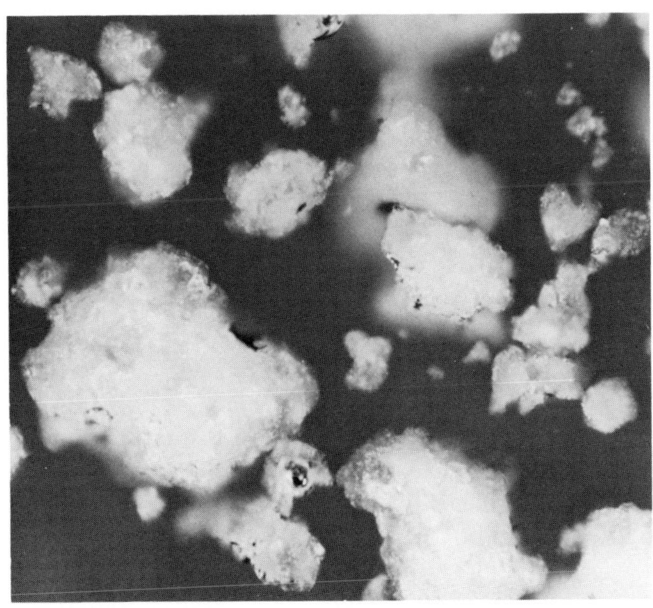

Figure 32. Ethyl cellulose, top light as well as transmitted light

Obviously, the particle analyst must use top lighting to fully characterize and identify any particle. What are the various approaches to this problem and which should he use?

There are two general procedures for illuminating the top surfaces of particles. One makes use of built-in or attachable light sources to obtain either incident brightfield illumination or incident darkfield illumination (Figure 33). This may be called vertical illumination, although, strictly, only the incident brightfield uses vertical light rays. The second general procedure for top lighting makes use of auxiliary illuminators; this is generally called unilateral illumination.

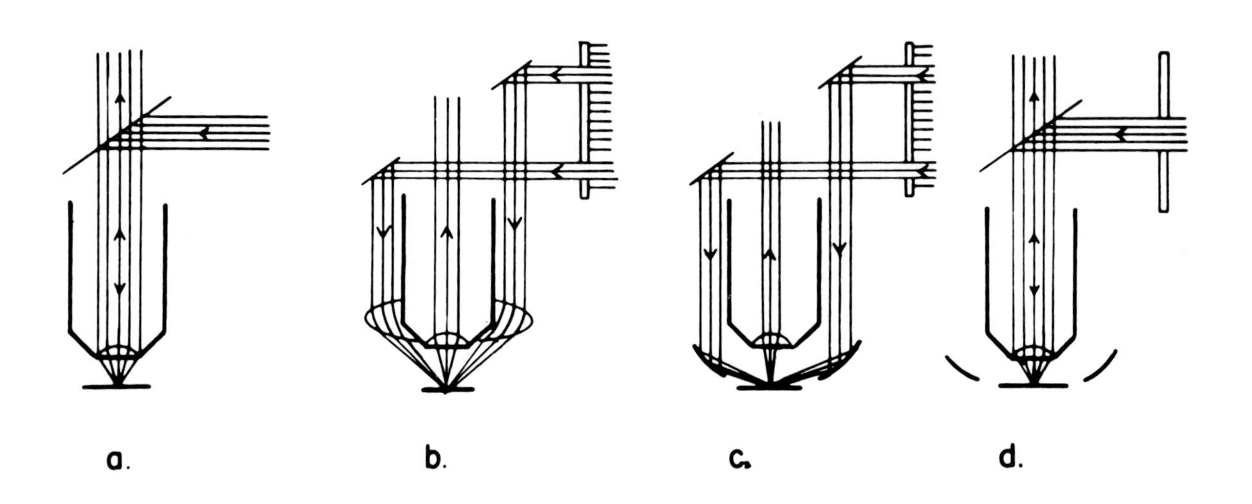

Figure 33. Vertical illuminators

a. Unilateral illumination

An auxiliary light source may be added to any microscope since one or more external illuminators may be placed at the side with the light beams directed downward from above the microscope stage, past the objective, onto the top of the particles. Such lamps must deliver a concentrated spot of light and they must be used at as high an angle to the horizontal as possible. The light, in any case, undergoes some reflection from the upper surface of the coverslip. Less light is lost if the angle of incidence is high. There is, of course, no such problem if the prep is uncovered. Particles on a membrane filter, or spread loosely on a slide, can be examined best by top lighting. A very dilute collodion "paint" (ordinary three-percent collodion solution diluted 100 times with ethyl or amyl acetate) of the particles may be prepared, spread on a slide and allowed to dry. The particles are thus attached to the slide but no visible collodion film remains. This permanent dry prep may be stored or it may be examined at once. The collodion does not interfere with examination of the particles mounted in various refractive index media. Any desired liquid is added, a coverslip is placed over the prep and the sample examined in transmitted light. The liquid can be washed away with benzene and the resulting dry prep is ready for a second liquid, dry examination with top lighting or for storage.

The objective may obscure the light path if the illuminator is set too high or if objectives with short working distances are used. Special objectives designed to permit passage of high-angle incident light are, however, now on the market. Usually, examination with a top light is limited to 16-mm 10X objectives or to those with lower power. This means a limiting total microscope magnification of about 200X. Anyone who wishes to use top lighting for particles too small to be observed properly at 200X has a choice:

1) Ignore the fact that the high-power objective obscures the direct light, use it anyway.

2) Use special long working distance objectives.

3) Use vertical illuminators incorporated into the microscope.

Each of these requires some further explanation, especially the first. It turns out that the light will enter the narrow slit between objective, e.g., 4-mm 45X, and coverslip , where it will be reflected back and forth. Enough light bounces around and finally finds its way to the particles so that the upper surfaces can be studied. It helps to use two auxiliary illuminators and it may be necessary to turn off the transmitted light beam in order to see the reflected light.

A variation of this technique that does not require an auxiliary illuminator involves intense, high NA, substage illumination with crossed polars. The objective reflects enough light to illuminate the tops of opaque particles. If the field itself is bright, the intensity of illumination on the particles is too low to register on the retina but if the field is darkened by crossing the polars, the color and shape of opaque particles can be seen with the 4-mm 45X objective.

Why, then, use any auxiliary or special vertical illuminators? The answer is twofold: top lighting is less intense with the multiple reflection method, and this method is less convenient. Properly arranged top lighting of sufficient intensity permits simultaneous observations with transmitted polarized and top lighting and thus saves time.

b. Long working distance objectives

Several companies manufacture long working distance objectives. These are usually either 8-mm 20X or 4-mm 40-45X. In every case, the working distance is equal to or greater than the working distance of a standard 16-mm 10X objective. It is therefore possible to obtain excellent top lighting using external light sources with these objectives.

c. Auxiliary illuminators

Many manufacturers produce suitable spotlight auxiliary illuminators. The essential requirement is a focusable spot image of the filament about 8-10 cm from the lamp. The spot size should not be more than 5-10 mm across.

An interesting possibility is the use of a light pipe, usually constructed from polymethylmethacrylate (Lucite, Plexiglass and Perspex are trade names). This material, in rod form of any desired diameter, can be used to focus the beam from an illuminator. When warmed to about 100° C, rods can be bent and drawn down to small diameter. The lamp can then be any distance away and located anywhere relative to the microscope. The light pipe gathers the light, transmits it to the microscope and "focuses" it on the specimen.

d. Vertical illuminators

Many manufacturers furnish attachments for the microscope fitting between the objective and bodytube so that light can enter the bodytube from one side and be reflected and focused through or around the objective. In some, such as most metallurgical objectives, a plane glass plate reflects the light straight down through the objective for incident brightfield illumination (Figure 33a). Others, such as the Leitz Ultropak (Figure 33b), have the light reflected around the outside of the objective and then refracted onto the specimen for incident darkfield illumination. The light is refracted by a lens with a hole cut in the center through which the objective protrudes. This condensing lens is focusable. Still other vertical illuminators allow either bright- or darkfield llumination, e.g., "Epi" type. These have some advantages over the other types in that either incident darkfield or incident brightfield are possible with a single objective. Only the reflecting mirror above the objective is changed (Figure 33 c, d).

All objectives intended for use with vertical illuminators are corrected for use without a coverslip, as they are intended primarily for uncovered, polished metal specimens. The incident darkfield objectives may be used with success with covered specimens, just as with unilateral illumination. Incident brightfield objectives, however, cannot be used with covered specimens, as the light coming from the objective will be reflected from the top of the coverslip filling the field with an intense glare.

Setting up illumination with incident brightfield objectives follows Köhler's method just as with transmitted light (Chapter IV. B. 2). The vertical illuminator will have two diaphrams, one a field diaphram and the other an aperture diaphram. When setting up the illumination it is best to use a plane, specular surface as a specimen. The field diaphram is closed somewhat and the specimen is brought into focus. It will be seen that the field diaphram is in sharp focus — the objective lens is also acting as the light condenser lens. The field diaphram is centered and then opened until it is just outside the field of view. The lamp filament is focused on the aperture stop (a ground glass will aid this step if the iris is not conveniently located). Using the Bertrand lens, or removing the ocular, center the aperture stop in relation to the objective aperture. Some vertical illuminators have provision for tilting the plane glass reflector. The specimen is now observed and the aperture stop set according to the requirements of the specimen (resolution, contrast).

For use in quantitative polarized light work, a right angle prism is used instead of the plane glass reflector. This gives more light and does not depolarize the beam but, because it occludes part of the objective aperture, it may reduce the resolution. With this reflector, the aperture stop is closed and centered on the prism.

Objectives for use with vertical illuminators, in addition to being corrected for use without a coverslip, are corrected for a longer tubelength because of the interposition of the reflecting elements. This distance may be 180-210 mm (185 mm for the Ultropak).

A crude but effective vertical illumination system can be quickly set up by arranging about half an 18-mm coverslip in a bit of modeling clay and fixing it at a 45° angle to the edge of the objective (Figure 34). The coverslip may also be fixed with clay to the slide rather than the objective. A very small segment of a coverslip will fit beneath 10X objectives.

Another interesting top light is a one-inch diameter ring of "neon" tubing. Aristo Grid Lamp Products, Inc.* can furnish one of these with transformer or you can have a signmaker make one for you. It can be arranged around the objective so that light strikes the particles from all sides and is reflected back into the objective. Such an illuminator gives a bright image but without shadows. This "flattens" the particles and may not give good shape interpretation. Request the signmaker to give you a green glow rather than the red neon glow. Along the same lines, some microscopists[4] use half a ping-pong ball lying on the stage over the preparation, with an opening in the top through which the objective peers. One or more auxiliary spot illuminators can be focused on the outer surface of

---

* Aristo Grid Lamp Products, Inc., 65 Harbor Road, Port Washington, L.I., NY 11050.

4. Albertson, C.E., "A light-tent for photomicrography," Microscope 14 (7), 253 (1964).

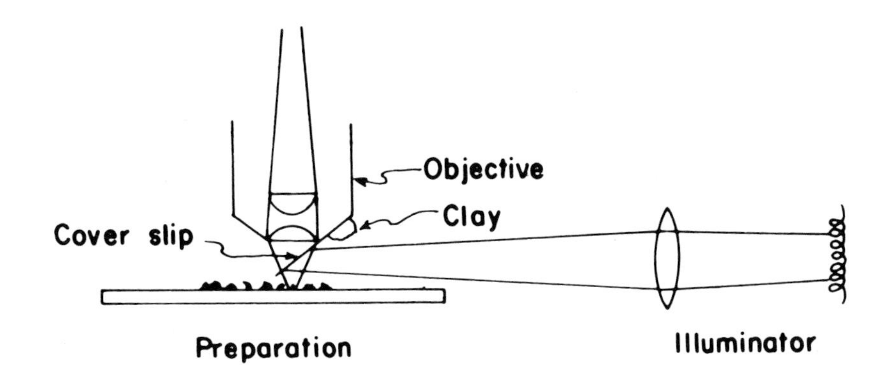

Figure 34. A makeshift illuminator for low power

the ping-pong ball to illuminate the specimen. All shadowing is, of course, lost and the particles are flattened, much of the surface texture is also lost.

The choice of system depends, to some extent at least, on the size of the budget. Entirely satisfactory results can be obtained with either built-in vertical illumination or auxiliary illumination. Reflected light intensity is less with auxiliary illumination but this causes inconvenience only in photomicrography. Since the proper exposure for top light is usually longer than for transmitted light, a double exposure is usually required.

With incident illumination there should be no problem in differentiating paint spray from oil soot or metal shavings from insect parts. To fully identify small particles, it is necessary to use all obtainable information. Often the use of top lighting makes unnecessary the use of more sophisticated tools, such as electron microscopy, x-ray diffraction or electron microprobe analysis.

C. LIGHT FILTERS IN VISUAL MICROSCOPY

Light filters are used in microscopy to 1) improve resolution, 2) insure correct particle color identification, 3) increase contrast, 4) reduce glare and discomfort to the eyes, 5) reduce light intensity, 6) cool the light beam and 7) alter or control spectral emission for a variety of other visual and photographic purposes.

The principles and requirements of light filtration are different in visual microscopy than in photomicrography because the human eye and photographic film respond differently to the same kind and quality of light, even though the purpose of filtration may be the same for both.

1. Daylight Filters

The most important light filter for visual microscopy is the blue, or so-called daylight filter. This is usually supplied with microscopes but it is remarkable how many microscopists ignore it. This filter is intended to absorb excess red and yellow from tungsten sources, giving the light an approximate daylight quality. Use of a tungsten source without this filter is excusable in the biomedical microscopist who depends upon morphology (in the strictest sense) only for identifying tissue elements. Even here, though, prolonged use of the microscope with this quality of illumination is fatiguing at best.

Tissue sections are stained either for visibility or for differentiation; precise color differentiation is almost never an important criterion in tissue identification. But for the particle analyst, colors — even very subtle colors — are extremely important, so it is essential that they not be masked by excessive red and yellow from unfiltered tungsten sources. The pale yellow of some catalysts and food products, for example, is undetectable in unfiltered tungsten light. This applies to top lighting also; daylight quality illumination should come from top lights. It is, at the same time, possible to get light that is too blue. The light should therefore be filtered so that the illumination approximates daylight, neither too yellow-red, nor too blue, but this is complicated by the human eye which acclimatizes to different hues.

Try this experiment. Using a low-voltage, coil-filament tungsten light source, or a ribbon filament lamp set to suggested operating voltage, introduce Kodak Wratten 80A filter in the light path and view a general air pollution sample. It will appear either perfectly correct or somewhat blue. View the specimen for a couple of minutes, then change to an 80B filter. You will notice immediately that, regardless of the original evaluation, the light is now definitely too yellow. This new impression becomes somewhat doubtful after a couple of minutes, and if the 80A is now reintroduced, the light is seen to be definitely too blue. If the 78A replaces the 80A, the light is seen to be definitely too red but, again, after a couple of minutes, it seems to be perfectly adequate and both the 80A and 80B are too blue! In short, blue is not blue; there are many blues and the human eye will adjust to almost any of them. Actually, with a given light source operated at a standard voltage, a blue filter can be found which, by repeated comparison with other blues, will be found ideal, even if the exact reason for selecting it over the others cannot be stated in concrete terms.

How does one go about selecting the correct blue ? To begin with, manufacturers supply blue glasses, but there is little similarity between the blues of different manufacturers, or even between the blues of the same manufacturer purchased at different times. (Incidentally, the blue filters supplied with microscopes are usually not intended as conversion filters for photomicrography with Daylight Type Color Film though some are satisfactory.) It can, however, be said that most of them are too blue. Besides this simple blue glass, some manufacturers offer at additional cost a daylight filter. This is a Corning glass filter, C.S. 1-62 (5900). With most light sources, this filter will be found too blue if on the thicker side of stock thickness (4.2-5.9 mm). See the Corning catalog, Glass Color Filters (Bulletin CF-1, Corning Glass Works, Optical Sales Department, Corning, New York) and Figure 35 for transmission curves. Corning makes other blue filters but this is by far the best for visual microscopy.

Kodak Wratten filters are very useful to the microscopist. The Kodak booklet on filters for scientific and technical use, which gives the transmission curves for all of the filters, is essential for the microscopist's library. The blue 78AA is recommended by Kodak as the visual filter to convert the color quality of tungsten sources to the visual equivalent of daylight. It is a photometric filter designed originally for visual use with a photometer using a standard lamp of approximately 2360°K. We recommend it for viewing colored specimens with what is generally accepted to be their standard daylight appearance. For visual purposes, it can be considered comparable to the Corning 1-62 (5900°K), the 78AA peaks at about 440 nm, the 1-62 peaks closer to 410 nm but both are about the same in the yellow and red regions (see Figure 35). The 78A, also a photometric filter, has higher transmission in both the blue and the red but may profitably be tried along with the 78AA. The 80A and 80B, used photographically with tungsten sources for daylight film (see Chapter VI. D) can also be used for visual purposes. The 80A may be found somewhat blue, depending upon the light source. The 80B is comparable to the 78A in transmission characteristics (see Figure 35). The 80 series are light balancing filters. The 82 (bluish) series are also light balancing filters, frequently used with the 80 series. Visually, they are not as useful as the 80 series as they effect changes of color temperature in steps of about 100°K. The color compensating filters are of little moment in visual microscopy. It must be emphasized here that the preceding statements do not apply to photomicrography, only to visual microscopy.

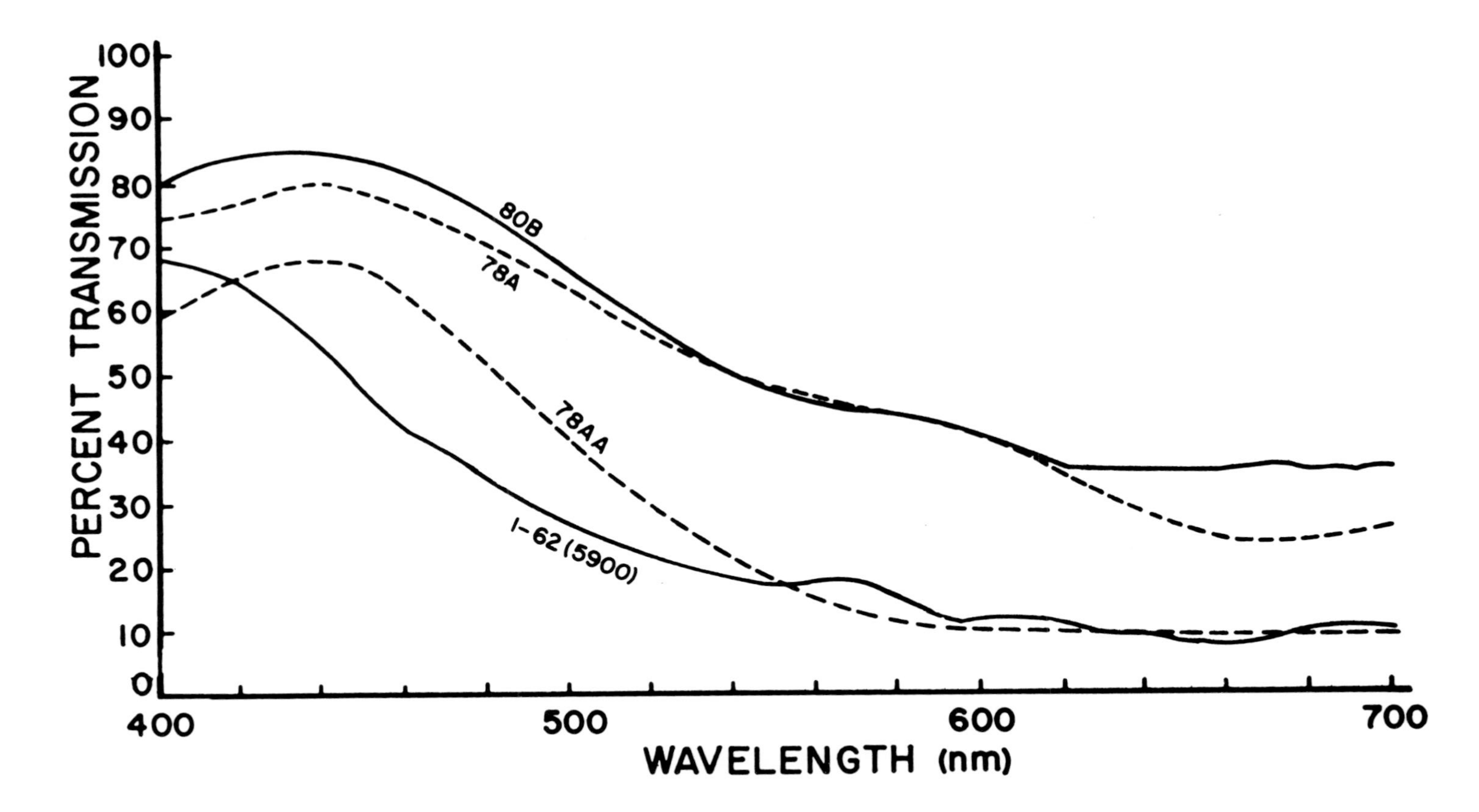

Figure 35. Transmission characteristics of four blue filters

In summary, illumination of daylight quality is essential in visual particle microscopy so that very pale colors will be true and not missed. The background with polars slightly uncrossed, i.e., about 15°, should be a pleasing neutral gray that will not cause discomfort after daily, prolonged viewing. Since tungsten light sources vary in color quality, depending on lamp age, voltage settings etc., it is not possible to recommend a particular blue filter. Besides, we are dealing with a psychophysical condition, so that much depends on individual judgment. The actual daylight blue filter used should be selected from a) the glasses supplied by microscope manufacturers (exact duplication should not be expected if breakage occurs), b) Corning C.S. 1-62 (5900°K) in the lower stock thicknesses, c) the photometric 78A and 78AA and d) the photographic filters in the 80 series — say, 80A and 80B.

2. Neutral Density Filters

After the blue, the most important filter in visual microscopy is the neutral density filter. Controlling light intensity with neutral density filters is preferred to reducing the voltage because the proper filters do not change the light color temperature. (The substage iris should never be used to control light intensity.) The neutral density filter should reduce light intensity without altering the color balance. Therefore, the filter should have as flat an absorption curve as possible, neither absorbing nor transmitting selectively anywhere in the visible spectrum, and it should also be available in several degrees of density. Among the finest filters of this type for both visual and photomicrographic use are the Inconel® neutral density filters supplied by Bausch and Lomb, Inc., Rochester, New York 14602. There are three neutral filters with densities of 0.7, 1.0 and 1.3 (20%, 10% and 5% transmission, respectively), available in 2-in. or 3-in. diameter. The more versatile are the 2-in. squares of which there are four: 0.3, 0.6, 0.9 and 1.2 (50%, 25%, 12.5% and 6.3% transmission, respectively). These may be used singly or combined to afford excellent control over light intensity. The four filters, together with a daylight filter, are available as a set (Bausch and Lomb 31-34-38-01). All other kinds of neutral density filters, i.e., those containing particles suspended in the glass, are suitable for visual work but many impart some color, and most are not recommended for color photomicrography. Inconel® filters have the advantage of reducing the number of filters necessary, as they are superior for both visual and photographic microscopy. In an emergency, the microscopist can reduce light source intensity by interposing, in the beam, screens (wire gauzes) of various meshes or stacks of plate glass or he can make a graded series of filters by photographing a white surface with black-and-white film at different exposure settings.

3. Contrast and Resolution Filters

Probably the next most important filters in visual microscopy are the contrast and resolution filters. Here, again, Kodak Wratten Filters are very convenient, useful and inexpensive. It is well known that resolution in microscopy depends largely on the wavelength of light used, resolution increases as wavelength decreases toward the blue end of the spectrum. The lowest practical wavelength is actually in the blue-green and Kodak Wratten Filter 45A was designed specifically for the highest resolving power in visual microscopy. Because it absorbs most of the visible spectrum, a strong light source is needed. The 45 has a somewhat narrower band, but higher transmission, than the 45A and is also good for resolution.

Belling[5] used the yellow-greens (Wratten series 66, 56, 57A, 58 and 61) for almost all of his work, preferring them to the light blue-greens and even the blues. Advantages of yellow-green light cited by Belling include a) restfulness to the eyes, b) improvement of corrections of objective and condenser, c) increased contrast of many stained specimens, d) regulation of light intensity, e) increased visual aperture because the red, blue and violet are cut out and f) improved acuity of vision because the chromatic errors of the eye itself are eliminated. For particle identification, con-

---

5. Belling, J., The Use of the Microscope, McGraw-Hill Book Company, New York, 1930.

tinued use of yellow-greens is undesirable but there may be occasional use for these filters, especially in connection with biomedical specimens.

Wratten filters recommended for contrast in visual microscopy include 38A (blue) for apparent contrast with faint yellow or orange specimens, 66 (light green) and 58 (green) for apparent contrast with pink and red specimens, 15 (deep yellow) and 22 (yellow-orange) for increased contrast in specimens stained with methyl green, methylene blue etc. In selecting contrast filters, choose a filter complementary in color to the object or stain. Thus, for pollens stained pink with basic fuchsin, a good choice of contrast filter is 58 (green).

4. Sodium Light Filters

A filter essential to every microscopist is one which isolates the sodium line, i.e., 589 nm. This filter is necessary for precise refractive index determination. Sodium vapor lamps may be used but filters are cheaper and entirely adequate. Orange glasses should be avoided because of their relatively broad transmission bands (see Figure 36). Gelatin filters are rather inconvenient here because, used singly, they have a rather broad band pass, they are heat sensitive and, unless cemented between glass, are inconvenient to handle. Interference filters are by far the best choice. One has to be careful, however, of pass bands which may be present in addition to the desired one. In second-order interference filters, pass bands occur wherever the optical spacing of the metallic films is any whole multiple of a half wavelength. For example, a nominal second-order 720 nm filter will have a third-order pass band in the blue at about 482 nm and a fourth-order pass band at about 368 nm. The transmission of the 720 nm band is about 38%; the transmission of the 482 nm band is about 58%. Such a 720 nm filter visually appears blue (see Figure 37). With a first-order filter, there are no pass bands on the long wavelength side of the nominal pass band, but there will be one at about one-half the wavelength of the nominal pass band. Some manufacturers use clear glass cover plates with these filters to extend the range of applications. Other manufacturers deliberately use colored glass cover plates to absorb all pass bands other than the one desired. Hence, it is better to obtain an interference filter of 589 nm in which the additional filter is an integral part of the entire filter. Such a two-inch square filter can be purchased from Optics Technology, Inc.* or one made by Schott Optical Glass, Inc.** can be obtained through Zeiss.

5. Filters for Dispersion Measurement (see Chapter X)

For dispersion measurement the microscopist may require — besides the 589 nm (D) filter — a 486 nm (F) filter and a 656 nm (C) filter. These interference filters together with integral gelatin filters to cut out unwanted pass bands are also available from Optics Technology. The three filters are used for determining refractive indices at their respective wavelengths. Alternatively, the microscopist may wish to purchase an interference wedge (gradation filter), made by Schott. This wedge covers the entire visible spectrum and is, in many ways, more convenient to use than separate filters but it is also more costly. The interference wedge is obtainable either from Zeiss or Leitz. An annular wedge interference filter as mentioned by Jones[6] may be used for dispersion and refractive index determinations.

---

* Optics Technology, Inc., 901 California Avenue, Palo Alto, CA 94304.

** Schott Optical Glass, Inc., Duryea, PA 18642.

6. Jones, F.T., "Dispersion and refractive index determinations by means of an annular wedge interference filter," Microscope 14 (11), 440-7 (July-August 1965).

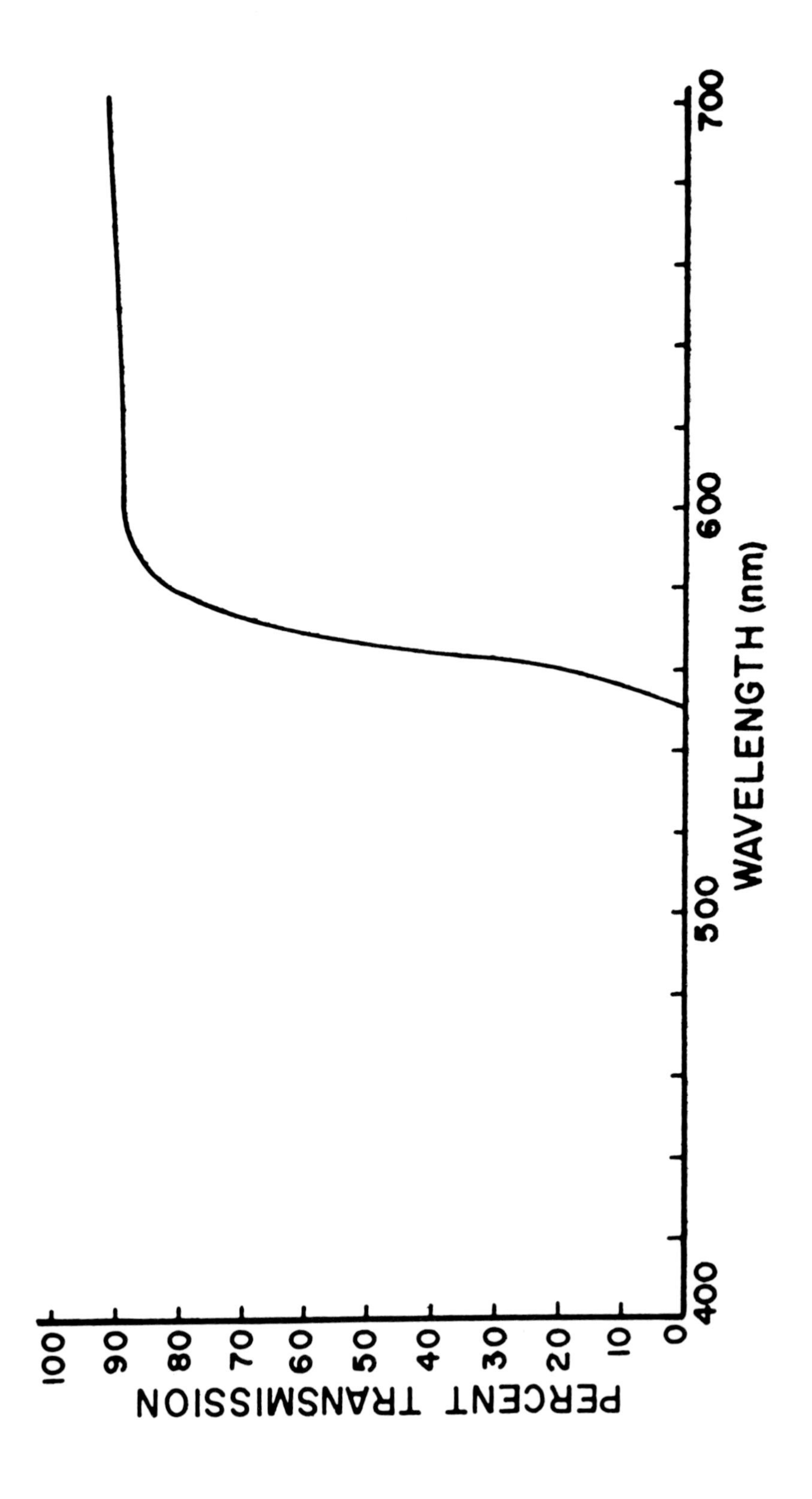

Figure 36. Transmission characteristics of an orange glass filter

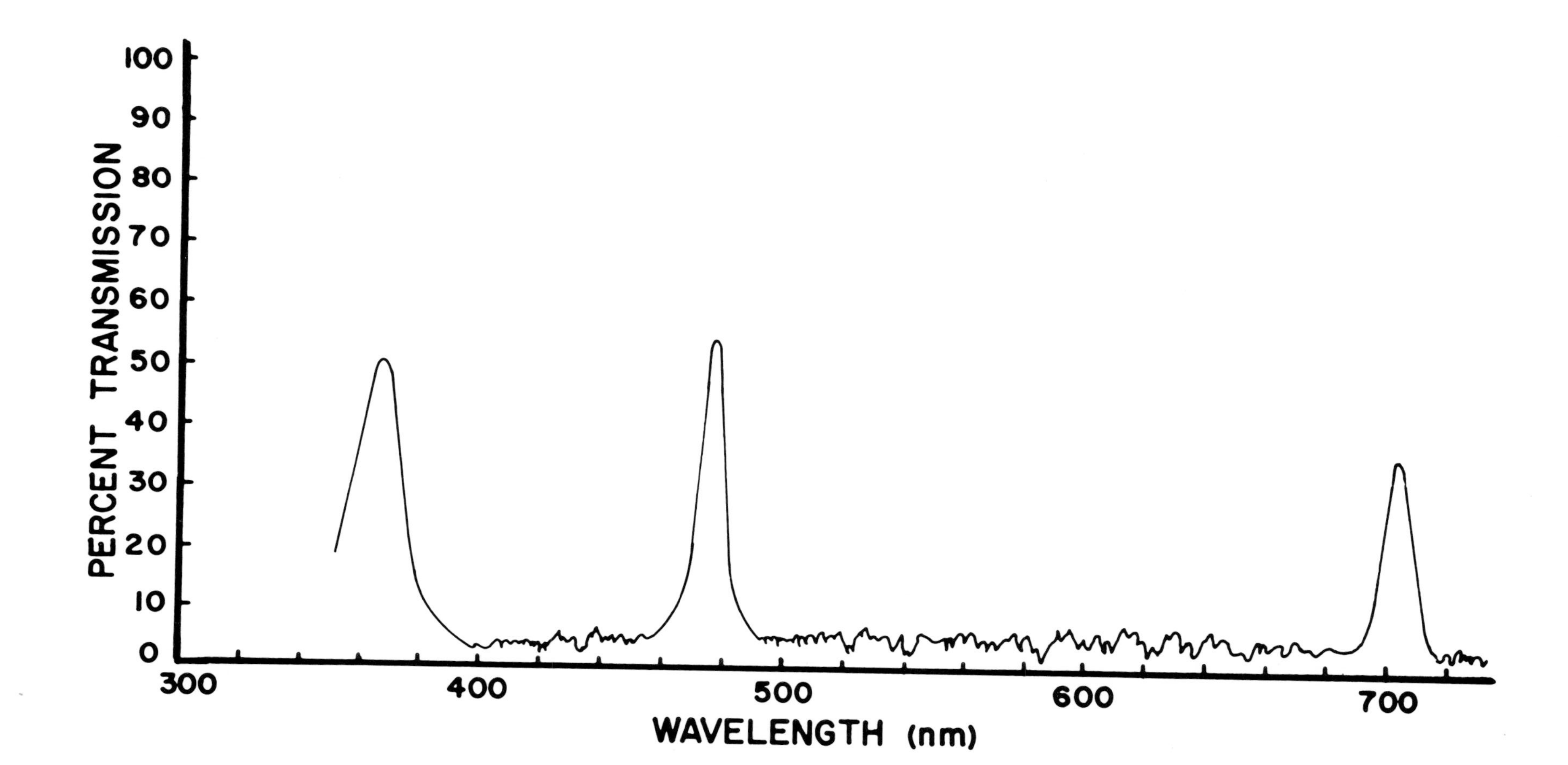

Figure 37. Transmission characteristics of a nominal 720 nm second-order interference filter

6. Glare Reduction Filters

One filter, very useful in visual microscopy but often overlooked, is the analyzer of the polarizing microscope used alone on top-lighted specimens. In some given orientation, determined by rotating either the analyzer or the specimen, the analyzer used alone will cut out objectionable glare from highly reflecting surfaces.

7. Didymium Filter

The didymium filter, so widely used for photomicrography of specimens stained with hematoxylin and eosin, can also be used to advantage visually. It is suggested[7] that biomedical specimens stained by the Ziehl-Nielson method are viewed to better advantage with the didymium and, that in darkfield illumination by the stop method, the filter darkens the background by absorbing scattered yellow light.

In addition to all of the above, special filters for visual use are required for phase contrast microscopy, fluorescence microscopy and for focusing in infrared photomicrography but these will be described in each of the respective sections.

8. Gelatin vs Glass Filters

Experiments with filters for visual microscopy should be made, if possible, with inexpensive gelatin filters. If the desired effect is achieved, either the same filters may be ordered, cemented in 2 x 2-in. B glass, or solid glass filters with the same or similar transmission characteristics, made by Corning[8], Chance[9] or Schott[10], should be purchased.

Filters should be kept away from excess heat or, alternatively, placed on the microscope side of a heat filter. They should be kept clean so that, if they are placed near the field diaphram, out-of-best-focus dirt particles will not mar the image. Finally, filters should be stored in filter safes or plush boxes when not in use.

7. Crossmon, G., Bausch & Lomb Focus 37 (2), 28 (Fall 1966).

8. Corning Glass Works, Optical Sales Department, Corning, NY.

9. Chance-Pilkington, St. Asaph, Flintshire, England. In U.S.: Alfa American Corporation, 5420 Walker Avenue, Rockford, IL 61111.

10. Schott Optical Glass, Inc., Duryea, PA 18642.

V. SPECIALIZED MICROSCOPE ILLUMINATION

A. POLARIZED LIGHT

Some of the most valuable observations possible with the light microscope are those made with polarized light. Ordinary unpolarized light consists of a bundle of rays having a common propagation direction but different vibration directions. Polarized light, on the other hand, has a single vibration direction. It is very important to remember that the vibration direction is always perpendicular to the propagation direction.

Polarized light is produced from ordinary light by reflection, by double refraction with a suitable crystal or by absorption, again with a suitable crystal.

1. Polarization by Reflection

Reflected light is usually almost completely polarized. The completeness depends on the composition (more directly, on the optical properties) of the reflecting surface and on the angle of incidence. Metallic mirrors are not good polarizers, conversely transparent substances at a characteristic angle of incidence may be nearly 100% effective as polarizers. For each transparent substance, the particular angle giving maximum polarization is the Brewster angle. The reflected and refracted rays are perpendicular to each other (Figure 38) and the mathematical expression is known as Brewster's law:

$$n = \frac{\sin i}{\sin r} = \frac{\sin i}{\sin (90-i)} = \tan i \tag{14}$$

where n is the refractive index of the medium of refraction; air is assumed to be the medium for the incident ray

i is the angle of incidence

r is the angle of refraction.

The angle for maximum polarization, i, is about 57° for most common optical glasses with refractive indices of about 1.520. The angle increases slowly as the refractive index increases, becoming about 60° at 1.75.

2. Polarization by Absorption

Crystals of most colored substances show different colors in different directions. That is, if one examines such crystals with polarized light, the transmitted light color observed due to absorption of light will vary with the position of the vibration direction (Figure 39). Such substances are called pleochroic. A pleochroic compound showing very strong absorption in one direction and very weak absorption perpendicular to that direction would transmit polarized white light. The plane of vibration corresponds to the weak absorption direction (Figure 39). Early polarizing sheets were composed of such crystals, all parallel. Herapathite (iodoquinine sulfate) is one such substance. It should be remembered that polarization by absorption depends on all visible wavelengths being absorbed but strong polarization in the visible range may not necessarily mean strong polarization in the ultraviolet and infrared ranges.

3. Polarization by Double Refraction

All crystals except those in the cubic system show double refraction. This means that a light ray incident on such a crystal will split into two components traveling at different velocities and having mutually perpendicular vibration directions. Each component is polarized and travels through the crystal at a characteristic velocity and with a vibration direction perpendicular to that of the other component. If, by some means other than internal absorption, one component can be diverted, the remaining transmitted beam will be plane polarized.

It is easiest to do this with a substance having a very large difference in its refractive indices. Calcite is usually used, cut in such a way that one component is totally reflected into the black mounting and absorbed (Figure 40). Such a device was first used by William Nicol in 1828.

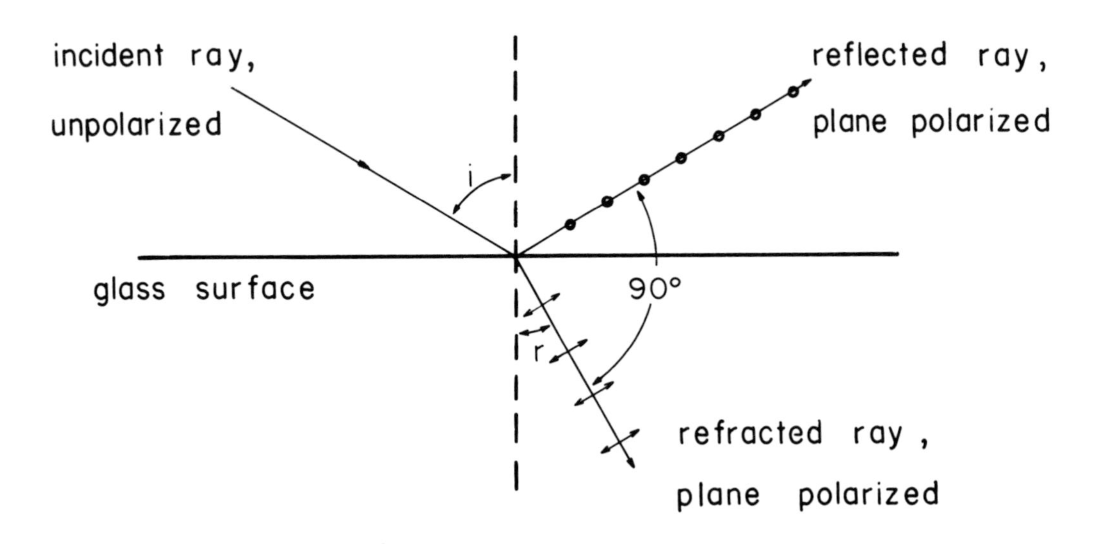

Figure 38. Polarization by reflection at the Brewster angle

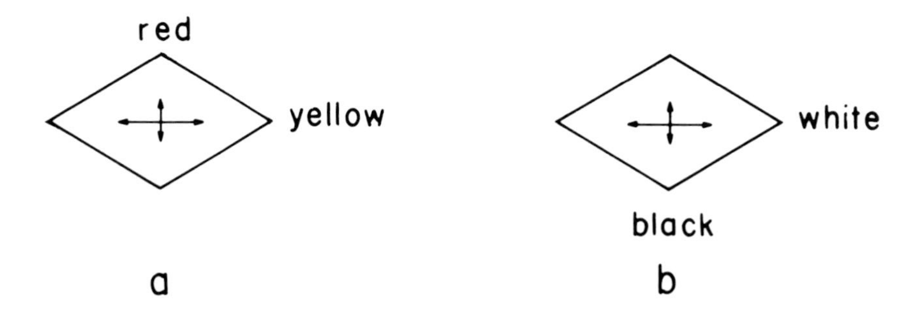

Figure 39. Pleochroism; absorption of white light by a crystal:
a) moderate difference in absorption, b) strong difference

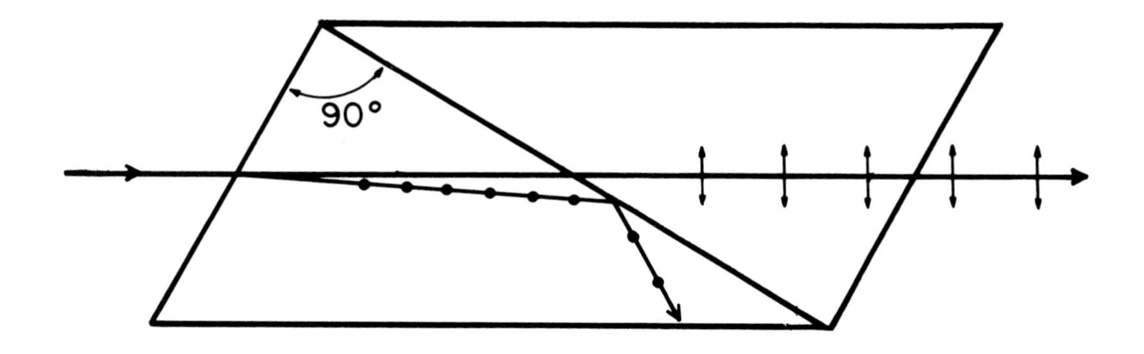

Figure 40. Schematic view of a Nicol prism showing transmission of plane polarized light after reflection and absorption on one compound

B. INTERFERENCE

In the same way that we considered diffraction (Chapter IV. A above) as a more general phenomenon in microscopy, we should also consider interference. Besides the obvious relationship with interference contrast and quantitative interference microscopy, the phenomenon of interference of light rays is the basis for image formation in general as well as polarized light microscopy and Zernike phase contrast.

Figure 41 shows a highly schematic representation of interference of light beams. The beam splitter may be diffraction as in microscope resolution or phase contrast microscopy or it may be double refraction as in polarized light microscopy and many forms of interference microscopy. The two beams must travel along paths of different optical length before recombination for interference to occur. In phase contrast the direct beams in an annular cone pass through a phase plate in the objective back focal plane. This phase plate introduces a path difference in the direct beams as

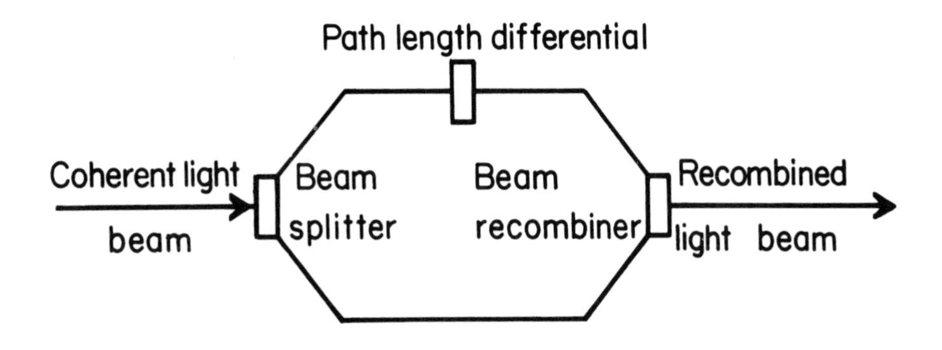

Figure 41. Requirements for interference between light beams

compared with the diffracted beams. In polarized light microscopy the doubly refractive specimen itself splits the light beam into two beams traveling with two different velocities through the specimen, thus the optical paths are different. Similarly, in interference microscopy, the beam splitter is a doubly refractive crystal plate below the specimen and the specimen detail introduces the phase differences.

Recombination is by image formation in phase contrast, by the analyzer in polarized light microscopy and by a second crystal plate in interference microscopy. Phase contrast is covered in Chapter V.C., polarized light microscopy in Chapter IX.

Interference phenomena are best understood by referring to the wave characteristics of light. Two light rays originating from a single source but which travel by different paths and arrive simultaneously at a point, give rise to interference effects. The intensity at that point will vary from dark to bright, depending on the phase differences between the two light rays.

1. Destructive Interference

The first requirement for interference is that the light, though split into any number of paths, must come from a single coherent source. Two light rays from a coherent source arriving at a point in phase agreement will reinforce each other, i.e., constructive interference (Figure 42). If they arrive completely out of phase they show destructive interference, i.e., they will cancel each other (Figure 43). Two light rays from a single source can interfere destructively by various mechanisms. One is reflection of a light ray by a thin, transparent film. One ray is reflected from the top surface, one from the bottom surface. The distance traveled by the latter ray in excess of the first is twice the thickness of the film, assuming normal incidence to the film (Figure 44). A graph expressing the thickness of a thin film in terms of refractive index of the film and the interference color is given in Particle Atlas Two, Volume I, Part 3, Section III. F.

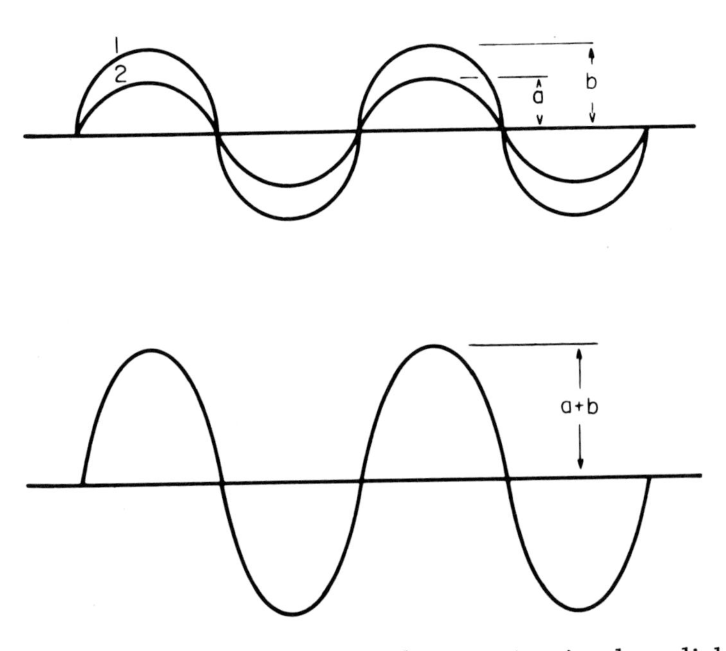

Figure 42. Constructive interference between two in-phase light beams

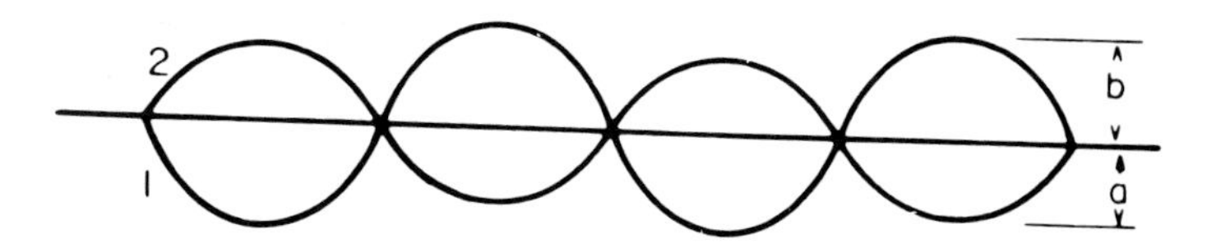

Figure 43. Destructive interference between two out-of-phase light beams

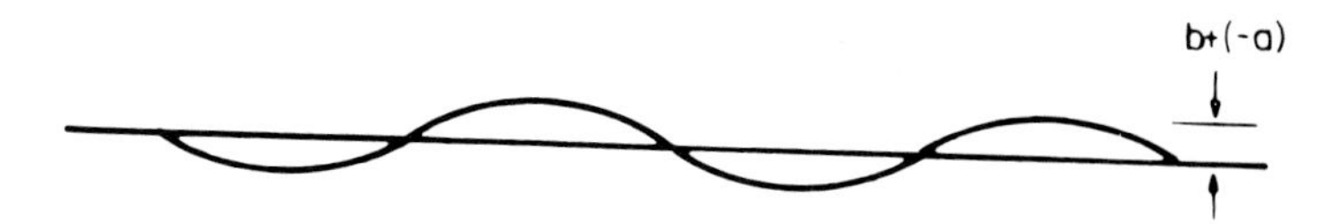

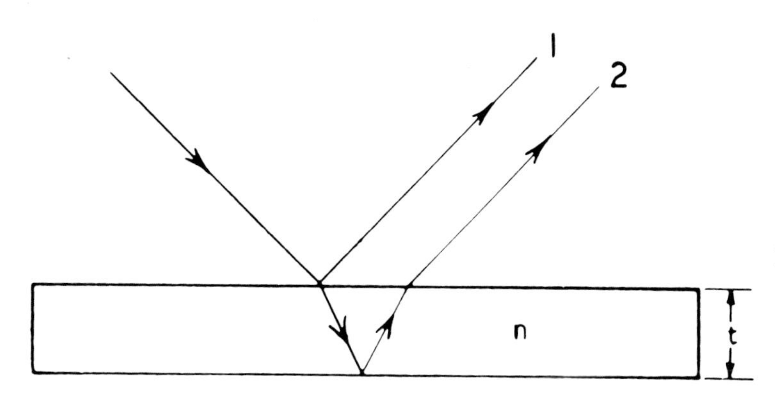

Figure 44. Thin film interference

An abrupt half-wavelength shift occurs on reflection from a denser medium back into a rarer medium. It occurs at those interfaces where the light approaches from the low refractive index side. On reflection from a film having a low refractive index, the second ray will undergo a half-wavelength shift upon reflection at the bottom surface and now reinforcement occurs when the optical thickness nt is $m\lambda/2$. The λ for which destructive interference occurs is given by:

$$\lambda = 2nt/m$$

where λ is the wavelength, n the refractive index of the film, t the thickness and m any integer.

A 175 nm film having an index of 1.5 will show destructive interference for $\lambda = 2(1.5)(175)/1 = 525$ nm or green. Since white light minus green is red the film will show a red interference color.

2. Simple Microscope Interferometer

A simple interferometer can be made by partially silvering a microscope slide and coverslip. A preparation between the two surfaces will show interference fringes when viewed with monochromatic light, either transmitted or from a vertical illuminator. The fringes will be close together with a wedge-shaped preparation and will reflect refractive index differences due to temperature variations, concentration differences, different solid phases etc. The method has been used to measure quantitatively the solute concentration around a growing crystal [11] (Figure 45).

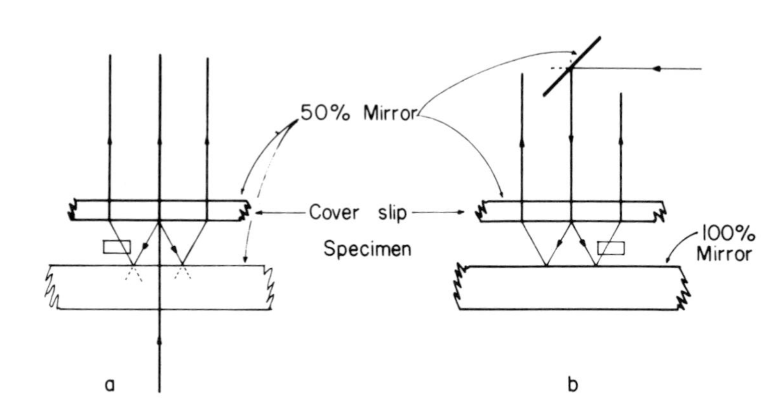

Figure 45. Microscopical method of viewing interference images by transmitted light (a) and by reflected light (b)

Each dark band represents a retardation equal to an odd number of half wavelengths. Conversely, each bright band is the result of an even number of half wavelengths of retardation.

With interference illumination, the effect of a transparent object of refractive index different from the medium in the microscope field is:

a. a change of light intensity of the object if the background is uniformly illuminated (parallel coverslip), or

b. a shift of the interference bands within the object if the background consists of bands (tilted coverslip).

The interference phenomena discussed here have been used as the basis for constructing special microscopes for the qualitative and quantitative examination of specimens possessing little contrast. Some of the microscopes utilizing interference phenomena are the phase contrast, differential interference contrast (Nomarski), transmission interference (Jamin-Lebedeff) and others (Dyson, Smith, Tolansky, Mach-Zehnder, Mirau).

3. Nomarski Differential Interference Contrast

The differential interference contrast (DIC) microscope after Nomarski (Figure 46) is a qualitative instrument like phase contrast microscopy for making more visible those phase specimens lacking contrast with ordinary brightfield. DIC microscopy does not replace phase contrast but complements it. Phase contrast is suitable for phase specimens with optical path differences of up to $\lambda/2$ but works best with specimens of $\lambda/10$ retardation. Greater path differences yield a more

11. Bunn, C.W., "Crystal growth from solution," Discussions of the Faraday Society No. 5, 132, Gurney and Jackson, London (1949).

pronounced halo which makes image interpretation difficult. DIC microscopy is best with specimens of path difference between λ/10 and 1λ. Thus, it is intermediate between phase contrast and conventional brightfield.

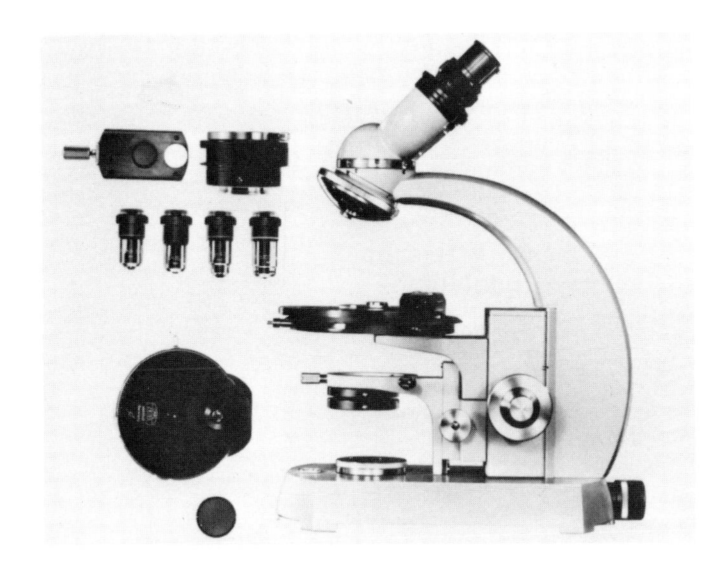

Figure 46. Zeiss Nomarski interference contrast optics for transmitted light (Photograph courtesy of Carl Zeiss, New York)

The interference contrast image is characterized by a distinct relief effect and shallow depth of field. The apparent three-dimensional image arises from the shearing effect of the Nomarski prisms. The shear is of the same order of magnitude as the objective resolution or even less. The recombination and interference of the resulting two beams creates a shadow effect. The full NA of the objective is utilized so field depth is small. This shallow field depth adds to the illusion of the relief image and makes possible optical sectioning without interference from structures below and above the plane of focus.

One must be careful in interpreting the DIC image, as it is tempting to believe the 3-D image is a true and accurate representation of the specimen. This may not be so; a "hill" may not really be a hill. It may, in fact, only be an optical discontinuity in a flat film. One observes a bright side and a dark side to each structure. These represent the phase gradients and it is possible to determine if a structure has a refractive index greater or lesser than its surround by observing this effect and comparing with a known. There is, furthermore, a "preferred orientation" in that, phase gradients will be visible only if they are aligned in the splitting direction of the system. Therefore, a rotating stage is necessary so that each structure may be optimally oriented, else adjacent structures with phase gradients at right angles to each other will not be accurately represented.

The differential interference contrast microscope is a two-beam instrument like all of the other interference microscopes. It is not quantitative because there is no visible separation of what would be its "reference beam" and "specimen beam". In a Michelson-type interference microscope, for example, we have complete beam separation and the reference beam is not affected by the specimen. In the Jamin-Lebedeff interference microscope the reference and specimen beams appear in the field of view at the same time and the beam separation ranges from about 56 μm (100X objective) to 560 μm (10X objective). The shear in the plane of the object in the DIC microscope ranges from 0.22 μm (100X objective) to 1.32 μm (16X objective). Since the shear is less than the resolution of the objective, the beams cannot be referenced to each other for quantitative purposes.

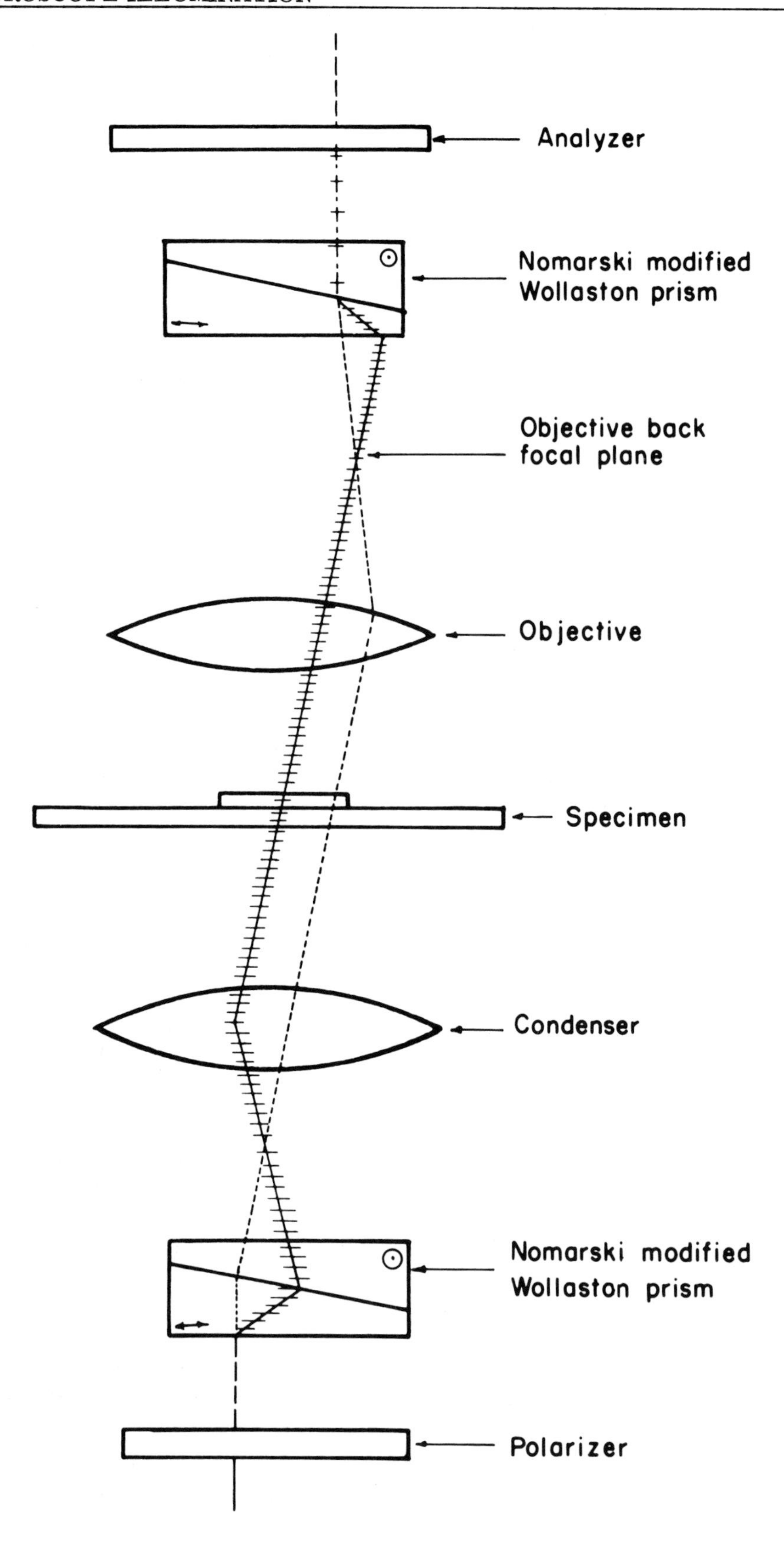

Figure 47. Differential interference contrast (Nomarski) microscope; showing exaggerated light path

The Nomarski differential interference microscope was developed during the middle 1950's. In construction it requires, first of all, a polarizer in the optical path (Figure 47). The plane polarized light thus produced is incident on a Wollaston prism modified by Nomarski. This prism (quartz prisms cut with vibration direction oriented as indicated in Figure 47), which is located beneath the condenser and at 45° to the plane of polarized light, splits the incident beam into two mutually perpendicular components. The two beams travel parallel to each other from the condenser and specimen to the objective. A second Nomarski-modified Wollaston prism behind the objective recombines the two components. The analyzer next makes the vibration directions parallel so that the light rays can interfere at the intermediate image plane. The close spacing of the separated beams from the lower prisms is what causes the image to appear in relief. The degree of relief effect depends on the optical path differences at the interfaces of different structures.

The lower prism is computed to give optimum contrast with a given objective. A different objective, therefore, requires a different prism. In practice, three are supplied in a special rotating condenser mount for the 10-16X, 40X and 100X objectives. The upper prism is used for all three lower prisms. The condenser mount holding the three DIC prisms is usually also provided with a brightfield position and one or two phase contrast annuli, so that it is possible to do brightfield, darkfield, phase contrast and DIC microscopy with the same condenser. The upper prism may have the analyzer incorporated with it on the same interference contrast slide where the entire unit replaces the analyzer of a polarizing microscope (Zeiss) or the slide may bear the prism only when the analyzer is located elsewhere above the prism slot (Leitz).

The condensers used in DIC microscopy must be strain-free and they are also usually achromatic-aplanatic. The objectives are planachromatic (Zeiss) which may also be strain-free (Leitz).

The DIC system is available for opaque specimens using incident brightfield objectives. Only one Nomarski prism is used in the objective mounts, as the same prism serves to separate the beams and recombine them.

To set up the Nomarski differential interference contrast microscope for transmitted light, one starts by crossing polars with the upper prism in place and with no objective, condenser or ocular. The polar is slowly rotated until the interference fringes in the back focal plane are sharp and the dark fringe lies in the center. The ocular, objective and condenser are inserted and the whole is set up for Köhler illumination. The lower prism for the appropriate objective is inserted and the upper prism is displaced by the knurled adjustment knob. A pronounced relief image of the specimen will be seen. It is possible to reverse the bright and dark sides of the image by displacing the upper prism one way or another. Which side the shadow lies on is arbitrary but it is customary to make the adjustment so that the shadows appear to be cast by the "setting sun", i.e., the upper side lighted, and the shadows cast on the lower right side. This will usually mean the phase-retarding structures appear in high relief (higher index than the medium) and the phase-advancing structures in low relief. The reverse is usually the case when Aroclor® (n = 1.66) is used as the mounting medium. If this is not done, most subjects will appear as depressions. Rotation of the adjustment knob to the extremes (maximum and minimum displacement or bias of upper prism) will introduce the colors of the Newton series and color contrast may be obtained.

4. Quantitative Interference Microscopy

The transmission interference microscope is a two-beam instrument developed on the ideas of Jamin (1868) and Lebedeff (1930). It provides qualitative information through contrast enhancement of phase specimens as does phase contrast and Nomarski DIC. In addition, however, the transmission interference microscope after Jamin-Lebedeff provides quantitative information about a specimen: phase difference, path difference, refractive index, thickness and volume and mass of particle.

Quantitative information about a specimen is possible because the separated reference and measuring beams can be seen in the field of view at the same time. The measuring beam only can be made to pass through the specimen. The system uses polarized light so that the measuring possibilities here (compensators etc.) are applicable to the interference image.

A polarizer (see Figure 48) provides plane polarized light to a beam splitter which in turn produces two beams due to its birefringence. The two perpendicular beams then pass through a $\lambda/2$ plate where their vibration directions are rotated 90°. This is done so that when recombined by an identical beam splitter above, they will interfere. The two beams pass parallel into the objective. One is a measuring beam passing through the center third of the field of view and the other is a comparison or reference beam passing through another third of the field of view. (The last third of the field of view will contain a ghost image with anisotropic specimens). If a specimen is placed in the measuring beam, its components will retard the light variably. The now retarded measuring beam and the reference beam are recombined by a beam combiner identical to the beam splitter. The interfering beams pass to the analyzer. The specimen and background will be seen as various colors when using white light, depending upon the tilt position of the condenser beam splitter.

The $\lambda/2$ plate and the beam splitter are mounted above the top lens of the condenser and may be tilted with the aid of a knurled knob adjustment. Tilting the beam splitter causes Newton's series of colors to pass through the field of view when using white light. The beam combiner is mounted directly in front of the objective front lens. It may be rotated about 5° for initial adjustment. Since each objective requires a different beam combiner, each objective requires either a different condenser (Zeiss) or a different beam splitter (Leitz).

For qualitative work, the stage is rotated and the beam splitter is tilted until the most favorable color contrast for the specimen is obtained. Areas in the specimen of the same color are the same refractive index if the thickness is the same, and vice versa. Path differences due to thickness or refractive index variations are indicated as differences in color.

The differences in refractive index and thickness which are seen as different colors may be quantitatively determined either separately, knowing one or the other, or simultaneously. This is possible because the measuring beam passing through the specimen and the reference beam passing outside of the specimen are sufficiently separated so that both appear in the same field of view and their colors can be compared. It will be recalled that in the Nomarski DIC microscope the beams are separated by a distance less than the resolving power of the objective. In the Zeiss Jamin-Lebedeff system, the maximum center-to-center beam separations are: 10X, 0.546 mm; 40X, 0.175 mm; 100X, 0.054 mm.

To make a quantitative measurement monochromatic light is needed, as the specific wavelength serves as the measurement standard. A wavelength of 546 nm is usually chosen because the $\lambda/2$ plate is corrected for that wavelength. Phase differences or path differences between the light passing through the specimen and light in the reference beam are measured with the aid of compensators. With the microscope in its initial position and the compensator in zero position, the background of the specimen (reference beam) is brought to maximum extinction by tilting the beam splitter while observing the specimen in monochromatic light. Maximum extinction may be determined by eye, by the half-shadow method, or by photometric means. The compensator (or analyzer when using Sénarmont compensation) is rotated until the specimen area of interest goes to maximum extinction (the background will become lighter). The angular rotation of the analyzer in Sénarmont compensation is related to the phase difference. Twice the reading of the analyzer is the phase difference. Thus, if the analyzer is rotated 30° for maximum extinction of the specimen area, the phase difference is 60°.

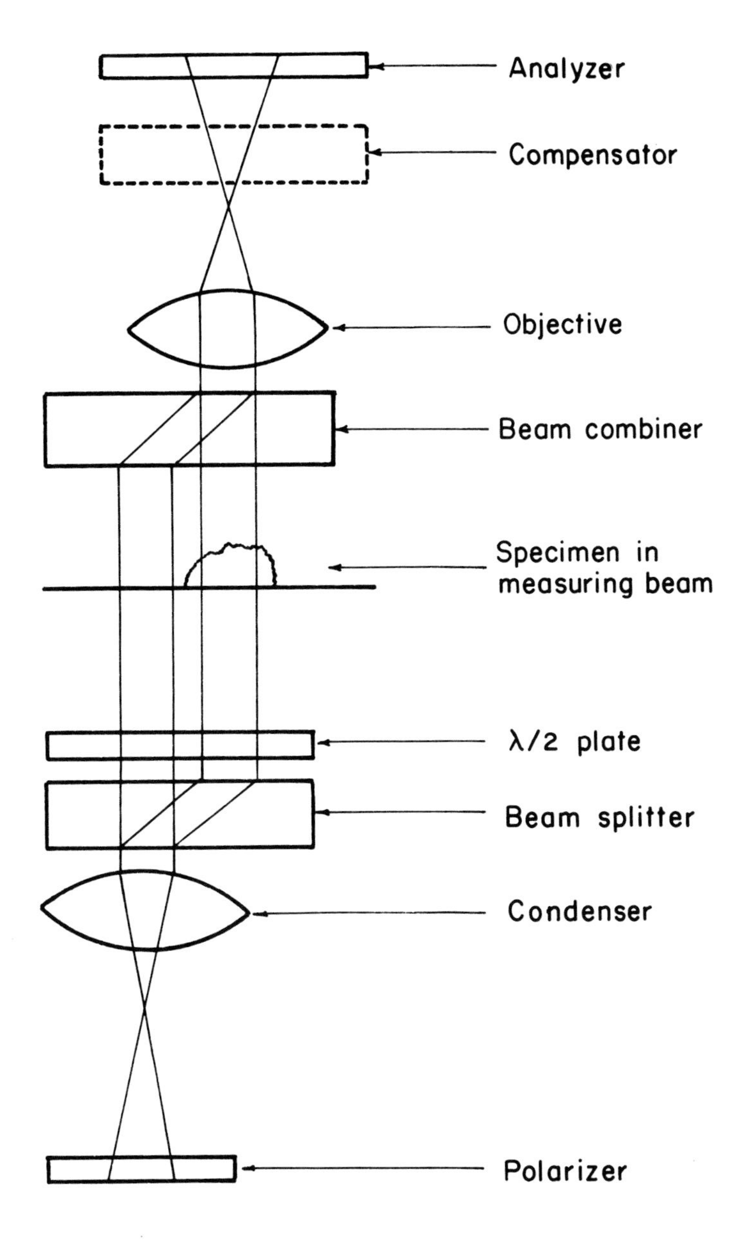

Figure 48. Jamin-Lebedeff interference microscope showing ray paths of measuring and reference beams

Before one makes a measurement, the magnitude of path difference is estimated from a Michel-Lévy chart. Path differences up to one wavelength are determined with a Sénarmont compensator, a $\lambda/4$ plate with lower index perpendicular to the transmission direction of the polarizer and rotating the analyzer. Path differences of this order may also be determined with a $\lambda/4$ rotary mica compensator. Measuring path differences less than one wavelength requires one of the Brace-Köhler rotating mica compensators: up to $\lambda/10$, $\lambda/20$ or $\lambda/30$. Path differences greater than one wavelength may be determined with a quartz wedge up to three or four orders or to more than 122 orders with an Ehringhaus rotary compensator. The number of whole wavelengths is determined by the color chart or with one of the compensators and then the fraction above the last whole wavelength is determined in monochromatic light by the Sénarmont method (Sénarmont $\lambda/4$ compensator and rotation of analyzer).

The mechanics of setting up each Jamin-Lebedeff system differ with each manufacturer's equipment. Practice in the manipulation and application of each system is necessary to achieve meaningful results.

Some of the basic equations involved in transmission interference (Jamin-Lebedeff) microscopy will clarify the role of the analyzer or compensator reading and the wavelength of the light used in obtaining quantitative information. First of all, the path difference, $\Delta\delta$, is the difference between the product of the thickness, t_s, and refractive index, n_s, of the specimen, and the thickness, t_m, and refractive index, n_m, of the medium:

$$\Delta\delta = (n_s t_s - n_m t_m) \tag{16}$$

The phase difference, θ, in degrees is:

$$\theta = (n_s t_s - n_m t_m)\frac{360}{\lambda} \tag{17}$$

or

$$\theta = \Delta\delta \frac{360}{\lambda} \tag{18}$$

where λ is the wavelength of the monochromatic light used — 546 nm in this system.

Practically speaking, this phase difference, θ in degrees, is simply twice the analyzer reading.

By transposition,

$$\Delta\delta = \frac{\theta\lambda}{360} \tag{19}$$

In other words, the path difference is about 1.516 times twice the analyzer reading. Example: Suppose a sample has been placed in the center of the field of view (measuring beam), the 546 nm filter has been placed in the light path, and the background has been brought to extinction by tilting the lower beam splitter. Suppose, further, that the Sénarmont compensator is placed in the compensator slot, and sample goes to extinction when the analyzer is rotated to 20°. Then the phase difference is:

$$\theta = 2 \times 20° = 40°$$

The path difference is:

$$\Delta\delta = \frac{\theta\lambda}{360} = \frac{40(546)}{360} = 60.6 \text{ nm}$$

Note that if the specimen thickness and medium thickness are the same, then

$$\Delta\delta = t\,(n_s - n_m) \tag{20}$$

This sets the basis for refractive index determination because if the thickness of the specimen is determined and the refractive index of the medium is known,

$$n_s = \frac{\Delta\delta}{t} + n_m \tag{21}$$

or, if two different media are used,

$$n_s = \frac{\Delta\delta_2 n_1 - \Delta\delta_1 n_2}{\Delta\delta_2 - \Delta\delta_1} \tag{22}$$

where $\Delta\delta_1$ is the path difference as measured in the medium with refractive index n_1 and $\Delta\delta_2$ is the path difference as measured in a second medium with refractive index n_2 (one medium may be air, $n_1 = 1$). Wavelength and temperature variation methods may also be used.[2]

Equation (20) can also be converted to:

$$t = \frac{\Delta\delta}{n_s - n_m} \tag{23}$$

in which case, thickness of the specimen may be determined. With unknown specimen refractive index, a second mounting in a different liquid is required,

$$t = \frac{\Delta\delta_1 - \Delta\delta_2}{n_2 - n_1} \tag{24}$$

Transmission interference microscopy has already been successfully applied to volume and mass determinations in biology and medicine and in textile fiber problems. The potential in particle characterization and identification and general industrial microscopy is very great.

The Mach-Zehnder transmitted light interference microscope and the incident light interference systems (Mirau, Tolansky, Michelson etc.) have not yet found meaningful application to particle characterization.

C. PHASE CONTRAST

Usually the microscopist studying microscopic particles can choose an appropriate mounting liquid to give ample contrast between particle and medium. The phase contrast microscope is a qualitative instrumental means of enhancing contrast, hence it is usually not helpful in particle studies. The microscopist who needs the phase microscope is one who does not have freedom of choice of his mounting liquid and whose combination of sample and mountant gives low contrast. Usually only the biologist is in this unhappy position. It is generally far better to enhance the inherent contrast in a specimen by choosing a liquid mountant with a refractive index very different from the specimen than to resort to a phase contrast microscope.

There are one or two situations, however, in which phase contrast may be helpful. For example, one might have very tiny particles for which the refractive index must be determined. If the particles are less than 1-2 μm in diameter, the Becke line may not be definitely indicative of the relative indices. In this case, use of phase contrast will permit easy observations of indices of the particles relative to the medium. With positive phase contrast (once called "dark"), particles having an index higher than the liquid will appear dark on a light field. Use of negative phase (once called "bright") or reversing the relative indices will result in light particles on a darker field.

1. Low Contrast Specimens

In addition to biological specimens and refractive index determination, the phase contrast microscope is useful for the examination of any specimen where the two refractive indices of particle and mountant are very close together. It is especially good for specimens which exhibit optical path differences of less than $\lambda/10$, although specimens with optical path differences up to $\lambda/2$ also benefit from phase contrast examination. Specimens with optical path differences greater than $\lambda/2$ are best examined with differential interference contrast (Nomarski) microscopy.

We recall that light travels as a sine-wave with a characteristic amplitude and wavelength. The human eye detects changes in amplitude (brightness) and wavelength (color). If such a wave passes through, say, a piece of glass, it will be slowed down, or retarded, with respect to a wave which does not pass through the glass and thus be out of phase with the nonretarded wave. Changes in phase such as this are not detected by the human eye. Most biomedical specimens can be termed phase specimens, as opposed to amplitude specimens, in that they consist of thin, transparent structures inside other thin, transparent structures. The refractive indices and thicknesses of the various structures do not differ by very much and the phase changes which take place with transmitted light are not detected. Lacking color and absorbance and not being able to detect the phase shifts, we find these specimens very difficult to examine. Staining generally kills and/or alters the specimen and stopping down the substage iris yields a poor image surrounded by diffraction lines.

2. Zernike's Invention

In 1932, F. Zernike, a Dutch physicist, discovered the phase contrast principle which enables the phase differences in a specimen to be converted to amplitude differences (light and dark), thus yielding a visible image of good contrast. The principle is based on the Abbe theory of microscopical image formation; a phase difference of $\lambda/4$ is introduced for the diffracted beams relative to the direct beam. Interference between these image forming beams on recombination in the final image is responsible for the enhanced contrast. After passing through a phase specimen (or any specimen for that matter) light consists of central, undeviated rays which form the zero-order diffraction spectrum in the back focal plane and diffracted, deviated rays which form the first, second and highest order diffraction spectra in the back focal plane (Figure 49). The light diffracted by a specimen is retarded $\lambda/4$ in addition to whatever retardation occurs due to the passage of the light through the sample. If the retardation could be increased by $\lambda/2$, the light passing directly through the specimen will destructively interfere with the diffracted light (Figure 49), producing amplitude differences. The phase contrast microscope increases the $\lambda/4$ retardation due to the specimen to $\lambda/2$ if the sample is sufficiently thin so that the path retardation may be neglected.

The Zernike method consists of an annular stop in the condenser and a phase plate in the objective (Figure 49). The annular stop in the condenser is focused by the objective to form an image of the annular ring of light in the back focal plane. There is nothing special about the condenser stop. It is simply a ring cut out of black paper or a disc of glass painted black except for a clear ring. The shape is not critical. Early stops were slits, and one manufacturer makes these in the form of a cross, but the ring annulus is the most generally satisfactory. The purpose of this annulus is to separate the direct rays from the diffracted rays.

The phase plate in the objective has a ring of absorbant material, such as evaporated metal. The ring is of such a size as to correspond with the image of the annulus. The direct light from the condenser (or background light) is thus much reduced in intensity — as much as 75% or more. In addition, a transparent phase retarding film is applied, usually by evaporation, over either the metal ring or all areas except this ring. The purpose of the transparent film is to retard either the direct beam or the diffracted beam relative to the other by $\lambda/4$. The additional retardation by the phase plate results in an approximate $\lambda/2$ retardation, the conditions for destructive interference between the direct and diffracted beam at the intermediate image plane. This renders the hitherto undetectable phase differences into easily visible light and dark. Note that all of the direct light is confined to the absorbing ring and almost all of the diffracted light is confined to other areas of the plate. Some, however, passes through the ring itself and this light causes the halos in the phase contrast image. If the absorbing ring were completely opaque, a darkfield image would result. As it is, the

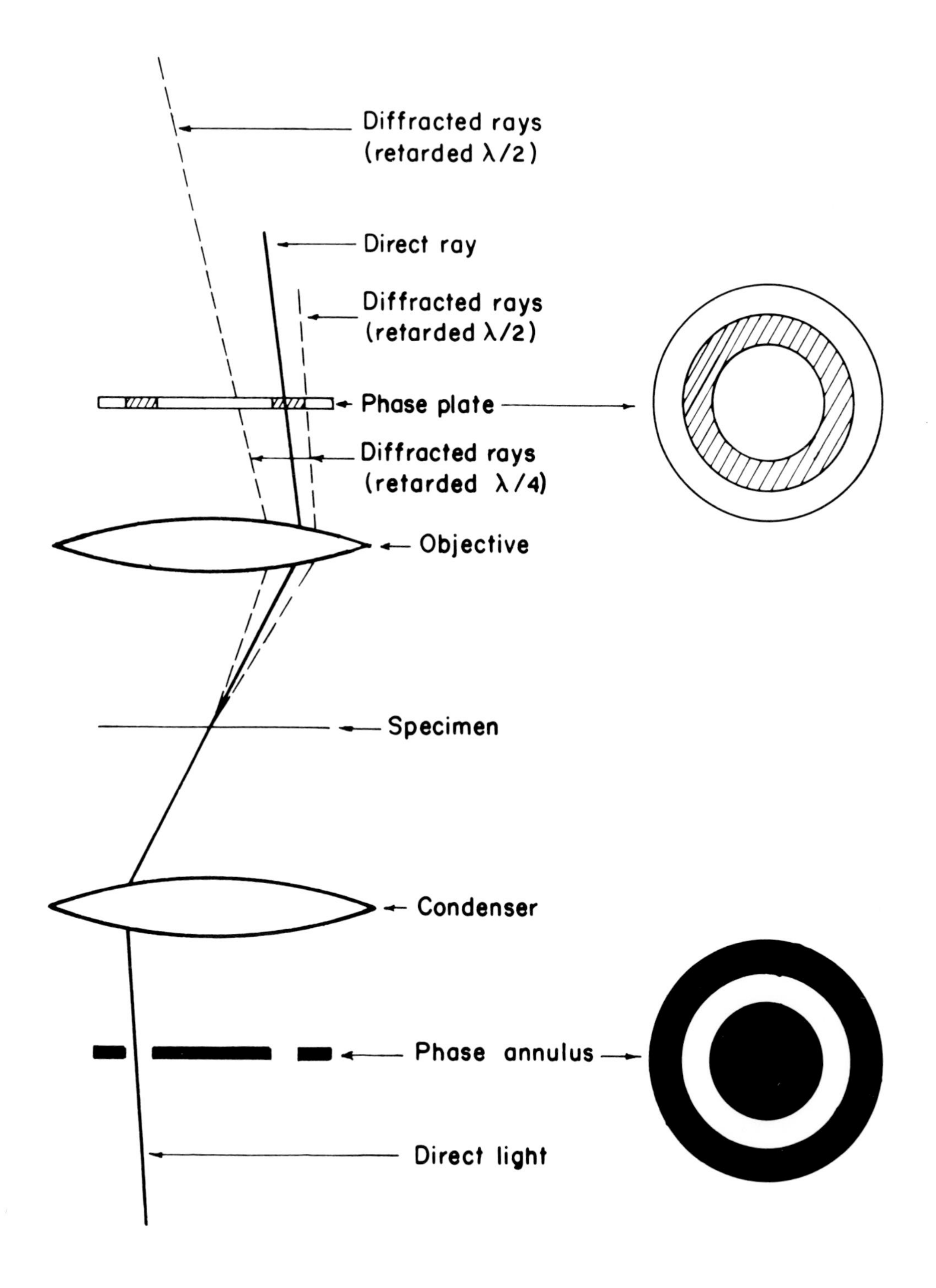

Figure 49. Phase contrast microscope, showing path rays of direct (background) light and diffracted light

absorbing ring simply attenuates the direct light to the point where interference involving the much less intense diffracted light can be seen.

Objectives used for phase contrast are identical to objectives for brightfield , but, in addition, have a phase plate (Figure 50). Phase objectives may be used for ordinary brightfield work but the resulting images lose some contrast. In most systems, a different annulus is required for each objective or numerical aperture so that several annuli are mounted on a rotating disc at the lower focal plane of the condenser. Other systems use just one annulus but have a focusing condenser system which projects a different sized image of the annulus for each objective.

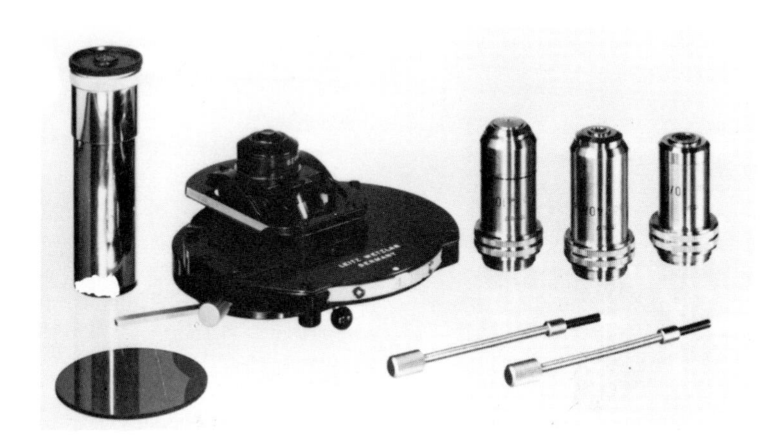

Figure 50. Zernike phase contrast accessories (photograph courtesy of E. Leitz, NY)

3. Proper Use

In setting up a phase contrast microscope, one starts with no annulus in place and Köhler illumination. The specimen is focused and the substage iris is opened. The annulus appropriate to the objective is rotated into position, and the objective back focal plane is observed — either with a phase telescope, Bertrand lens, or by removing the ocular and looking down the tube. The phase plate with its ring will be seen, together with the (usually off-center) annulus of light. The image of the annulus is centered so that it is precisely superimposed on the ring of the objective phase plate. The ocular is then replaced, or the Bertrand withdrawn, and the phase contrast image is observed.

A green filter is frequently used for maximum contrast.

D. FLUORESCENCE

Fluorescence microscopy is frequently confused with ultraviolet microscopy. In ultraviolet microscopy, radiation of about 254 nm or 275 nm (short UV or far UV) forms images invisible to the human eye. The images, formed through absorption differences within the object, must be recorded on photographic film or converted to visible yellow-green on a fluorescent screen as in RCA's electronic image converter, Ultrascope. Quartz or reflection optics are necessary for ultraviolet microscopy — or, more accurately, ultraviolet micrography — since glass systems absorb radiations below about 300 nm.

Fluorescence microscopy, on the other hand, is based on the phenomenon that occurs when radiation of one wavelength falls on a substance and is changed to a longer wavelength (Figure 51). Fluorescence microscopy most commonly uses invisible radiation of 365 nm (long UV or near UV) to incite visible fluorescence, i.e., to cause certain substances to emit rays in the longer, visible portion of the spectrum. Visible blue-violet rays can also cause certain substances to fluoresce at longer

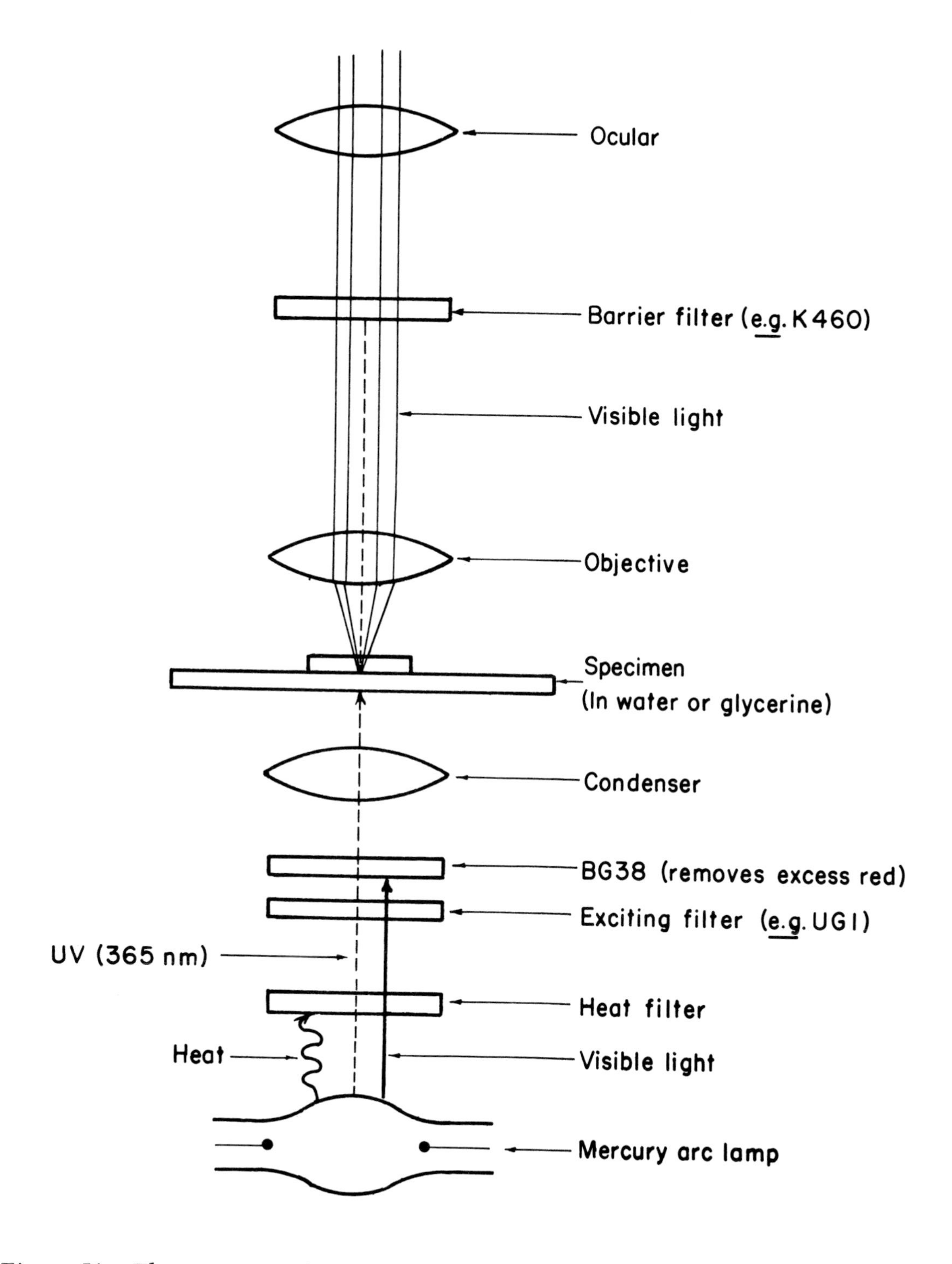

Figure 51. Fluorescence microscope. Schematic showing production of UV, excitation of specimen, adsorption of UV and transmission of visible light to the eye

wavelengths in the green, yellow and red. Visible UV fluorescence is excited by radiation of 365 nm, and blue-violet fluorescence is excited by radiation peaking at 400 nm. Glass systems are adequate for near UV or blue-violet fluorescence microscopy.

Fluorescence is primary or secondary. Primary fluorescence, sometimes called autofluorescence, is due to the object itself. Objects which ordinarily do not autofluoresce can be impregnated with certain dyes, called fluorochromes, which will fluoresce when excited by UV radiation. This induced fluorescence is secondary fluorescence. The biomedical microscopist is interested in secondary fluorescence; the particle analyst, in primary fluorescence.

Fluorescence microscopy can be used in conjunction with tracer techniques, in which fluorescent substances are deliberately used to tag and trace effluent etc.

1. Light Sources

Probably the least expensive way to produce "black light", or radiation of 365 nm, is to purchase a fluorescent lamp in the BLB series. This is basically a germicidal lamp, emitting 253.7 nm radiation which strikes the lamp phosphor coating, causing it to emit the longer 365 nm. This lamp, available in a variety of lengths and wattages, can be used in an ordinary fluorescent desk lamp. The disadvantage is the relatively low intensity. Better sources are the stronger lamps made for mineralogists. The most intense sources are high-tension sparks between magnesium electrodes, iron arcs, 100 W and 250 W mercury vapor arc lamps, water-cooled mercury lamps and the various quartz mercury lamps. Ideal lamps for particle analysis are available from the microscope manufacturers. These frequently employ high-pressure mercury arc lamps, such as the HBO 200.

Very effective use has been made of the 12 V, 100 W quartz iodine lamp for blue-violet fluorescence [12].

2. Filters

The filters which are used to isolate the exciting radiation from the light source are called exciting, or primary filters. Filters used to absorb unwanted exciting radiation are called barrier, or secondary, filters.

Exciting filters which produce 365 nm radiation include the Kodak Wratten filter 18A (glass) and the Corning Violet Ultra Glass 5860. If the Corning filter is ordered thicker than five millimeters, no visible radiation will pass through it. Other filters for 365 nm include Corning's Red Purple Ultra Glass 5874 and 5970. Needham[13] recommends still others. Again, ideally the microscope manufacturers who supply UV sources will provide the filters. Examples of these are the Schott filters UG1 and UG5.

Barrier filters, which absorb scattered and residual UV, are placed in or over the oculars or in the intermediate tube anywhere above the objective. They are usually a very pale straw color. The barrier filter for UV can be the Kodak Wratten 2A, which does not fluoresce and absorbs all wavelengths less than 400 nm, or the 2B or 2E can be used. Alternatively, the barrier filters supplied by the manufacturers as UV-absorbing can be used, e.g., K430, K460, K490.

Blue-violet radiation can be produced with Corning glass filter 5850, four millimeters thick (the radiation ranges from about 30 nm to 540 nm, peaking at about 400 nm) or the Schott BG12, Kodak Wratten 49 or 50 can be used as the exciting filter.

---

12. Young, M.R., and J.A. Armstrong, "Fluorescence microscopy with the quartz-iodine lamp," Nature 213, 649-50 (1967).

13. Needham, G., The Practical Use of the Microscope, Charles C. Thomas, Springfield, IL, 1958.

Barrier filters for blue-violet fluorescence are orange. They absorb excess blue-violet radiation so that the fluorescence colors may be seen. Barrier filters for blue-violet may be the Kodak Wratten no. 15, G, the Ilford 107, or one provided by microscope manufacturers as blue-absorbing, e.g., K530.

In addition to exciting and barrier filters, heat absorbing filters are frequently necessary. Also, a filter is sometimes necessary to absorb the visible red light passed by some UV filters. A Chance ON 13 glass filter may be used to absorb heat and red from a quartz iodine lamp[12]. With mercury arc lamps, solutions of copper sulfate will remove both heat and the red. A cell two inches thick requires a 4% solution, for example, a quarter-inch cell filled with a 32% copper sulfate solution. A few drops of sulfuric acid should be added. A BG38 will also remove excess red, and is more convenient to use.

3. Setup and Alignment

A fluorescence microscope (Figure 52) is set up in essentially the same manner as a conventional microscope except that uranium glass is used to trace the UV rays and aid in alignment. Uranium glass can be ordered from Corning as Fluorescent Canary, or a uranium glass microscope slide can be supplied by Klinger Scientific Apparatus Corporation (83-45 Parsons Boulevard, Jamaica, New York 11432). The bright blue bits that one sees on the mirror in setting up a fluorescence microscope are textile and paper fibers impregnated with "optical whiteners".

One should place barrier filters in the oculars or intermediate tube before looking at a preparation. Remove the polarizer and analyzer from the light path (polars are destroyed by the heat from strong light sources). The most contrasty images and those against the blackest background are obtained only when a darkfield cardioid condenser is used. This is not essential unless extremely weak fluorescence is anticipated. Either the room itself should be darkened or eye caps should be used with reduced room light.

To insure that alignment is correct, and before looking at an unknown, it is a good idea to look at a preparation that is known to fluoresce. A good standard is a slide preparation of the mineral willemite which fluoresces a strong yellow-green.

4. Sample Preparation

Unfortunately, the glass and mounting media most commonly used in particle analysis fluoresce. The fluorescence of microscope slides and coverslips is slight and not objectionable for routine purposes. For complete absence of fluorescence, one must use quartz. A quartz slide can be purchased from A. H. Thomas Company, Philadelphia, as Transparent Silica Micro Slide, Catalog no. 7036-S. Quartz microscope slides and coverslips can be purchased from A. D. Jones Optical Works, Inc. (2400 Massachusetts Avenue, Cambridge, Massachusetts 02440) or from Esco Products (Oak Ridge Road, Oak Ridge, New Jersey 07438).

Aroclor® (Monsanto), unfortunately, not only fluoresces in UV, but is "burned". All of the resinous mounting media fluoresce as do the most common immersion oils. The most useful mounting in fluorescence microscopy is pure glycerin ($n = 1.47$). Water and Fluormount® are also satisfactory mountants.

It is a little difficult to assess the application of fluorescence microscopy to particle characterization and identification. Because fluorescence usually seems to depend more on trace elements in a sample than on the major component, we can say that fluorescence will not be as generally useful as, say, refractive index. Furthermore, many substances, with or without trace elements, fluoresce so faintly that their fluorescence cannot be used analytically.

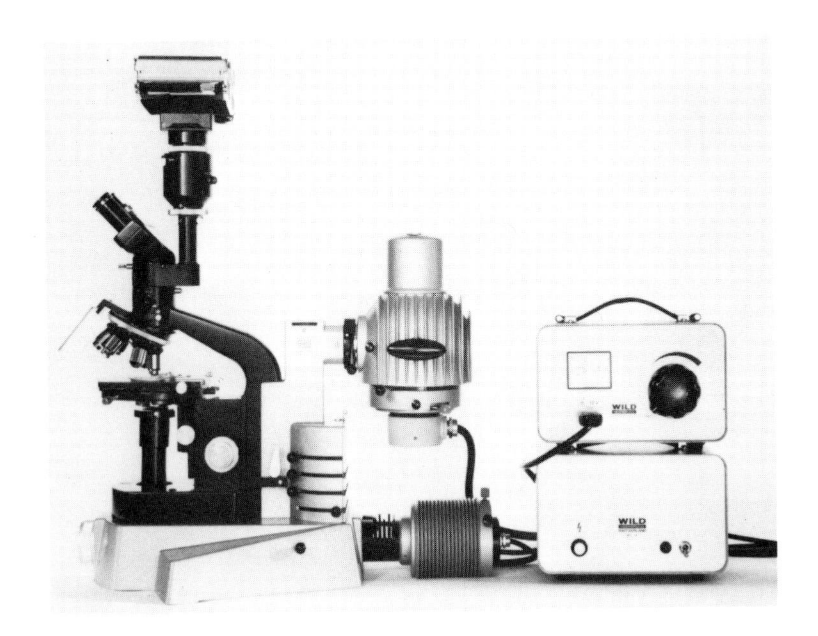

Figure 52. Fluorescence microscope (photograph courtesy of Wild Heerbrugg)

In situations where the fluorescent mineral willemite is used for tracer studies, fluorescence becomes an important analytical adjunct. Single small particles of natural willemite can be recognized and counted because of their characteristic green fluorescence when they are irradiated with near ultraviolet radiation. Willemite is seldom used today as a tracer because a microscopist can, and should, use normal stack effluent as its own tracer.

We can say that fluorescence effects provide one sensitive characteristic of a given source of a particular substance. Two different samples of a given mineral, e.g, calcite, will usually fluoresce differently; they will show either different colors or different intensities or both. Conversely, two samples of a given substance from the same source will usually show the same fluorescence properties. This does make fluorescence a particularly good tool for insuring that two given samples came from the same source or from different sources. This kind of analysis, particularly useful in criminalistics, can also be applied to biological materials as well as industrial chemicals and minerals.

A classification of particles by fluorescence appears in Chapter VI, The Particle Atlas, Edition Two.

VI. PHOTOMICROGRAPHY

The object of this section is to describe the optics and to discuss the principles and methods of photomicrography.

There are two general approaches to photomicrography: one uses any camera with its own lens; the other uses the microscope to project an image onto a photographic film.

Current practice in the industrial research laboratory favors the use of film rather than plates and, usually 35 mm or Polaroid®. The latter is used because of the "instant picture" feature and 35 mm is used because it is fast, easy to use and the dark room techniques are rapid and simple. Only a few laboratories still use larger film negatives. There is little loss of quality with 35-mm films since either the black and white or color film can be enlarged considerably before lack of definition becomes apparent. In other words, a photomicrograph can be taken at, say, 100X and subsequently enlarged to 1000X or even more without "empty magnification". One should think in terms of objective NA and a final magnification of 1000 NA.

Many 35-mm camera systems are so constructed that photomicrographs with a 100X, 1.30 NA objective and a 10X ocular will show magnifications of only about 350X on the negative (rather than 1000X). At 1000 NA, however, the magnification should be 1300X, hence the negative is simply enlarged 4X photographically. The final print should be as good as a photomicrograph taken originally at a negative magnification of 1300X and then printed 1:1.

A. DEPTH OF FIELD

In addition to determining the final print magnification, the numerical aperture partly determines the depth of field. Photographic depth of field (d) is determined by the wavelength (λ) of the light used, the refractive index (n) of the space between coverslip and objective, and the numerical aperture (NA):

$$d = \frac{\lambda \sqrt{n^2 - (NA)^2}}{(NA)^2} \tag{25}$$

Note that depth of field is increased if the wavelength is <u>increased</u> toward the red (but resolution suffers) and if immersion objectives are used but it is <u>decreased</u> with increased numerical aperture. The equation can be plotted graphically, taking 550 nm as average white light (Figure 53).

Photomicrographs can be taken with any camera if both microscope and camera are focused at infinity and if the camera can be held firmly over the ocular during exposure. Focusing the microscope at infinity involves only normal focusing with a relaxed eye capable of focusing easily on distant objects. Many camera manufacturers, especially of 35-mm cameras, furnish adapters to hold their cameras on the microscope. Obviously a reflex camera will be easier to use or, at least, easier to use with confidence that the focus is correct.

B. MAGNIFICATION <u>VS</u> PROJECTION DISTANCE

The microscope can be used as a microprojector with a photographic film in place of a projection screen to record the image. To understand the effect of microprojection on image magnification, see Figure 54. The projection distance, D, in cm, determines the magnification, M, of the projected image. The relationship is:

$$M = \frac{D}{25} \times OM \times EM \tag{26}$$

where OM and EM are the magnifications of objective and ocular, respectively.

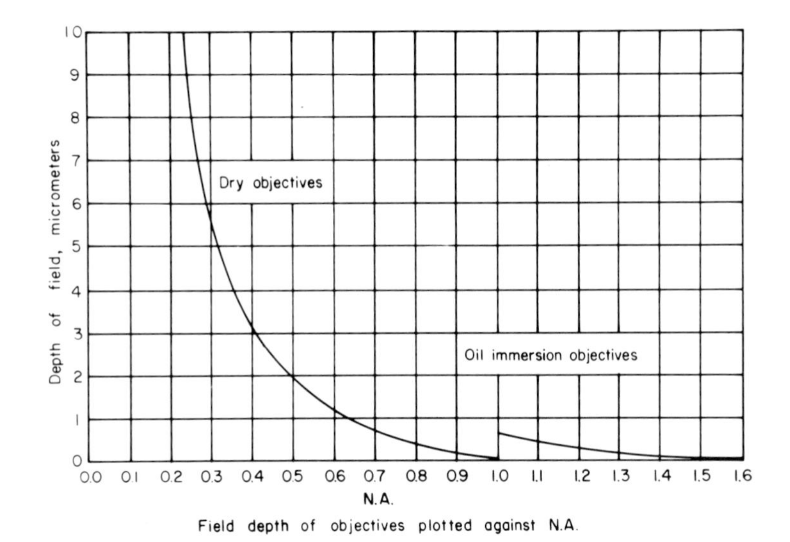

Figure 53. Field depth of objectives as a function of NA

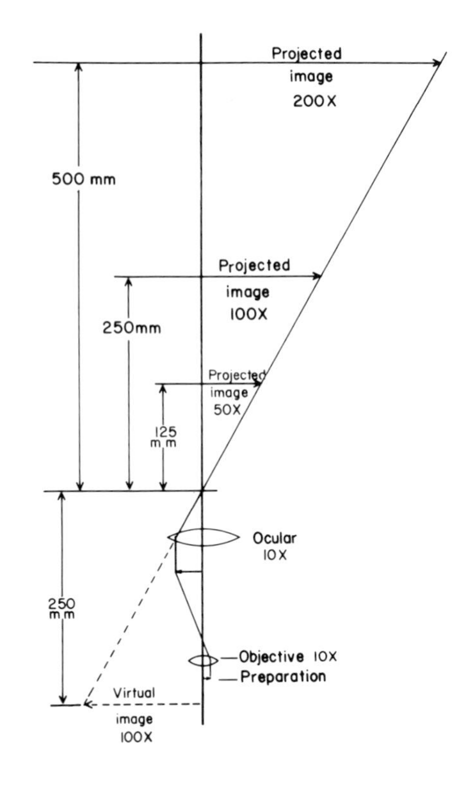

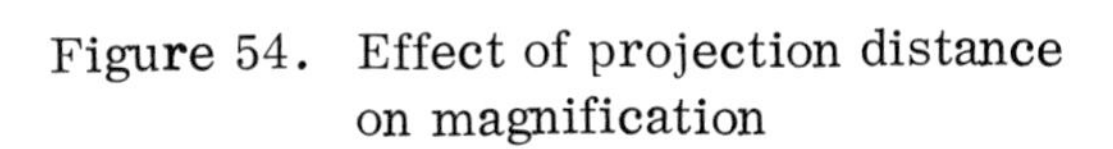
Figure 54. Effect of projection distance on magnification

C. EXPOSURE DETERMINATION

Automatic cameras and exposure controls are gaining wide acceptance in photomicrography. This is probably due to the microscopist's unwillingness to spend a good deal of time learning photomicrographic procedures — especially exposure control. This section attempts to explain the major factors affecting accurate exposure measurement and offers a technique for measuring exposure for polarized light, phase contrast, dispersion staining and other nonbrightfield microscopical systems.

If the occasional photomicrographer does not have a light meter, he has only his best guess as a guide to exposure. Probably most of these "gamblers" use Polaroid[14] film. It is fairly common in some quarters to use two or three frames of Polaroid® to gauge the correct exposure. This process is often repeated for changes in magnification or illuminating systems such as phase, polarized light etc. Actually, once the correct exposure has been obtained for a certain microscopical and illuminating system, it is possible to calculate, with reasonable accuracy, the exposure for changes in: magnification and numerical aperture (NA), illuminator intensity and film speed (ASA).

1. Factors Affecting Exposure

If the magnification is changed only by changing the ocular magnification, or the intermediate magnification as with the Zeiss Optovar[15], Bausch & Lomb Dynazoom[16] or Leitz Vario-Tube[17], the new exposure can be calculated from:

$$\text{new exposure (sec)} = \left[\frac{\text{new magnification}}{\text{old magnification}}\right]^2 \text{ x old exposure (sec)} \tag{27}$$

This equation is derived from the fact that a change in magnification results in a proportional change in the area of field of view. This equation is only an approximation since the cited ocular magnifications are nominal. This equation should have cumulative errors below 20% and is therefore useful for exposure estimates.

The same equation applies to change in projection distance which results, of course, in a change of magnification

$$\text{new exposure (sec)} = \left[\frac{\text{new film distance}}{\text{old film distance}}\right]^2 \text{ x old exposure (sec)} \tag{28}$$

The film distance is measured from the film plane to the ocular eyepoint. Some manufacturers offer 1/3X, 1/2X or 1X adapter cones with fixed but different film plane distances from the eyepoint. The change in magnification which corresponds to these film plane changes is listed on the adapter and can be used to calculate the correct exposure by equation (27).

The magnification is usually changed with the objective rather than the ocular or projection distance. The numerical aperture (NA), in addition to the magnifications, must now be considered in calculating exposures. The exposure varies inversely as the square of the NA. Since the substage iris is rarely opened to the full NA of the objective, the working NA is used to calculate the new exposure.

$$\text{working NA} = \frac{\text{objective NA + condenser NA}}{2} \tag{29}$$

A fair approximation of the condenser NA can be read directly from the scale (if present) below the slot for the arm of the substage iris diaphram. If the scale is present, it may be given directly in NA or in mm of the substage iris aperture or in arbitrary units of substage iris aperture. Another and more accurate way to determine the condenser NA is with the Bertrand lens of a polarizing microscope and an ocular micrometer. The diameter of the full and reduced (with the substage iris) back

---

14. Polaroid Corporation, Cambridge, MA 12139.

15. Carl Zeiss, Inc., 444 Fifth Avenue, New York, NY 10018.

16. Bausch & Lomb, Inc., Rochester, NY 14602.

17. E. Leitz, Inc., Link Drive, Rockleigh, NJ 07647.

focal plane are measured in ocular micrometer scale divisions (osd). There is a direct relationship between osd and NA. For example, a 10X/0.30 NA objective back focal plane measures 14 osd. The aperture of the substage iris measures 10 osd. The NA of the condenser is:

$$\frac{10 \text{ osd}}{14 \text{ osd}} \times 0.30 = 0.21$$

The working NA is therefore:

$$\frac{0.30 + 0.21}{2} = 0.25$$

The working NA is now used along with the new magnification to calculate the new exposure:

$$\text{new exposure (sec)} = \left[\frac{\text{new magnification}}{\text{old magnification}}\right]^2 \times \left[\frac{\text{old NA}}{\text{new NA}}\right]^2 \times \text{old exposure (sec)} \qquad (30)$$

Without a photometer the correct exposure for changes in illuminator intensity can be obtained in several ways. With black and white film the light intensity can be controlled by varying the illuminator voltage or current. A plot of exposure vs transformer settings can be made from an exposure test series. With color film the illuminator voltage must give as nearly as possible the correct color temperature for that film and the intensity must then be controlled with neutral density filters. Accurate and neutral (gray) filters are sold by Bausch and Lomb, Inc. Filters with transmissions of 50, 25, 12.5 and 6.3% are available in 2-in. coated glass squares, while those of 20, 10 and 5% are available in 3-in. diameter glass.

When colored filters are used for contrast control, increased resolution etc., the exposure can be corrected by the filter factor usually supplied with the filters. The exposure is simply multiplied by the filter factor which increases exposure time. We usually assume that the illuminator lamp has a constant light flux for a given setting. Unfortunately, as nonhalogen tungsten lamps age, their output can drop to 80% of their initial intensity and they change color temperature. These effects do not occur with tungsten-halogen lamps.

The final major consideration in calculating exposure is film speed. When film of a different ASA value is used, the new exposure is calculated from:

$$\text{new exposure (sec)} = \frac{\text{old ASA}}{\text{new ASA}} \times \text{old exposure (sec)} \qquad (31)$$

The general equation which incorporates all the exposure factors is:

$$\text{new exposure (sec)} = \left[\frac{\text{new magnification}}{\text{old magnification}}\right]^2 \times \left[\frac{\text{old NA}}{\text{new NA}}\right]^2 \times \frac{\text{old ASA}}{\text{new ASA}} \times \text{old exposure (sec)} \qquad (32)$$

It should be obvious that the ultimate answer to exposure control is test exposures. In equations (27)-(32), the old exposure is the "best by test".

It should also be obvious that test exposures have to be run for each major change in microscope system, i.e., polarized light, dark field, phase etc.

2. Visual Exposure Meters

A visual meter consists of a step-graduated negative incorporated in a viewing ocular[18,19]. The steps vary geometrically in density, e.g., 1, 2, 4, 8 etc. The entire range of stepped densities is superimposed on the field of view. In use, the microscope ocular is removed and replaced with the visual meter. The exposure is determined by observing the greatest density step whose identifying mark can still be read.

A simple visual meter can be made by exposing a number of frames of 35-mm black and white film (negative) with the photomicrographic attachment. This film is placed between the microscope illuminator and the substage condenser so that it covers the field of view. The exposure is now determined by extinction of the brightest portion of the field. The exact reference frame is obtained when the particle or section can just no longer be seen. These test strips are actually superior to the ocular visual meter because the full range of subject contrast can be examined. It is also an excellent way to determine exposure for darkfield , phase, Rheinberg, dispersion staining and polarization colors. This method eliminates the need to estimate the area percent occupied by the sample.

3. Photometers

There are three main types of photoelectric cells that can be used for photomicrography. They are, in order of increasing sensitivity, a) photovoltaic cells; b) photoconductive cells such as cadmium sulfide and cadmium selenide; and c) photomultiplier tubes. The photovoltaic cell is commonly used for macrophotographs , photomacrography and bright field photomicrography. Exposure meter adaptors for microscopes (over the ocular types) are sold by the major photometer manufacturers or can be easily constructed. The low sensitivity of the photovoltaic cell makes them unsuitable for nonbright field photomicrography. Because of their low sensitivity, these meters can only be used to measure full-field intensity instead of selected areas. Photoconductive cells (CdS and CdSe), from 100X to 1000X more sensitive than photovoltaic cells, are most frequently used in photomicrography with manual or automatic cameras. These cells can be tailored to a wide variety of sensitivities, spectral response and rise and decay times.

A number of schematics[20,21] for photometers have been published for do-it-yourself enthusiasts. Science and Mechanics markets a supersensitive CdS darkroom meter[22], Model A-3, available assembled or in kit form. It is an excellent photometer for photomicrography. The four linear sensitivity ranges vary by factors of 10; the fourth range has a sensitivity that gives realistic exposure readings up to 8 hours. In addition, this meter can be used for darkroom printing and densitometry, is portable (13 cm x 18 cm x 6 cm) and can be used for macro photography. Science and Mechanics also offers a smaller version that does not have a variable resistor for darkroom printing.

---

18. White, G.W., "The visual determination of exposure time in photomicrography," Microscope 14, 1, 34-6: 2, 55-8 (1963).

19. One is sold by Practos Products, Hamburg-Grossflottbek, Germany.

20. Schneider, G.T., and C.W. Powell, "An amplifier for microscope photometers," J. Biol. Photo. Assoc. 28 (1), 35-6 (1960).

21. International Rectifier Solar Cell and Photocell Handbook, International Rectifier Corporation, El Segundo, CA, 1964.

22. Science and Mechanics Kit and Product Division, 505 Park Avenue, New York, NY 10022.

Photoemissive cells coupled to photomultipliers are about 20X more sensitive than CdS and CdSe cells. They require a high voltage source and stabilizer which increases their cost. These photomultiplier exposure meters are incorporated mainly in automatic photomicrographic units which divert only 10% of the total light to the photometer instead of the 25-80% measured in manual units. This 10% is reflected from a beam divider in the optical path to the photomultiplier. The photomultiplier amplifies the current to charge a capacitor while the shutter is open and when it discharges, at a preset level, the shutter is tripped. The main advantage of the automatic cameras is instant bright field photomicrography for the novice, hands-free work for the expert, and the ability to photograph moving subjects which haven't a variable light intensity. In spite of a price tag that is 4 to 6 times higher than manual photomicrographic units, they cannot be used to measure exposure directly for nonbrightfield microscopy. "Fudge" factors for the specific microscopical system and another for the percent of field occupied by the sample must be used. The factor for the technique will be a constant and can be electronically compensated by the automatic unit. The sample area factor must be estimated for each prep which results in using the automatic camera in its manual override mode.

4. Brightfield Photomicrography

Ideally a correctly exposed black and white brightfield photomicrograph will show a just fully exposed white background. Any subject detail will be represented as an attenuation of this full exposure, i.e., shades of gray. The same general idea holds for color photomicrography. The proper way to measure the exposure meter reading is to measure the background either in an empty portion of the field or full field after removing the preparation. Often a full-field reading taken with the preparation in place may be used if the attenuation of the light by the detail is not excessive; that is, if the amount of subject detail in the field of view is not excessive. A field of view containing 50% by area of coal particles would show a meter reading reduced by 50% over the full-field reading. Of course, the correct exposure is independent of the number of coal particles in the field.

A test strip is then taken and the best exposure is chosen, i.e., white background with proper attenuation of light for all detail. The meter reading (EMR) and best exposure are then related as follows:

$$\text{best exposure (sec)} = K/EMR \tag{33}$$

As an example, assume the best exposure is 0.1 sec with an EMR of 40; K is therefore 4. The proper exposure for any subsequent brightfield photomicrograph can quickly be determined using equation (33) with K = 4. Obviously, a film having the same ASA rating must be used and the EMR must be measured in the same way. Changes in magnification, lamp intensity and substage aperture will all be taken into account by the meter. You can measure EMR with the beam-splitting prism (to the side-viewing tube) in and take the picture with it out just as long as you did the same thing when you took the test strip and determined K. It is best to remove the prism for the exposure if only because it may collect dirt which can appear out of focus in the final picture. With some cameras, the prism can be removed or is automatically removed during the exposure.

Light measurements should be made as close to the film plane as possible. With external photometers, this usually means taking intensity readings at the focusing telescope or focusing screen. Readings can also be made over the visual ocular (whose eye lens must be kept scrupulously clean) or by removing the ocular and holding the photometer in place. Factors for contrast or color-correcting filters are not used if the filters are in the light path during intensity measurement.

5. Nonbrightfield Photomicrography

The technique described below applies to darkfield , dispersion staining and polarized light photomicrography. The greatest error in determining the exposure directly for nonbrightfield methods occurs when the sample does not fill the field. Petrographic thin sections which fill the field between crossed polars can be measured reasonably accurately but how can crushed mineral grains which occupy an unknown area of the field be accurately exposed? Also, what is the effect of slightly uncrossing the polars on exposure? The only way to get an accurate exposure with nonbrightfield systems is to make the measuring technique independent of the amount of sample covering the field. This is especially true when photographing a dynamically changing field such as recrystallization from a melt, in polarized light. The exposure in polarized light (dispersion staining etc.) for any sample should be the same whether there is one or a thousand particles in the field of view or whether one or one-hundred percent of the field is occupied by the sample. The exposure is governed by the brightness of the brightest particles and this is reasonably independent of color (± 30% over the entire spectrum). That is, the correct exposures for a second-order blue, a first-order white or a third-order green are within the latitude of the film. In a randomly oriented preparation, some particles will be close to, or at, extinction in polarized light. It is again important that the exposure reading is not affected by these particles. By exposing for the most intense particles and choosing films with the widest latitude, all particles will show their correct film density. The correct exposure is determined by switching from nonbrightfield to a standard type of brightfield illumination, e.g., to plane polarized light by removing the analyzer and measuring the brightfield intensity. Brightfield illumination always maintains constant ratios to illumination with crossed polars, dispersion staining with the central stop etc.

The EMR is measured with the corresponding brightfield but the exposure test strip is taken with nonbrightfield. With dispersion staining, the corresponding brightfield could be the annular stop or no stop at all; with darkfield a higher NA objective or removal of the darkfield stop.

The proper exposure for slightly uncrossed polars is determined by an exposure test strip after measuring the EMR with no analyzer. The value of K calculated as in equation (33) can then be used subsequently. The proper exposure is usually almost the same for slightly uncrossed or crossed polars.

The exposure for other film speeds is obtained by multiplying the exposure by the ratio of the two ASA ratings, e.g., for an ASA 50 film, the correct exposure was 1/10 sec at a meter reading of 48. At the same meter reading, an ASA 25 film should be exposed 1/5 sec.

$$\frac{50 \text{ ASA}}{25 \text{ ASA}} \times 1/10 \text{ sec} = 1/5 \text{ sec}$$

Rather than going through each of the foregoing equations for each individual specimen photograph, a graph may be constructed for a particular film. Figure 55 shows the relationship between best possible exposure (BPE) and exposure meter reading (EMR) using a Science and Mechanics exposure meter Model A-3 and a film having an ASA speed of 50. This figure is completely general (except for the restriction to a film of ASA 50) and can be used by any photomicrographer. The appropriate value of K must be determined for each kind of photomicrography, i.e., crossed polars, phase, interference, dispersion staining, darkfield etc. Each of these requires measurement of the EMR without a prep and with an easily standardized brightfield corresponding to the nonbrightfield to be photographed. The following are examples:

Conditions for: Exposure	Meter reading	K (ASA) = 50)
1. brightfield, no polars	brightfield, no polars	200
2. brightfield, one polar	brightfield, one polar	200
3. slightly uncrossed polars	brightfield, one polar	1,500
4. crossed polars	brightfield, one polar	1,500
5. darkfield	standard prep, ground glass surface	1-3000
6. central stop, dispersion staining	annular stop	6,000
7. central stop, dispersion staining	no stops, full objective aperture	6,000
8. negative phase	remove condenser annulus	4,500

*depends on specimen contrast; K = 30 for low and 10 for high contrast specimens.

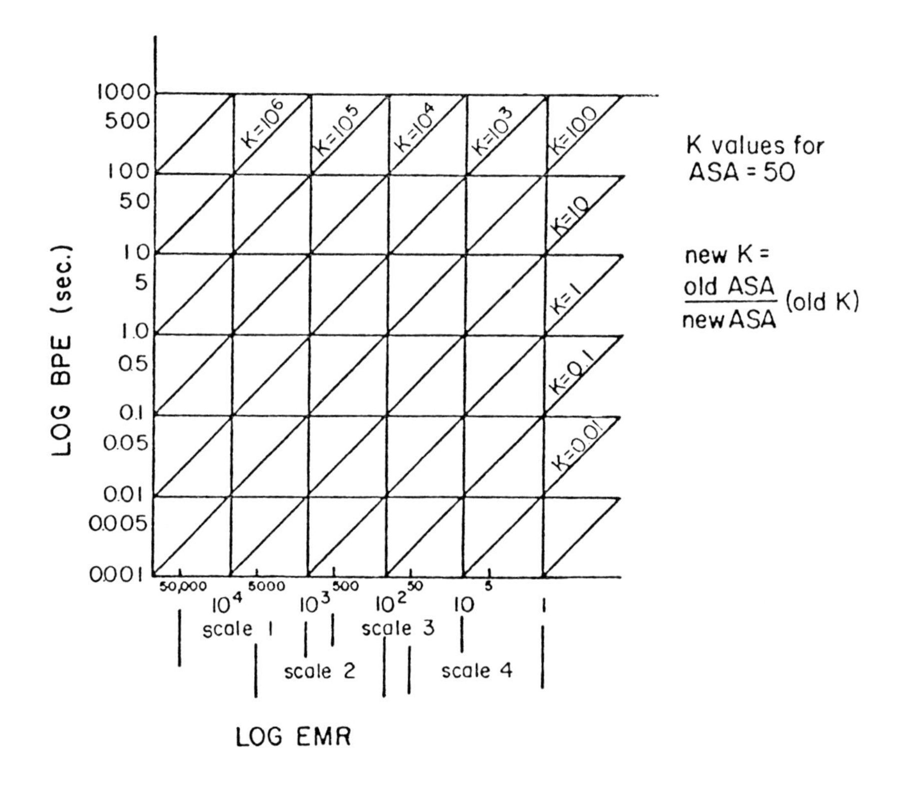

Figure 55. Best possible exposure time (BPE) as a function of exposure meter reading (EMR) using a Science and Mechanics meter

6. Top Light

It may be helpful here to describe one procedure for determining the best exposure for top-lighted particles. First, use any of the suggested systems (Chapter IV. B. 4) to arrange the lighting so that the particles are top-lighted. Second, replace the preparation with a piece of white 3 x 5-in. file card (or any convenient white surface kept for the purpose) and read the exposure meter. The latter can be a photometer held over the side viewing tube of the camera, over the ocular, on the ground glass of a reflex camera system or anywhere in the light path from the microscope. It must be placed in the same spot every time, however. Third, replace and focus the preparation and take a test series of exposures. Fourth, after development, choose the best exposure and substitute the data in the formula:

$$K = (EMR)(BPE) \quad (34)$$

K then becomes a constant for that setup and film as described above. The formula and the value of K thus determined can be used in all future photomicrography if the meter reading is always measured on a piece of 3 x 5-in. file card or equivalent standard. If the type of film is changed, K also changes; K is indirectly related to the ASA rating of the film. Double the ASA rating and you halve the exposure time. It may be possible to expose for transmitted light and reflected light simultaneously if the two light intensities are properly adjusted to give equal exposure times.

7. Film Packs

Polaroid® film packs are used in some photomicrographic backs without a dark slide. The question then arises as to how to determine exposure with a minimum waste of film. The answer is simple if Köhler illumination is used. With the 4 x 5-in. back, the film envelope is pushed in by steps to produce the familiar exposure test strips. With the film pack back, this is not possible. The simple expedient is to expose the entire frame at the least expected shutter speed. Next, close down the field iris until it cuts into about one-sixth of the field of view; repeat the same least expected shutter speed. Close down the field iris about one-sixth more, expose at next slowest shutter speed. Continue to close down the field iris and expose with successively slower shutter speeds. An example of an exposure series would be: 1/120, 1/120, 1/60, 1/30, 1/15 and 1/8 sec. This will give a regular series of exposures from 1/4 - 1/120 sec as shown in Table XI. The resulting test exposure print will not be of the usual stepped variety, but, instead, will show a series of concentric rings of exposure. This works best with a well-corrected condenser which produces a sharp black image of the field diaphram. However, even with the two-lens Abbe condenser, this technique will put you in the ball park, and reduce film waste.

8. Vibration

Camera shutters can cause fuzziness by vibrating the film plane. No matter what photomicrographic equipment you use, the shutter is very likely to introduce some fuzziness. Some shutters are worse than others and of at least equal importance is the rigidity of the entire microscope-camera arrangement. As an example of this effect, we took a set of photomicrographs of a test diatom using a Bausch and Lomb LC polarizing microscope fitted with an IPSCAPHOT with a Leica back. Both the IPSCAPHOT and the Leica back have shutters. A series of exposures was made using only the IPSCAPHOT shutter. An identical series was made using only the Leica shutter. The results are illustrated in Figure 56. Obviously that particular setup cannot be used for photomicrography with exposure times faster than about 1/10 sec. The slowest speeds prove best as they allow the initial vibration to cease. In addition to this series, a shutter was mounted externally on the lamp, and an exposure at the worst shutter speed was taken. This last exposure is the sharpest of all. Generally, an external shutter is best although they are not generally used because of the inconvenience. Note, too, that the results are somewhat better using the Leica focal plane shutters rather than the IPSCAPHOT shutter.

Table XI. Total exposure in circular areas of film exposed by successive closing of lamp iris

Area	Exposure Increments	Total Exposure
Center ring	1/120 + 1/120 + 2/120 + 4/120 + 8/120 + 16/120	32/120 ≅ 1/4
Center + 1	1/120 + 1/120 + 2/120 + 4/120 + 8/120	16/120 ≅ 1/8
Center + 2	1/120 + 1/120 + 2/120 + 4/120	8/120 = 1/15
Center + 3	1/120 + 1/120 + 2/120	4/120 = 1/30
Center + 4	1/120 + 1/120	2/120 = 1/60
Outer ring	1/120	1/120 = 1/120

You can easily check whether your camera shutter causes this trouble by taking a series with different exposure times, making corresponding light intensity changes with each exposure. Be sure to focus very carefully for each shot.

If your short exposure pictures are fuzzy and the degree of fuzziness is progressively worse as the exposure is shortened, then shutter vibration is a problem. The possible solutions are: a) an external shutter in the light path; b) control of light intensity to give longer exposure times; and c) increasing the rigidity of the microscope-camera system.

9. Photomicrographic Records

Successful photomicrography depends on critical evaluation of test rolls and specimen photographs. The photomicrographic record included here (Figure 57) was designed for 35-mm photomicrography. With slight modifications, such as bellows extension etc., the form can be used for any photomicrographic system.

D. FILTERS FOR PHOTOMICROGRAPHY

1. Color Contrast Filters

The light filters one uses for photomicrography depend upon whether one is using black and white or color film.

When shooting black and white film, the only filters of any consequence are the color contrast filters. Intensity can be controlled by changing the setting of a variable voltage transformer placed in the circuit. The resulting changes in color temperature are unimportant in black and white photomicrography.

Passable photomicrographs may be taken with black and white film without any filtration at all. However, it is highly recommended that a Wratten 58 (green) filter be placed in the light path during all black and white photomicrography. The reason for this lies in the optics, not the film. Both the condenser and objective, regardless of correction, retain various degrees of spherical and chromatic aberrations. Chromatic aberration results in several sizes of images in the same image plane, a different size for each different wavelength in white light. The final black and white image is not sharp, no matter how carefully the microscope is focused, because each different wavelength-size image is individually recorded on the film with fidelity. By inserting a green filter in the light path, all wavelengths of white light are removed except those for which the condenser and objective are pri-

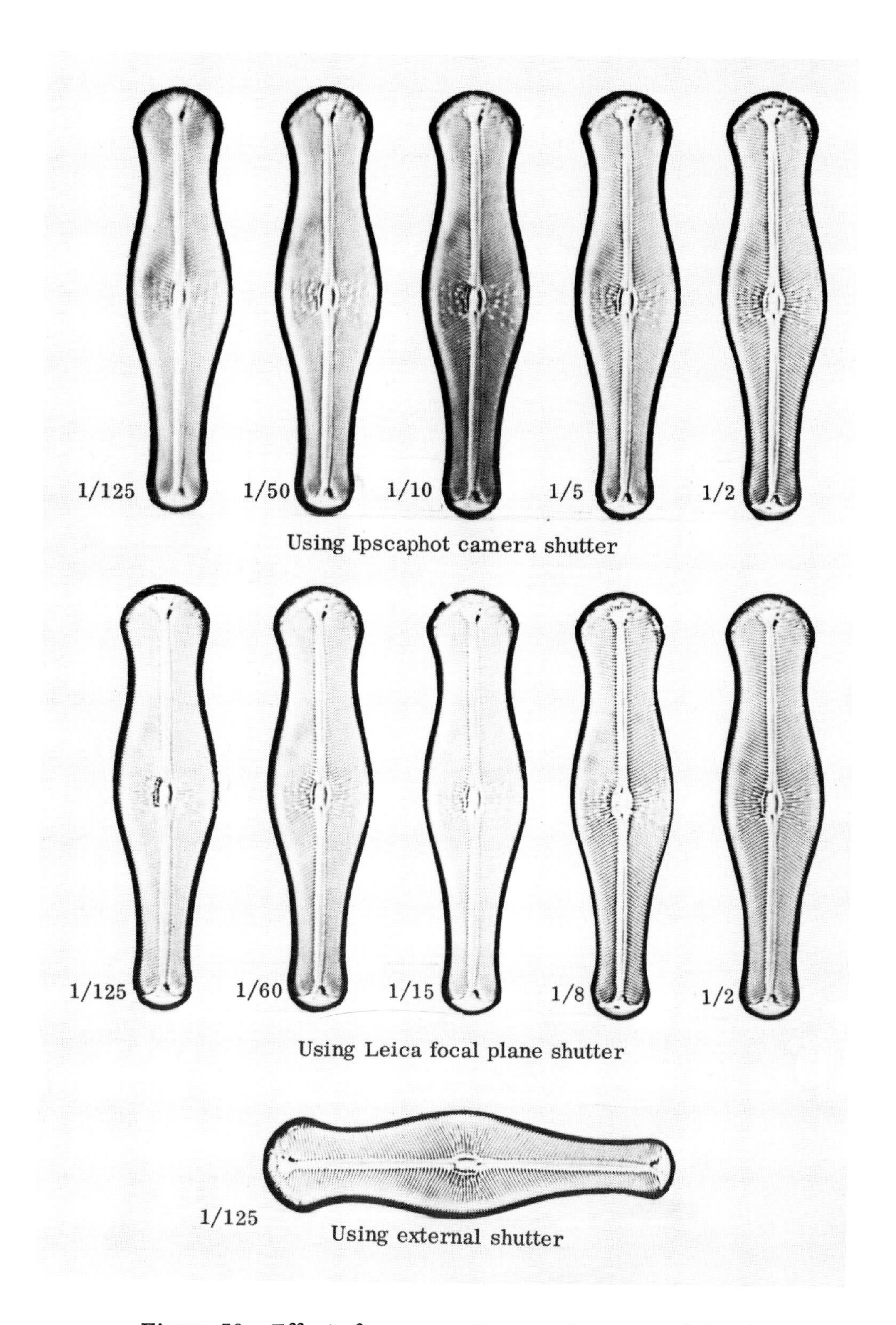

Figure 56. Effect of exposure time on sharpness of detail

PHOTOMICROGRAPHIC RECORD

Film: ASA: Date:

Filter(s): Camera: Tube Length:

Illum.: Spec. Apparatus:

Other:

Exp.	Subject	Ocul.	Obj.	EMR	Time sec.	Notes and Remarks	Evaluation
1							
2							
3							

Figure 57. Photomicrographic record form

marily corrected. Only one image size reaches the film and the resulting photomicrograph is crisp and sharp. The green filter is a "must" for black and white photomicrography.

In black and white photomicrography color contrast filters are frequently desirable and sometimes absolutely necessary. These filters are used when a colored subject either lacks or has too much contrast. The principle behind filter selection is, then, to choose the complementary color of the subject if more contrast is desired and the same color if contrast is to be reduced. For example, if a pollen grain is stained with acid fuchsin (a red dye) and the stain is very weak resulting in a pale pink coloration, the resulting black and white photomicrograph will lack contrast. In this case we would select a Wratten 58 (green) or Wratten 66 (light green) to put in the light path. The contrast on the black and white film will be increased. On the other hand, if we had a green particle which was very dark green, we might lose some detail because of too much contrast. We would again choose the greens, this time to reduce the contrast in the final image. The basic contrast filters for black and white photomicrography include, besides the 58 (green), the 66 (light green), the 22 (yellow-orange), the 15 (yellow) and the 38A (blue).

2. Color Balancing Filters

The most important filters in color photomicrography are the color balancing filters. Photographic films are made to be used with three principal types of light sources which differ from each other by their color temperature (°K). Type B color film will have correct color temperature balance when used with light sources of 3200°K; for example, some household lamps and most tungsten light sources used in microscopy. Type A color film requires a light source with a color temperature of 3400°K; for example a studio floodlight. Daylight type film, of course, requires daylight for correct color temperature, of 5500°K equivalent to "midday" daylight. Practically speaking, the higher the color temperature of the light source, the bluer the light; the lower the color temperature, the yellower the light. Therefore, if a Type A or Type B film is exposed with daylight, the excessive blue in the source will result in overly blue transparencies. If a daylight type film is exposed with the usual low voltage, tungsten lamp used in microscopy, the resulting transparencies will be overly yellow.

Differences in color temperature between light sources and films can be eliminated; that is, the color temperatures can be made to match by inserting yellowish or bluish filters in the light path; blue filters to raise the color temperature, and yellow filters to lower the color temperature. Thus, a daylight film (5500°K) can be used with the common microscope illuminator (ca.3200°K) by inserting sufficient blue filters in the light beam to raise the color temperature from 3200°K to 5500°K (Table XII).

Table XII. Color balancing filters

Filter Color	WRATTEN Number	To obtain 3200 K from:	To obtain 3400 K from:
	82C + 82C	2490 K	2610 K
	82C + 82B	2570 K	2700 K
	82C + 82A	2650 K	2780 K
Bluish	82C + 82	2720 K	2870 K
	82C	2800 K	2950 K
	82B	2900 K	3060 K
	82A	3000 K	3180 K
	82	3100 K	3290 K
	No Filter Necessary	3200 K	3400 K
	81	3300 K	3510 K
	81A	3400 K	3630 K
Yellowish	81B	3500 K	3740 K
	81C	3600 K	3850 K
	81D	3700 K	3970 K
	81EF	3850 K	4140 K
	80A	3200 to 5500	
Blue	80B	3400 to 5500	
	80C	3800 to 5500	
	80D	4200 to 5500	
	85C	5500 to 3800	
Orange	85	5500 to 3400	
	85B	5500 to 3200	

The 81 and 85 filters are seldom needed in microscopy. The exception, which seldom occurs, is when a xenon light source is used with Type A or B film. Most low voltage light sources in microscopy have a color temperature of 2800 °K to 3100 °K, although 3200 °K can be obtained by over-volting the lamp but that considerably shortens its expected life. The most important filters, therefore, turn out to be those which effect color temperature changes of + 100 °K to + 400 °K; that is, color balancing filters 82, 82A and 82B. The equivalent of 82C can be obtained by combining 82 and 82B. If a daylight film is being used, one raises the color temperature to 3200 °K or 3400 °K first, and then either the balancing filter 80A or the color balancing filter 80B is added to take the color temperature in one step directly up to 5500 °K. The 80A changes the color temperature from 3200 °K to 5500 °K, the 80B changes the color temperature from 3400 °K to 5500 °K.

Figure 58 shows a useful graph for determining the proper color conversion and balancing filters given the color temperatures of source and film. When the source color temperature is only approximately known, e.g., incandescent lamps, the choice is given of filters to try experimentally. One reads 3100 °K on the abscissa as the color temperature of a new 12 V tungsten lamp. Reading vertically to a value of 3200 gives a diagonal line corresponding to filter 82. With that filter, type B film (3200 °K) should give good color balance. In a similar fashion, 82A or 82B filters should be tried experimentally with type A film (3400 °K).

The color temperature of the microscope lamp must be established through exposing a test roll of film. The lamp should be operated at the recommended voltage or amperage and a series of micrographs taken with no filter, the 82, the 82A, the 82B, the 82 and 82B combined, and the 82A and 82B combined (the 80A or 80B should also be used for daylight type film). The resulting transparencies should be evaluated by projection or by using a light source of daylight quality to determine what filtration should be used with successive rolls of film at the same lamp voltage setting.

3. Neutral Density Filters

Good neutral density filters are extremely important in color photomicrography, as once the conditions for correct color temperature are established, the intensity of the light must be changed by inserting neutral density filters, not by altering the voltage setting. Most neutral density filters are unsuitable for color photomicrography, especially those which are dyed in the glass, as they impart a slight color cast to the final transparency. Neutral density filters made by evaporating stainless steel and other metals on glass do not impart these undesirable color tinges. Among the finest filters of this type for both visual and photomicrographic use are the Inconel® neutral density filters supplied by Bausch and Lomb, Inc., Rochester, New York 14602. There are three neutral filters with densities of 0.7, 1.0 and 1.3 (20%, 10% and 5% transmission, respectively), available in 2-in. or 3-in. diameter. The more versatile are the 2-in. square ones, of which there are four: 0.3, 0.6, 0.9 and 1.2 (50%, 25%, 12.5% and 6.3% transmission, respectively). These may be used singly or combined to afford excellent control over light intensity.

Filters should be kept away from excess heat or placed on the microscope side of a heat filter. They should be kept clean so that if they are placed near the field diaphram, out-of-best-focus dirt particles will not mar the image or photomicrograph. Finally, filters should be stored in filter safes or plush boxes when not in use.

E. FLUORESCENCE PHOTOMICROGRAPHY

Fluorescent objects are self-luminous and, because of the relatively low intensity of fluorescence, the exposures are much longer than for normal photomicrography and extend over a larger range. The intensity, which determines the exposure, varies with the a) intensity of the source, b) efficiency of the object in converting UV to visible, c) quantity of fluorescent material, and d) absorption properties of the optical elements and filters. A very sensitive exposure meter is

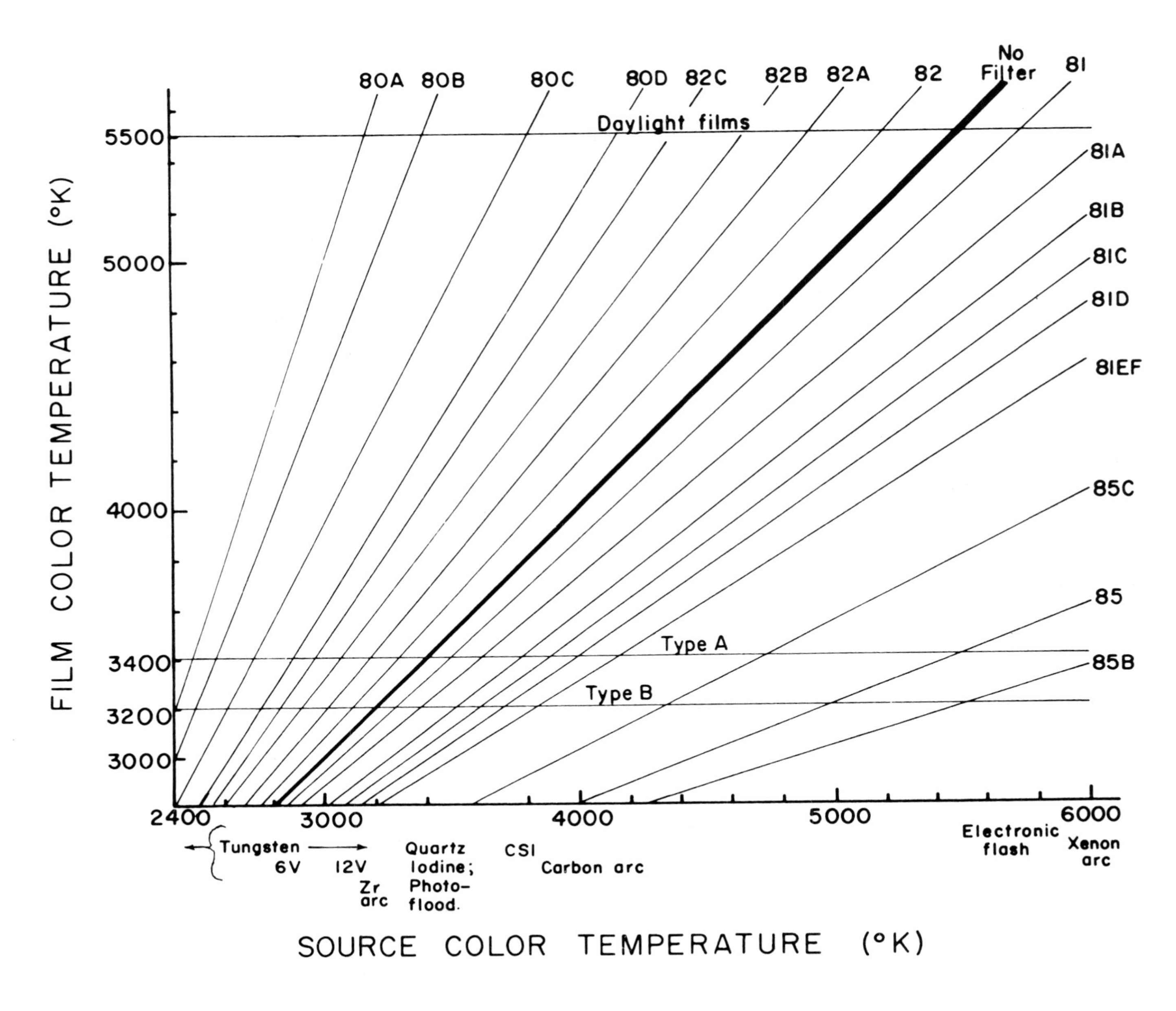

Figure 58. Color conversion and color balancing filters for various light sources and any color film

necessary to measure the fluorescent light. After practice, experience and evaluation of test rolls, the exposure can be estimated with bracketing by the photomicrographer. Comminuted willemite makes a good standard for the test roll of strongly fluorescing objects.

Because residual UV is recorded as blue on color film and degrades all fluorescent colors, the barrier filters must be in place during photomicrography. A cardioid darkfield condenser must be used for absolutely black backgrounds.

The best color film for fluorescence photomicrography is High Speed Ektachrome Film, Daylight Type (ASA 160), a 35-mm film. This size is best because the projected distance is short which cuts down the exposures necessary. Even with high speed 35-mm color film, however, exposures will be much longer than with the conditions usually used. Objects which ordinarily require 1/10 to 1/2 sec exposure on ASA 50 film with slightly uncrossed polars may require 8 to 24 sec exposure on ASA 160 film with UV illumination. Exposures of many minutes are often required. In any case, we are within the range of reciprocity failure of color films. Gelatin color correction filters cannot be used in the illuminating beam because they fluoresce. The color resulting without the corrections is surprisingly good and is adequate for most purposes. If correction is needed, a filter can be mounted with the final transparency.

Apart from fluorescence, incidentally, 365 nm radiation can also be used for differential absorption studies[23] or whenever the highest resolution obtainable with glass optics is required in photomicrography.

A paper by Royer and Maresh[24] discusses general considerations in fluorescence photography.

F. PROBLEMS IN PARTICLE COLOR PHOTOMICROGRAPHY

Particle identification depends on optical data (refractive index, birefringence, color etc.) as well as morphological data (shape, homogeneity, surface texture etc.) obtained with a polarizing microscope. It also depends on having accurate, unambiguous descriptions of standards readily available. One of the most accurate, least ambiguous methods for describing standards in permanent yet accessible records is color photomicrography in polarized light. But this technique presents all of the common problems of color photomicrography plus one — the "green disease". We will treat the problems related to the most frequently used 35-mm photomicrographic color film.

1. Slightly Uncrossed Polars

Almost all microscopical samples are mixtures of particles with one (isotropic) or more than one refractive index (anisotropic). Shape, size and absorption colors can be seen with conventional brightfield illumination (no polarizing filters), but under these conditions, isotropic glass particles are indistinguishable from anisotropic quartz. Crossed polarizing filters (polars) provide a dark background which renders isotropic particles invisible and anisotropic particles brightly colored. Slightly rotating one polar lightens the field and makes all particles visible. Colorless isotropic particles acquire the same color as the background, while anisotropic particles show nearly the same interference colors as with crossed polars.

---

23. Spinell, B. N., "Simplified λ 365 mμ photomicrography with improved results," J. Biol. Phot. Assoc. 29, 145-52 (1961).

24. Royer, G. L., and C. Maresh, "Photography of fluorescent phenomena," J. Biol. Phot. Assoc. 15, 107-19 (1947).

For photomicrography, the polars should be uncrossed until all isotropic particles and their absorption colors, if any, are visible. However, the interference colors of anisotropic particles are "lost" if the polars are uncrossed too much. How much they can be uncrossed depends principally on the intensity of the interference colors. The object is to obtain a gray background on which boundaries of isotropic particles and interference colors of anisotropic particles can be observed equally well. Most photomicrographs can be taken with the polars uncrossed 15°, but as little as 5° or as much as 25° can be used in some circumstances. Occasionally, two photomicrographs are necessary — one with crossed, one with uncrossed polars — especially for substances with faint interference colors such as wheat starch, acetate rayon or apatite. The easiest way to choose the proper degree of uncrossing is to observe a mixed preparation of isotropic and anisotropic particles in Aroclor® 5442 while slowly uncrossing the polars. They are properly uncrossed when both interference colors and morphology are easily seen (Figure 61). This position of slightly uncrossed polars is correct for most samples.

2. Color Temperature

Strict color temperature according to the principles outlined earlier must be maintained for each film.

3. Reciprocity Failure

The reciprocity law states that the exposure density is proportional to the intensity times the duration. Therefore, equivalent exposure densities should be obtained by proportionately increasing the exposure time with a lower light intensity. Reciprocity failure means that equivalent exposure densities are not obtained for a given intensity and exposure time. This failure generally occurs with exposures less than one one-thousandth of a second or greater than one second. Such short exposures are not usually encountered in photomicrography but long exposures are quite common. These require not only greater than calculated exposure times but also additional color-correcting filters. Most films require color-correction filters with exposure times greater than one second. When anticipated exposure times exceed 10 seconds, it is best to choose a more intense light source to avoid sample vibration and the lengthy procedure of selecting the proper color-correcting filters.

4. "Green Disease"

Photomicrographers face an uncommon problem when using slightly uncrossed polars. They get a bluish-green to green background (Figures 59-62) on the transparency, instead of the gray observed in the microscope field. The problem is related to reciprocity failure even though it occurs with normal exposures of 1/10 to 1/125 sec.

Most color films cannot accurately reproduce all colors if the light intensity ratio of the brightest to the darkest objects exceeds 3:1. If this ratio is exceeded, exposures correct for the bright objects will result in underexposed, "off-color" dark objects. With slightly uncrossed polars, the intensity ratio of interference colors to background is often as great as 15:1. Since our exposures are chosen to be correct for the interference colors of anisotropic particles, the background is underexposed and off-color-green instead of gray. Transparencies of colorless isotropic particles that appear green cannot be used as reference standards or as records of visual microscopy. The appearance of colored isotropic particles is also masked or altered. Anisotropic particles, however, show correct colors within the limits of film fidelity. These green backgrounds have been called the "green disease".

G. FILMS

Many of the films that give green backgrounds with slightly uncrossed polars are suitable for brightfield photomicrography. If the particles are correctly exposed under these conditions,

Figure 59. Metal shavings taken with slightly uncrossed polars using Agfachrome CK-20 film.

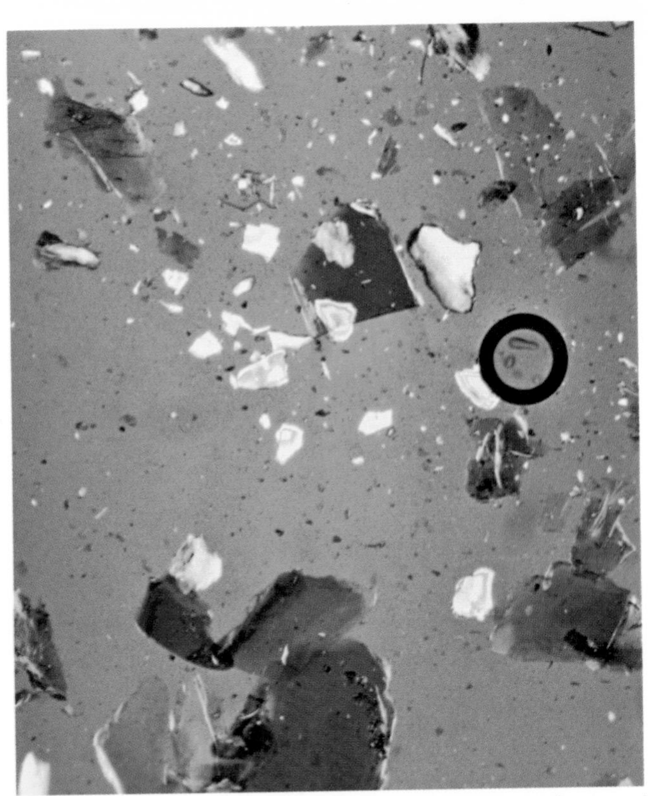

Figure 60. Biotite mica and silicate minerals taken with crossed quarter-wave plates and slightly uncrossed polars using Eastman Kodak Company Kodachrome II-A.

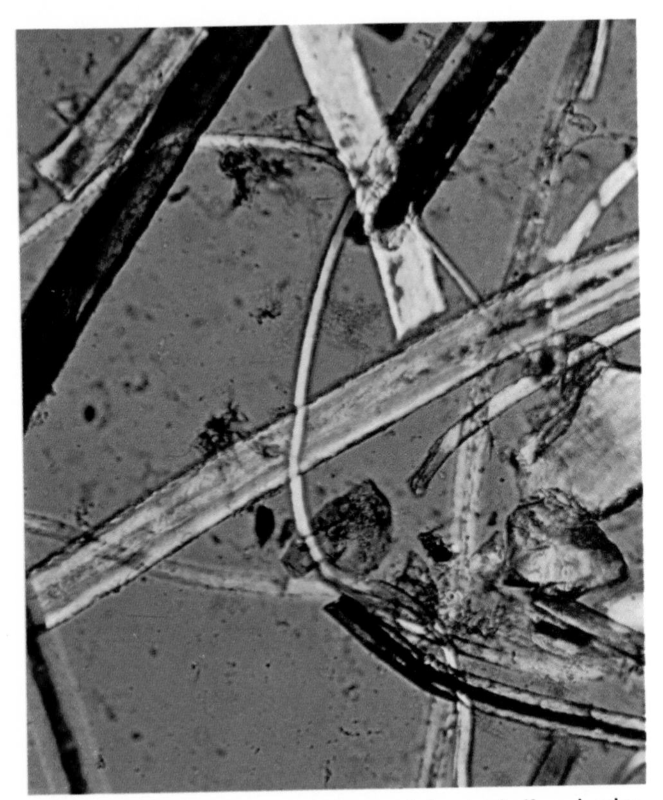

Figure 61. A mixture of wool, human hair, dandruff and other types of dirt from rug cleaning; slightly uncrossed polars using Brand X film.

Figure 62. Incinerator ash, gypsum, calcite, pollen and olivine; crossed quarter-wave plates and slightly uncrossed polars using Brand Y film.

Reprinted from
THE PARTICLE ATLAS
McCrone and Delly
Ann Arbor Science Publishers, Inc.

the background is fully exposed and therefore colorless. These films can also be used for photomicrography with crossed polars, since the background on the transparency is so underexposed that colors cannot normally be seen. Colored backgrounds can still be seen by holding the transparency close to the eye and viewing it against a bright light source.

During the preparation of the first edition of The Particle Atlas, an extensive film evaluation program was undertaken to find the film best suited to the requirements demanded by slightly uncrossed polars. At that time (1967), Agfachrome CK-20, a type B (3200°K) film, gave excellent gray backgrounds with slightly uncrossed polars. Only the slightest hint of magenta was detectable when the angle between the polars is 5°, too small an angle to be useful for particle photomicrography. With the polars at 15°, the film showed an exceptional gray latitude. Furthermore, the color fidelity of this film for normal exposure was excellent; even the most subtle absorption colors were reproduced. All photomicrographs in the first edition of The Particle Atlas were taken with Agfachrome CK-20; Eastman KR-135 (ASA 64) was used for Volume V of the second edition.

The particle photomicrographs in Volume II of the second edition were taken under different conditions: crossed quarter wave plates at 45° and slightly uncrossed polars. The resulting circularly polarized light shows no extinction for anisotropic particles. The background is variously gray to black depending on how well the quarter-wave plates match. Kodachrome II Professional, type A (3400°K) film, gave excellent results under these conditions.

A further modification was made for Volume V of the second edition because the slightly uncrossed polars used in Volume II left a slight retardation in the field of view which was subtracted from some particles and added to others (depending on their orientation). This became apparent when the subject also showed low retardation since then some particles, at least at thin edges, gave black as a polarization color. The effect is most apparent with apatite, number 118 on page 372 of Volume II; asbestos, number 121, page 373; microcline, number 169, page 391; quartz, number 183, page 398 and In Figure 61 in this book. This problem was eliminated in Volume V by John Delly. He maintained precisely crossed polars and precisely crossed matched quarter-wave plates precisely 45° to the polars and added a very short overall double exposure after removing one polar. With Eastman KR-135 film an excellent gray background results with good visibility of isotropic as well as anisotropic particles — and with none of the latter at extinction no matter how they are oriented.

In time, film manufacturers change their formulations so that it is necessary, periodically, to run new test rolls to determine which of the currently available films renders the visual image with the highest fidelity.

H. STEREOPHOTOMICROGRAPHY

Stereophotomicrography is a valuable interpretive technique for studying surface texture, e.g., distinguishing holes from bumps, observing particle shape and simply seeing the third dimension of objects. This section discusses techniques for obtaining stereophotomicrographs with a single objective microscope at magnifications from 50X to 1000X.

Stereograms, or stereo pairs, consist of two photographs (or drawings) that correspond to the left- and right-eye views of a three-dimensional object. If the object is viewed at 25 cm (the conventional image distance in microscopy) with an average interpupillary distance of 6.5 cm, the calculated convergence angle between the eyes and the object is 14.8°. This interpupillary convergence is simulated in stereophotomicrography with a single objective by a) using oblique (converging) transmitted light or b) tilting the specimen. Either of these systems gives full frame photographs and can be used with roll film or cut film.

There are at least three separate methods for obtaining stereophotomicrographs with oblique transmitted light: a) de-centering the substage diaphram; b) using split polarizing filters; c) placing a half-screen over the objective front lens. In each method light passes through the objective from only one side of the condenser for each stereo frame. Consequently this light is directional and corresponds to the same view seen by each eye.

Advantages of these methods are that samples remain flat, so that those in fluid mounts do not move and that, with a flatfield objective, the full field of view can be in focus. The disadvantages are loss of resolution (and occasionally excessive contrast) and reduced illumination which requires extended exposures.

1. De-Centered Diaphram

If the substage diaphram is reduced to half the diameter of the objective aperture and de-centered, the sample will be obliquely illuminated. Unfortunately, few microscopes are equipped with a de-centerable substage iris, but if yours is, you're all set. Observing the back focal plane, de-center the diaphram so that it just touches the edge of the objective aperture and the center of the field. The back focal plane can be seen through the Bertrand lens or through a low-power magnifier held in front of the telescopic ocular. If the microscope has a photographic tube, simply remove an ocular to see the back focal plane. Photograph the first stereo frame with the diaphram thus de-centered. For the second shot, move the diaphram to the opposite edge. If a line joining the centers of the diaphram positions is reasonably parallel to the film transport direction, the stereo pairs can be viewed stereoscopically without separating the frames. One can, of course, simulate the de-centered substage diaphram by first shading one half, then the other half of the substage aperture.

2. Split Polarizing Filters

A split polarizing filter, for oblique sample illumination, can be inserted in those microscopes not equipped with a de-centerable substage diaphram. Glue two Polaroid® filter half circles together so that they just overlap and their privileged (light-passing) directions are perpendicular. The glued edge should be parallel to one privileged direction and perpendicular to the other.

Place the filter in the substage filter holder or above the substage diaphram with the glued edges north-south. Place an uncut sheet of polarizing filter in front of the illuminator to complete the oblique illuminating system. Orient the illuminator-polarizer so that its vibration direction is parallel to half of the split polarizer. Light will pass through the parallel segment but will be blocked by the other half. This provides oblique illumination for the first frame. For the second frame, turn the illuminator-polarizer 90° so it is parallel to the other half of the split polarizing filter.

The split polarizing filter and the illuminator-polarizer should always be below the sample. If either were above the sample, anisotropic substances would show their complementary polarization colors. Instead, with both polarizing filters below the stage, the sample is illuminated with plane polarized light. Though this technique gives excellent stereo pairs, it cannot be used with crossed or slightly uncrossed polars. This disadvantage, added to the limitations of all oblique illumination methods, makes this technique slightly less valuable than the de-centerable diaphram or tilting stage methods.

3. Objective Half-Screening

A small paper cylinder is made to fit around the objective. A half circle of opaque paper is glued or taped to the bottom of the cylinder, and the entire cylinder is slid over the objective with the half circle stop against the front lens or its metal guard. Half of the front lens should be covered. This stop blocks half the light to the objective. The cylinder is simply rotated 180° after the first exposure to give the stereo pair. This method can be used only with 20X or lower magnification objectives. Because the front lens of 40X and higher magnification objectives is so small, the paper stop can be positioned neither easily nor accurately. The two halves of the objective aperture must be aligned parallel to the film transport direction.

4. Tilting Stage

The apparatus and procedure described below provide stereophotomicrographs in color with partially uncrossed polars, transmitted illumination and indirect or top lighting. One of the two views is taken with top lighting, the other without. When these two photomicrographs are fused, as in stereo viewing, the illusion of top lighting on both views is practically perfect, especially if the stereo pair is viewed using your stronger eye with the top-lighted photomicrograph.

At the same time, it is possible to differentiate between the appearances of particles under the two different types of illumination, indirect and transmitted, by closing one eye at a time and viewing the two photomicrographs alternately. The object of this rather complicated illumination system is to make possible differentiation between opaque and transparent particles, anisotropic and isotropic particles, and any combinations of these various types of particles. In addition, the surface luster and gloss as determined by the texture are also well-recorded under these conditions.

It is most convenient to use a 35-mm photomicrographic camera and color film. Any reflex camera arrangement may be used. If the camera itself is not reflex, then a side viewing tube below the camera should be used.

Our design for a tilting stage is shown in Figures 63 and 64. The drawing gives the critical dimensions and others to scale. Figure 64 shows the stage in place. It consists of a flat rectangular aluminum frame that is fastened to the circular stage of a polarizing microscope. Two holes in the frame align over holes tapped for the universal stage; two clamping screws hold the frame in place. (A rectangular microscope stage must be tapped to accept the thumbscrew or the tilting stage can be clamped in place.) A nickel plate with one short rod at each of its long ends rides in grooves cut in the frame; these grooves and rods form the tilt axis. This plate accepts a full 25 x 75-mm (1 x 3 in.) slide. A 12.5 mm hole is cut at the center of the plate for transmitted light. Glass slides are held to the magnetic nickel plate with small bar magnets.

The tilting stage should be aligned carefully.

1) Fasten the tilting stage loosely to the rotating microscope stage with the thumbscrews.

2) Rotate the microscope stage so that the tilt direction is parallel to the film transport direction. Roll film is usually moved from east to west (right to left). The stage therefore tilts to the east or west with its long axis north-south.

3) Place a particle slide preparation or stage micrometer on the tilting stage and add the bar magnets to hold the slide in place.

4) After the microscope is focused on the sample, tilt the stage alternately east and west and observe the line along the sample that stays in focus. Align this tilt axis parallel to the north-south cross line by rotating the microscope stage. Lock the stage to prevent accidental movement. Superimpose the tilt axis on the north-south cross line by sliding the tilting stage into position.

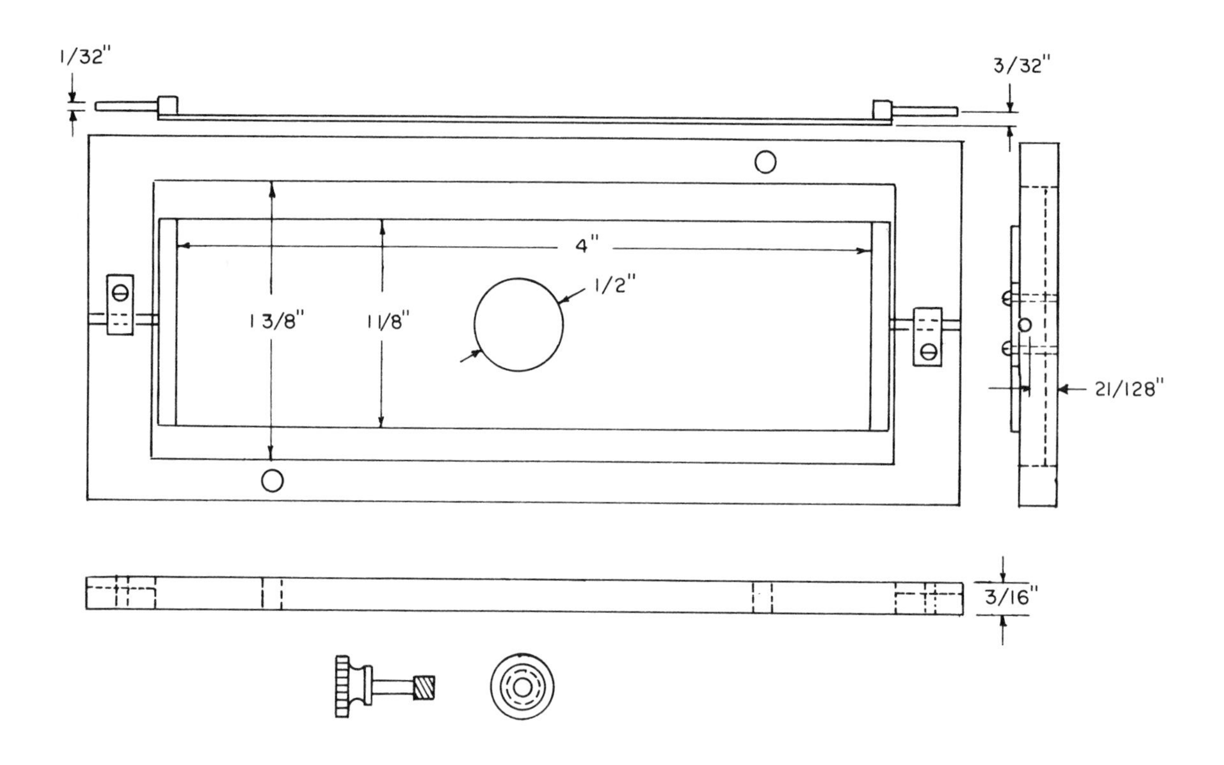

Figure 63. Working drawings for the tilting stage

Figure 64. Photograph of the tilting stage

(The holes in the frame of the tilting stage are oversize to permit this positioning.) After tightening the thumbscrews, the stage is aligned for any sample.

5) Position the photomicrographic attachment so that the film is transported from east to west. Visual alignment is sufficient to give satisfactory stereo pairs.

The proper angle of tilt is not, by any means, a unique angle which must be controlled within ± 5 min of arc. Any angle between 3 and 7° each way gives a good stereo effect. The tilting stage shown here gives 7° each way at maximum tilt and 4° in a controllable intermediate position. When first tested, a series of stereo pairs was taken with tilt angles of 2 x 1°, 2 x 2°, 2 x 3° etc. up to 2 x 7° (7° each way). On viewing these, it was apparent that the strongest stereo effect was shown by the 2 x 7° shots but that 2 x 3° was nearly as good. The apparent height difference of object points increases, naturally, with the angle of tilt; 4° each way is most realistic.

A reproducible tilt angle is obtained by alternately pushing each side down until it contacts the microscope stage. The angle can be decreased by placing coverslips beneath the tilt plate on the microscope stage. Each no. 2 coverslip reduces the tilt angle by approximately one degree. Similarly, the tilt angle can be increased by placing coverslips under the tilting stage frame, although this places more of the field out of focus. If one wishes to repeatedly use an angle less than 7°, one can solder or glue a pointer to the tilt plate and mount a fixed scale, reading in degrees of tilt, on the frame. Any angle up to 7° can then be selected.

Once the stage and camera are aligned, taking a stereo pair is as easy and quick as photographing the same sample field twice.

Samples for stereophotomicrography should be prepared in a viscous or nonflowing medium such as Aroclor® 5442. Particles in low-viscosity media, such as water, often move as the stage is tilted from side to side. The sample should be mounted at the center of a standard 25 x 75 mm slide. The slide can then be moved on the tilting stage so that any area under an 18-mm diameter coverslip can be photographed.

The sample field can be chosen with the magnets in place since the slide can be moved easily. Set the tilt plate horizontal while searching the slide; it will remain horizontal unless it is deliberately tilted. It is close enough to the microscope stage that Köhler illumination can be used with any condenser-objective combination. However, for oil immersion, it is sometimes necessary to remove the microscope stage insert (when a stage is so equipped) so that the condenser may be raised.

5. Exposure Control

Control of exposure is a twofold problem since both top and transmitted illumination must be used. Considering first the transmitted light, there is the problem of showing isotropic particles as well as anisotropic crystals. The background must, therefore, be gray so that the dark outlines of isotropic particles will be visible, as well as the polarization colors of anisotropic particles. The absolute light intensity is important but so is the angle between the vibration directions of the polarizer and analyzer. To help in making a choice, the light intensity was measured using a built-in photocell as a function of angle between these two vibration directions, with no preparation in the field and with the upper polar removed (Table XIII). Trial exposures have shown that polarization colors will be properly exposed when the polars are crossed about 75° so that the meter reading with crossed polars is about one-twentieth that with only one polar. The photomicrograph is then exposed so that the background is underexposed gray; all birefringent crystals should be fully exposed and all isotropic crystals will be visible as black outlines on the gray background.

For top illumination, the proper exposure should fully reveal any white or colored material and register grays or blacks as such. This is accomplished by using a small square of good quality bond paper in place of the preparation and measuring the exposure meter reading from this paper. The paper should be covered with a coverslip if the preparation will be a covered preparation. About 15% of the incident light is lost at the coverslip even if the incident angle of the top light is high. The angle of tilt makes a considerable difference; in one test, the readings were 8 for the preparation tilted 7° toward the lamp, 14 for 7° away from the lamp. The angle of incidence of the light beam should be at least 30° from the horizontal. Since the intensity of the top lighting is usually too low and since only one of the stereo pairs is to be top lighted, the illuminator should be placed at the left so that the "right eye" shot (left side of stage tilted down), which must be the second shot, can be the one with top lighting. If the stereo pairs are exposed in the reverse order, then the effect is reversed (bumps will become depressions and vice versa) unless the two are cut out and interchanged right for left.

Table XIII. Light intensity for crossed polars, parallel polars and intermediate positions

Angle	Exposure meter reading No preparation	
	Two polars	No upper polar
0° (parallel polars)	50	63
15	45	63
30	33	63
45	20	63
60	9	63
75	3	63
90 (crossed polars)	0	63

The two light intensities (transmitted and top) should be adjusted (with neutral density filters) so that both require the same exposure, otherwise a double exposure is required. Some stereo pairs are shown in Figures 65-67.

6. Viewing Stereo Pairs

To obtain the stereo effect, it is necessary to visually "fuse" the left and right eye images. This can be accomplished with a stereo viewer or with unaided eyes. Both viewing methods have the same limitation — the maximum usable print or frame width must not exceed about 65 mm. This is an average interpupillary distance and should be the maximum separation between identical points in the stereo pair. If these image points are farther apart than an individual's interpupillary distance, the pair cannot be fused with unaided eyes. (Large-format stereo viewers used for x-ray radiographs and aerial photographs are available.)

Any pair of image points separated by distances equal to or less than the interpupillary distance can be fused. This includes unmounted stereo transparencies in which image points are only 36 mm apart. The viewing techniques described below apply to 35-mm transparencies as well as any larger prints or transparencies that do not exceed interpupillary distance.

Realist Inc.[25] manufactures a stereo viewer that accepts stereo transparencies mounted 62 mm apart. The stereo transparency frames are only 20 x 23 mm, however, conforming to the once popular macro stereo camera format. This necessitates reducing the 35-mm transparency from 24 x 36 mm to 20 x 23 mm, which results in an unfortunate loss of field. Another drawback to these viewers is that Eastman Kodak Company[26] cannot mount 35-mm stereo transparencies that are on adjacent frames. Their automatic stereo mounting equipment accepts only macrostereo transparencies spaced at every other frame (with the smaller frame size). Therefore, 35-mm stereo transparencies must be mounted by hand which is time-consuming and often results in transposition of frames. This transposition gives the pseudoscopic effect of making the image seem to be set behind the film plane. The correct stereo view should cause the image to seem to be in front of the film plane.

It is easy to construct a stereo viewer for unmounted 35-mm transparencies from cardboard and two matched lenses. The lenses should have approximately 50-mm focal lengths (nominal 5X) and 30-mm diameters. The lenses are secured between two pieces of cardboard with the lens centers separated by the interpupillary distance. The lenses should be turned slightly towards each other so that they focus in the same plane, 36 mm apart. The focal plane of the lenses locates the film position. A cardboard slot for the transparencies is placed at this position. The stereo pairs need only be cut from the roll and dropped into the homemade viewer. The viewer is then held towards a bright neutral background for illumination.

A similar viewer can be constructed for small prints such as Polaroid® 3-1/4 x 4-1/4 in. film packs. (A half inch of the 3-1/4 in. dimension must be sacrificed.) The viewer lenses should be set parallel for these prints. Since the prints cannot be backlit either a strong incident light source or a well-lit room is needed.

The stereo pairs shown in the figures are mounted this same distance apart; hence they can be fused by holding them up so that you look over them at the far wall. As you look at the far wall, but conscious of the out-of-focus images of the stereo pairs, note that there are apparently three images, the center one being the coalesced doublet. Now transfer your attention and your eyes to this doublet and bring it into focus. It should "sharpen up" to give a 3-D image with one extraneous image on either side which should be ignored. If you lose the doublet repeat the above procedure a few times. Only two reasonably good eyes and some practice are needed before you can merely look at any stereo pair directly and by looking "through the paper" at an imaginary far wall, fuse the two images.

If you still have difficulty, one or more of the following simple tricks may help.

1) Bring the stereo figure up to the eyes and only an inch or less away. Note that each eye sees a blurred image; the left eye the left figure, the right eye the right figure. Now move the stereo figures away from the eyes until a clear image of each is formed by the eyes (at your normal reading distance) with the right eye still focused on the right figure and the left eye still focused on the left figure. A stereo effect should be obtained.

2) Hold the stereo pair at the proper viewing distance (your normal reading distance), then interpose a sheet of cardboard about 8 to 10 in. long between the tip of your nose and a line separating the two stereo figures. Your left eye cannot now see the right figure nor your right eye the left figure; however, each eye looking at its own figure should enable you to see a stereo effect.

3) Cut two 1/8 in. holes symmetrically in a 3 x 5 in. file card with a distance between them equal to your interpupillary distance. Look through these two holes simultaneously with both

---

25. Realist Inc., Menominee Falls, Wisconsin.

26. Eastman Kodak Company, Rochester, New York.

Figure 65. Stereophotomicrographs of table salt, slightly uncrossed polars, 70X

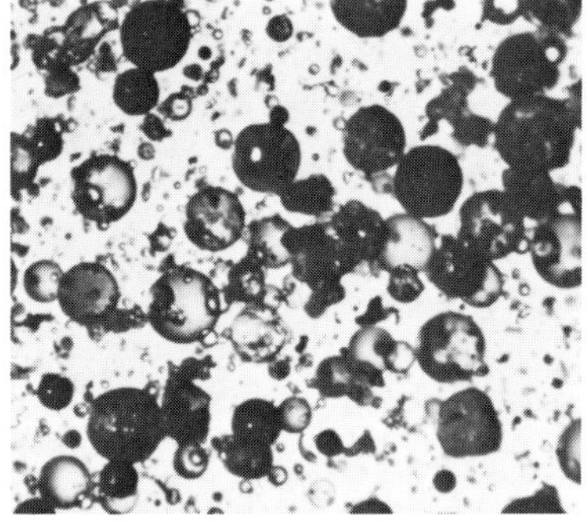
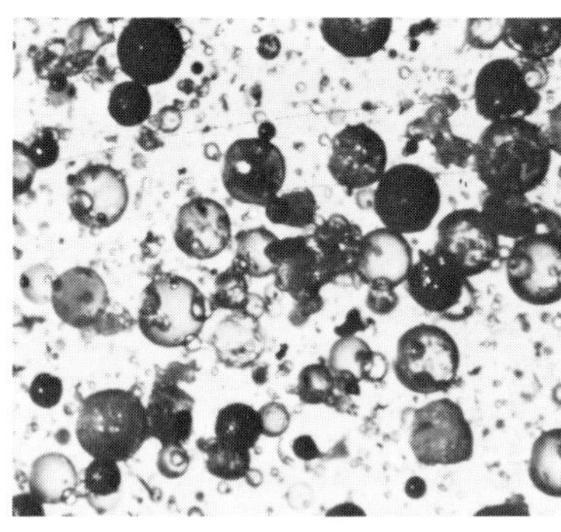

Figure 66. Stereophotomicrographs of power plant flyash top lighted for right eye, slightly uncrossed polars, 70X

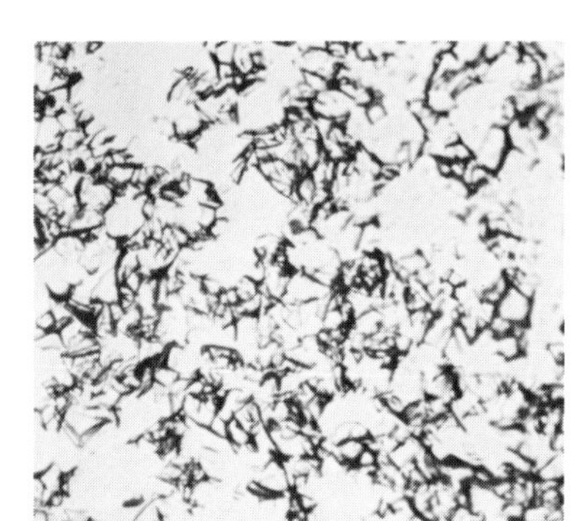
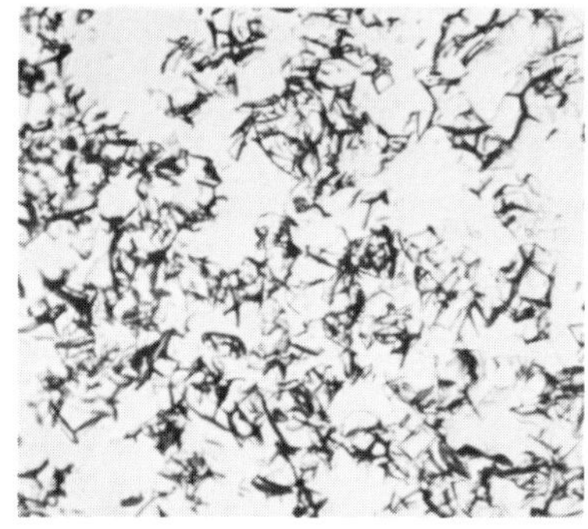

Figure 67. Stereophotomicrographs of perlite top lighted for right eye, slightly uncrossed polars, 70X

eyes at a stereo figure held about 10 in. away. Again, the left eye should be restricted to the left figure and the right eye to the right figure. The two figures should fuse to give a stereo effect.

Stereo pairs can be fused with the unaided eye, i. e., without the aid of the usual viewers, but this does require some practice.

First, hold one finger pointing vertically and about 10 in. before your open eyes. Now look past your finger at some distant object (across the room is far enough) and note that you are conscious of two images of your finger (one by each eye). Now hold another finger from the other hand pointing vertically alongside the first finger the same distance from your eyes and about 2 in. apart. Note that as you look at the far wall there are now four images of the two fingers. Note also that if you move the two fingers closer together or farther apart, the two pairs of blurred images move together or apart. At one position, with your fingers about 2 in. apart and your eyes still focused on the far wall, two of the four images can be made to coalesce (fuse).

VII. MICROMETRY

A. INTRODUCTION

Very small linear distances can be measured very accurately with the microscope. This permits particle size determination and the quantitative analysis of physical mixtures of very small particles. The usual unit of length for microscopical measurements is the micrometer (1 µm, 10^{-3} mm or about 4 x 10^{-5} in.). Measuring particles in electron microscopy requires an even smaller unit, the nanometer (1 nm, 10^{-3} µm or 10 A). Figure 68 shows the approximate average and weight of a few common airborne materials.

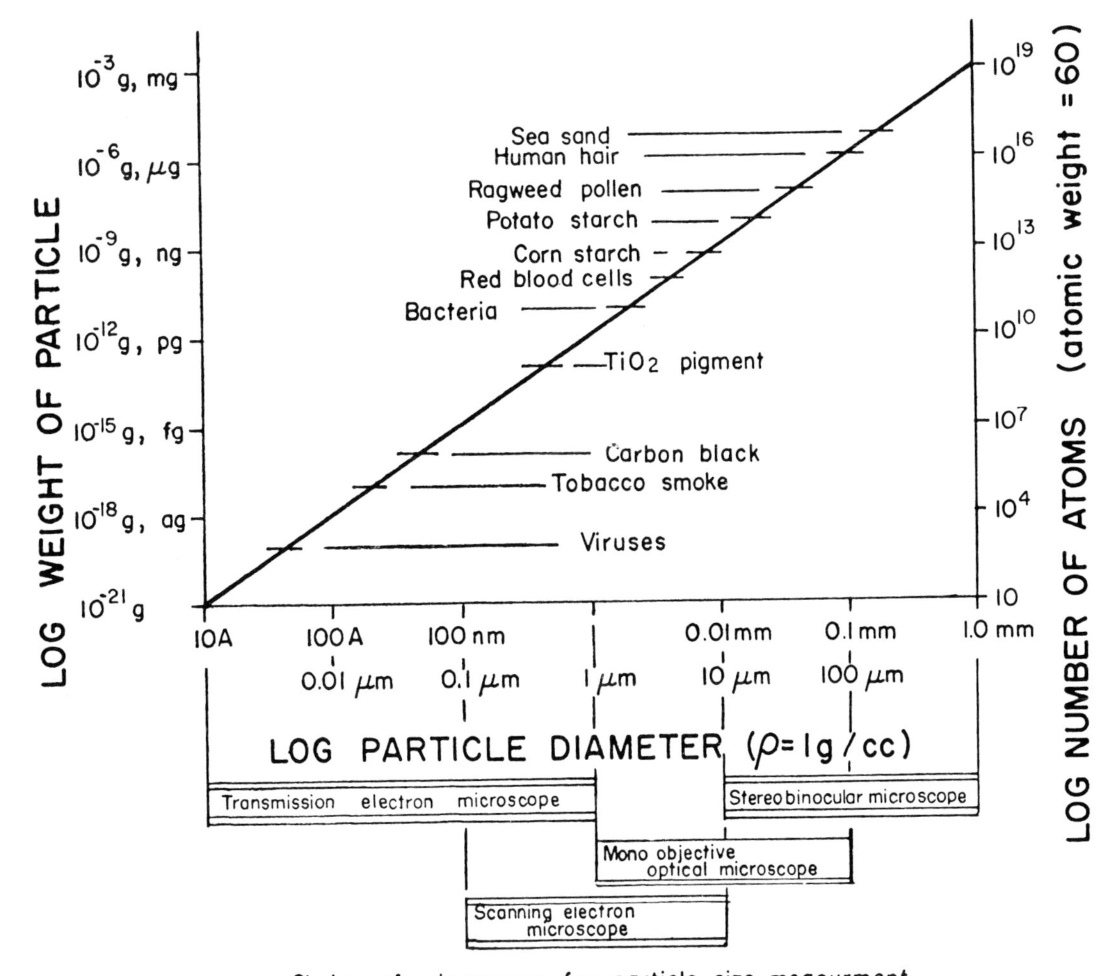

Figure 68. Relationship between particle size, weight, number of atoms and choice of microscope for measuring various sizes

The practical lower limit of accurate particle size measurement with the light microscope is about 0.5 µm. Measuring a particle smaller than this with the light microscope leads to errors which, under the best circumstances, may reach about ± 100% (usually +). Either the scanning electron microscope (SEM) or the transmission electron microscope (TEM) must be used to obtain accurate measurements if the particle size is below about 1 μm. Figure 68 illustrates the choice of microscope appropriate to each size range.

B. METHODS OF PARTICLE SIZE MEASUREMENT

One of the principal uses of high resolving power is the precise measurement of particle size. There are, however, a variety of approximate and useful procedures as well.

1. If the microscope magnification (product of the magnification of objective and ocular) is known, particle size can be estimated. For example, with a 10X ocular and a 16-mm (or 10X) objective, the total magnification is 100X. A particle that appears to your eye to be 10 mm in diameter in the microscope has an approximate actual size of 10 mm divided by 100 to give 0.10 mm or 100 μm. This is, in no sense, an accurate method, but it does permit quick estimation of particle size; the error in this estimate is usually 10-25%. It can be made more accurate, with a monocular microscope, by holding a millimeter rule ten inches before the left eye while looking through the microscope with the right eye. The images of rule and particle can then be superimposed and the particle image diameter can be measured. Dividing this size by the magnification of the microscope gives the particle diameter.

2. Another approximate method is also based on the use of known data. If we know approximately the diameter of the microscope field, we can estimate the percentage of the diameter occupied by the object to be measured and calculate the object's approximate size. The field size depends on both the objective and the ocular although the latter is a lesser influence. Field size should be determined with a stage micrometer scale for each objective and ocular. If this is done, estimates of sizes by comparison with the entire field diameter can be quite accurate, to 5-10%.

3. The movement of a graduated mechanical stage can also be used for rough measurement of diameter of large particles. Stages are usually graduated (with vernier) to read to 1/10 millimeter, or 100 micrometers. In practice, the leading edge of the particle is brought to one of the cross lines in the ocular and a reading is taken of the stage position. Then the particle is moved across the field by moving the mechanical stage in the appropriate direction until the trailing edge just touches the cross line. A second reading is taken and the difference in the two readings is the distance moved or the size of the particle. This method is especially useful when the particle is larger than the field, or when the optics give a distorted image near the edge of the field.

4. The above method can be extended to projection or photography. The image of the particles can be projected on a screen with a suitable light source or they may be photographed. The final linear magnification, M, on the projection surface (or film plane) is given approximately by

$$M = \frac{D}{25} \times OM \times EM \qquad (35)$$

where OM = objective magnification

EM = ocular magnification

D = projection distance from the in centimeters.

The image detail can then be measured in centimeters and the actual size computed by dividing by M. This method is usually accurate to within two to five percent depending on the size range of the detail measured.

5. Stated magnifications and/or focal lengths of microscope optics are nominal and vary a bit from objective to objective or ocular to ocular. For accuracy, a stage micrometer is used to calibrate each ocular / objective combination. The stage micrometer is a glass microscope slide that has, accurately engraved in the center, a scale, usually two millimeters long, divided into 200 parts; each part represents 0.01 mm (10 μm). When this scale is observed, projected or photographed, the exact image magnification can be determined. For example, if five divisions of the stage micrometer measure six millimeters when projected, the actual magnification is

$$\frac{6}{5\ (0.01)} = 120X$$

This magnification figure can be used to improve the accuracy of method 4.

6. The simplest procedure is based on the use of a micrometer ocular. Since the ocular magnifies a real image from the objective, one can place a transparent scale in the same plane as the image from the objective and thus superimpose the scale on the image. This is done by first placing an ocular micrometer scale graticule in the ocular. The ocular micrometer scale is arbitrary and must be calibrated with each objective used. Measurement methods using ocular scales are among the most accurate.

The accuracy of the measurement depends on recognition of the edge of the particle when observing an image which may be slightly out of focus and certainly involves errors introduced by the kind and quality of the illumination as well as errors involved in the optical system itself. Even the refractive index of the mounting medium relative to the index of the particle being measured has an effect on the apparent size. In all cases, the error will be minimized if the magnification of the optical system is sufficient to image the particle over at least ten ocular scale divisions. At worst, under these conditions, the edge of the particle can be measured accurately to within ± 0.25 division. Hence the overall error in measurement considering two sides of the particle could be ±0.5 division, or 10%, for a ten-division particle. This is quite conservative and one might expect to do better than this. He should not expect, however, to do better than ± 2% with the filar ocular scale. The error depends, of course, on the size and increases rapidly as the size decreases to the 1-μm range.

7. The filar micrometer ocular is one of the simplest and most direct ways of measuring the diameter of a single particle. This device has a scale in the ocular superimposed on the image of the object particle in the field of view. If this scale has been calibrated by means of a stage scale so that the number of micrometers (μm) per ocular scale division is known, particles can be measured directly by comparison with the ocular scale. Accurate measurement of the particle is facilitated in this case by having a movable line which can be set on opposite edges of the particle by a rotating drum which is itself divided into fractions of an ocular scale division. The difference between the drum readings after carefully setting the movable line on opposite edges of the particle can then be expressed in fractions of a scale division. The actual dimension in micrometers (μm) is then calculated from the calibration data for that magnification, i. e., the number of micrometers (μm) per ocular scale division.

8. The filar ocular is very simple to use and more accurate than measurement against an ocular scale without the filar movement. It is probably not quite as accurate as the image splitting or shearing ocular. The latter is a very clever arrangement involving a binocular system in which one observes two superimposed images of the particle, one of which is displaced by known micrometer (μm) distances from the other image. If the images are displaced to the point that the two images appear to be just in contact, with neither overlap nor gap between them, the displacement of the image measures precisely the diameter of the particle. Although setting the image displacement depends somewhat on illumination and optical imagery, the setting is more accurate than with the filar ocular. One can expect with such a device to be able to measure particle diameters to within ±1% over the range down to about 20 μm, perhaps ± 2% down 10 μm, ±5% down to 5 μm, ±10% down to 2 μm and no better than ±25% below 2 μm. The error goes up quickly to around ±100% below 1 μm.

C. CALIBRATION OF OCULAR SCALE

Each stage scale has divisions 100 μm (0.1 mm) apart; one or more of these are usually subdivided into 10 μm (0.01-mm) divisions. These form the standard against which the arbitrary divisions of the ocular scale are calibrated. One must calibrate each objective separately by noting the correspondence between the stage scale and the ocular scale. First focus the ocular scale with the eye lens of the measuring ocular then, starting with the lowest power objective, focus on the stage scale and arrange the two scales parallel (Figure 69). It should be possible to find the number of ocular divisions exacly equal to some whole number of divisions of the stage scale, expressed in micrometers (μm).

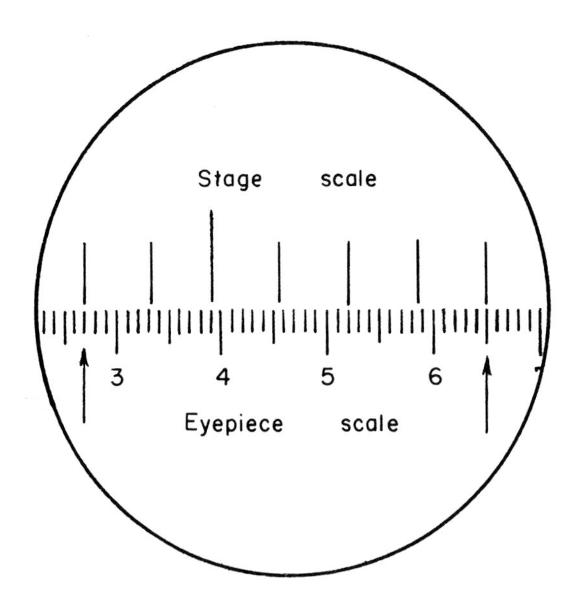

Figure 69. Comparison of stage scale with ocular scale

The calibration consists of calculating the number of micrometers (μm) per ocular scale division. To make the comparison as accurate as possible, a large part of each scale must be used. Let's assume that, with the 16-mm objective, six large divisions of the stage scale (ssd) equal 38 ocular scale divisions (osd). Hence:

$$38 \text{ osd} = 600\ \mu m$$

$$1 \text{ osd} = 600\ \mu m/38 = 15.8\ \mu m$$

Thus when that ocular and scale are used with that 16-mm objective at that microscope tube length each division of the ocular scale equals 15.8 μm, and the scale can be used to measure accurately any object on the microscope stage. A particle, for example, observed with the 16-mm objective and measuring 8.5 divisions on the ocular scale, is 8.5 x 15.8 or 134 μm in diameter.

Each objective on your microscope must be calibrated in this manner. Table XIV illustrates a convenient way to record the necessary data and to calculate μm/osd.

Table XIV. Example of calibration data for an ocular scale

Objective	Tube length	ssd/osd	μm/osd
32 mm (4X)	160 mm	18/44	1800/44 = 40.9
16 mm (10X)	160 mm	6/38	600/38 = 15.8
4 mm (45X)	160 mm	1/30	100/30 = 3.33

Linear distances on other styles of ocular graticules are also calibrated against the stage scale by similar procedures.

D. DETERMINATION OF PARTICLE SIZE DISTRIBUTION

The physical bulk properties of a particulate material usually depend on the particle size distribution — i.e., on the relative numbers of particles of each size that make up the material. The bulk behavior can be studied and predicted from the average sizes, which are calculated from the distribution data and from the size frequency data themselves.

The measurement of size distribution differs from the measurement of individual particles in several important aspects. First, it is essential that the sample be representative and that those particles counted be also representative. It is particularly important to adequately represent the larger sizes, since most of the weight of the sample could be in a relatively small number of large particles. It is essential that sufficient particles be counted for good sampling in all size ranges. This is one of the major limitations of the microscopical procedures for particle size distribution measurement. One must count far too many particles in the lower or middle size range in order to be sure to have a sufficient sample in the larger sizes.

Second, the measurement of any one particle diameter does not have to be as accurate as the one measurement of a single particle in section B, above. This is because any error due to rapid setting of the particle boundary will be offset by corresponding errors in the opposite direction with subsequent particles. This makes it possible to measure size distribution of particles in a given sample much more rapidly than would appear possible from the single particle measurement discussion. Whereas several minutes might be spent making sure of the diameter of the single particle sample, one might measure as many as 500 particles per hour in making a size distribution analysis.

The number of particles to be measured in a sample in order to obtain accurate and representative data depends on the variety of sizes and the diversity of shapes. If the shape is highly irregular or elongated and if there is a wide diversity of sizes, it may be necessary to count several thousand particles. On the other hand, if the samples are regular in shape and not too diverse in size, only 100 or so particles may be sufficient. The determination is, however, tedious and should only be undertaken as a calibration exercise for other automatic procedures or unless only a few such analyses are contemplated.

1. Statistical diameters of particles

The measurement of particle size varies in complexity depending on particle shape. The size of a sphere is defined by its diameter. The size of a cube may be expressed by the length of an edge or a diagonal, and the surface area, volume and weight (if the density is known) of cube or sphere can be calculated directly from such a dimension.* However, if the particles are irregular, weight and other properties cannot be calculated directly, and the particle "size" must include information about the shape of the particle. The expression of this shape takes a more complicated form.

An irregularly shaped particle has a number of different dimensions which might be measured as "diameters". Figure 70 shows three statistical diameters commonly used in determining

* for shape factors see: Horvath, H., Staub Reinhaltung der Luft (English) 34 197-202 (1974).

particle size distribution, and a circle equal in area to the particle.

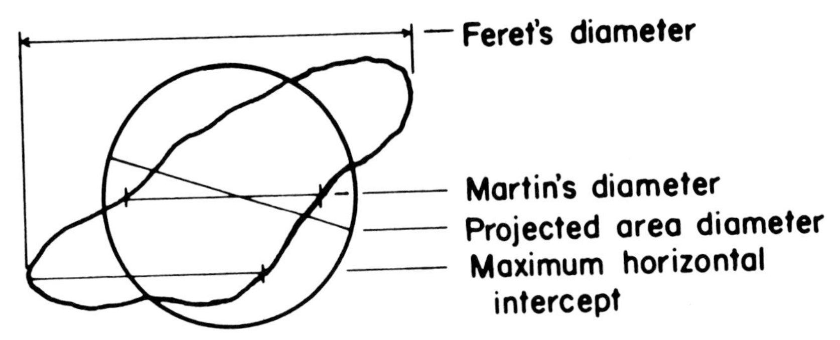

Figure 70. Various statistical diameters

Feret's diameter is the distance between imaginary parallel lines tangent to the particle profile and perpendicular to the ocular scale. It is usually measured with a filar ocular.

The maximum horizontal intercept is the longest diameter from edge to edge of the particle, parallel to the ocular reference line. This statistical diameter is the one obtained with image splitting or shearing oculars. It always falls between Feret's and Martin's diameters, or equals one of them.

Martin's diameter is one of the most often used for particle sizing. It is measured as the dimension (parallel to the ocular scale) that divides a randomly oriented particle into two equal projected areas (Figure 71).

Projected area diameters are found by comparing the projected area of the particle with the areas of reference circles on an ocular graticule. Two adjacent circles, one larger and one smaller in area than the particle, define a size class into which the particle fits. The statistical diameter of the size class (and particle) is the arithmetic mean between the diameters of the two circles.

2. Use of Martin's diameter

Martin's diameter is the simplest means of measuring and expressing the diameters of irregular particles, and is sufficiently accurate when averaged for a large number of particles.

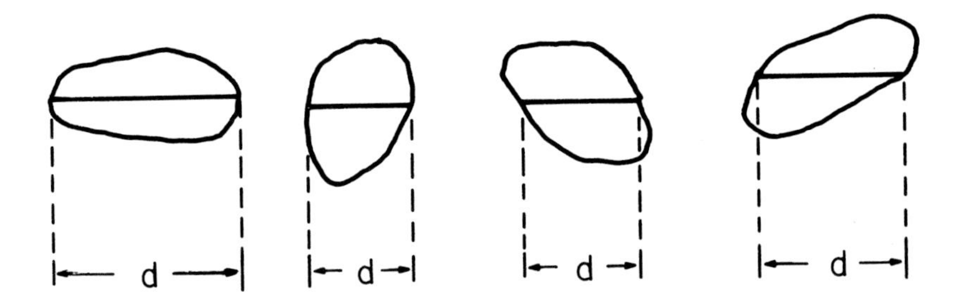

Figure 71. Martin's diameter

The more particles that are counted, the more accurate will be the average particle size. Platelike and needlelike particles should have a correction factor applied to account for the third dimension since all such particles are restricted in their orientation on the microscope slide. When particle size is reported, the general shape of the particles as well as the method used to determine the "diameter" should be noted.

The determination of particle size distribution is carried out routinely by moving a preparation of particles past an ocular scale in such a way that their Martin's diameter can be tallied. All particles whose centers fall within two fixed divisions on the scale are tallied. Movement of the preparation is usually accomplished by means of a mechanical stage but may be carried out by rotation of an off-center rotating stage. A sample tabulation appears in Table XV . The ocular and objective are chosen so that at least six, but not more than twelve, size classes are required and sufficient particles are counted to give a smooth curve. The actual number tallied, which may vary from 200 to 2,000, depends on the regularity of the particle shape and the range of sizes. The size tallied for each particle is that number of ocular scale divisions most closely approximating Martin's diameter for that particle.

Table XV. Particle size tally for a sample of starch grains

Size class (osd*)	Number of particles	Total
1	𝍸 𝍸 𝍸 I	16
2	𝍸 𝍸 𝍸 𝍸 𝍸 𝍸 𝍸 𝍸 𝍸 𝍸 𝍸 𝍸 𝍸 𝍸 𝍸 𝍸 𝍸 𝍸 𝍸 III	98
3	𝍸 𝍸	110
4	𝍸 II	107
5	𝍸 𝍸 𝍸 𝍸 𝍸 𝍸 𝍸 𝍸 𝍸 𝍸 𝍸 𝍸 𝍸 𝍸 I	71
6	𝍸 𝍸 𝍸 𝍸 𝍸 𝍸 𝍸 𝍸 𝍸	45
7	𝍸 𝍸 𝍸 𝍸 I	21
8	II	2
		470

* osd = ocular scale divisions

3. Calculation of size averages

The size data may be treated in a variety of ways; one simple straightforward treatment is shown in Table XVI. For a more complete discussion of the treatment of particle size data, see Chamot and Mason[27], page 436.

---

27. Chamot, E. M., and C. W. Mason, Handbook of Chemical Microscopy, vol. I, third ed. Wiley, New York, 1958.

Table XVI. Calculations for particle size averages

d. (Aver. diam. in osd)	n	nd	nd^2	nd^3	nd^4
1	16	16	16	16	16
2	98	196	392	784	1568
3	110	330	990	2970	8910
4	107	428	1712	6848	27392
5	71	355	1775	8875	44375
6	45	270	1620	9720	58320
7	21	147	1029	7203	50421
8	2	16	128	1024	8192
	470	1758	7662	37440	199194

The averages with respect to number, $\bar{d}_1$; surface, $\bar{d}_3$; and weight or volume, $\bar{d}_4$, are calculated as follows for the data in Table XVI:

$$\bar{d}_1 = \Sigma nd/\Sigma n = 1758/470 = 3.74 \text{ osd} \times 2.82^* = 10.5\ \mu m$$

$$\bar{d}_3 = \Sigma nd^3/\Sigma nd^2 = 37440/7662 = 4.89 \text{ osd} \times 2.82 = 13.8\ \mu m$$

$$\bar{d}_4 = \Sigma nd^4/\Sigma nd^3 = 199194/37440 = 5.32 \text{ osd} \times 2.82 = 15.0\ \mu m$$

* 2.82 micrometers (μm) per osd (determined by calibration of the ocular - objective combination used for the determination).

Cumulative percents by number, surface and weight (or volume) may also be plotted from the data in Table XVI. The percentage of the total weight or volume of the sample which is finer (or coarser) than any given diameter is calculated for each diameter, e.g.,

$$\left[\sum_{d=0}^{d=3} nd^4 \Big/ \sum_{d=0}^{d=8} nd^4 \right] \times 100 = \% \text{ finer than } d = 3 \quad (36)$$

The calculated percents for the cumulative weight or volume curve are plotted against d in μm.

Finally, the specific surface, S_m, in square meters per gram, may be calculated if the density, D, is known; the surface average, d_3, is used. If D = 1.1,

$$S_m = 6 / \bar{d}_3 D = 6 / 13.8\ (1.1) = 0.395\ m^2/g. \quad (37)$$

4. Use of projected area diameter (British Standard method)

The British Standard method[28] for determining particle size distribution utilizes the projected area diameter illustrated in Figure 72. Particles are sized by comparing their projected areas with the areas of reference circles on a standard graticule in the ocular. Each particle is assigned to a size class defined by two adjacent circles, which represent the size limits of that class. Thus the distribution of sizes is obtained in terms of the diameters of circles having the same projected areas as the particles. The British Standard graticule is shown in Figure 72.

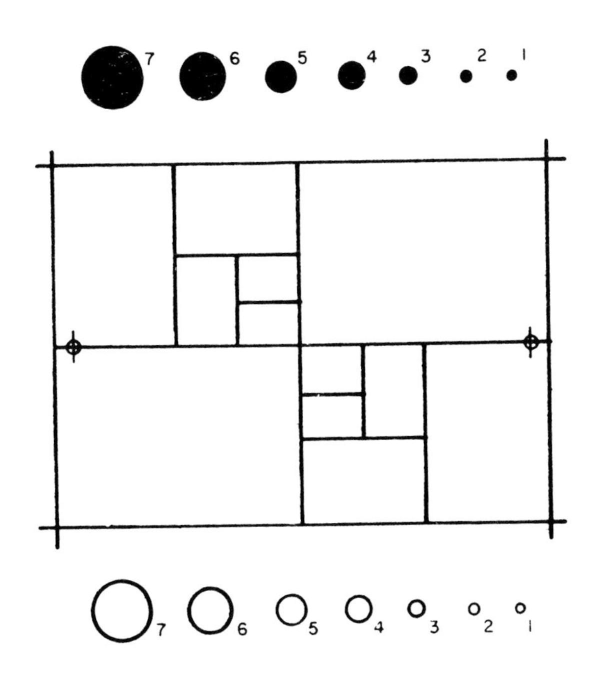

Figure 72. British Standard graticule

The circle areas double progressively, hence the diameters progress by $\sqrt{2}$, so that the size classes can form a continuation of the standard series of sieves for particle sizing. The microscope method covers particles between 0.8 and 150 μm, and it usually is necessary to change the objective magnification during a size determination. The rectangular grids of the graticule are used to interrelate the magnifications used and define the size and number of fields of particles to be counted and sized. The reference circles in the graticule are matched to recommended size classes by adjusting the microscope tubelength for each magnification used — if the tubelength cannot be adjusted other size classes must be calculated.

The size distributions with respect to number and to weight, (or volume, if the sample is not homogeneous), are determined separately. The number of particles that must be sized in order to get statistically valid results can be calculated after preliminary scans have been carried out. Final results are calculated as cumulative percents by number and weight.

For elongated (acicular) particles the Standard recommends a graticule on which rectangles replace the reference circles. Other graticules employing circles have been designed; the best

---

28. British Standard 3406: Part 4 (1963) "Optical microscope method for the determination of particle size of powders." British Standards Institution, 2 Park St., London W.1, England.

known are the original "globe and circle" by Patterson and Cawood[29], the Porton graticule[30], and several graticules designed by Fairs[31]. Cadle[32] discusses their use.

E. COUNTING ANALYSIS

Particulate mixtures can often be quantitatively analyzed by counting the total number of particles of each component in a representative sample. The calculations are, however, complicated by three factors, the average particle size, shape and density of each component. If all of the components were equivalent in particle size, shape and density, then the weight percentage would be identical to the number percentage. Usually, however, it is necessary to determine correction factors to account for the differences in size, shape and density.

When properly applied, this method can be accurate to within ± 1 percent and, in special cases, even better. It is often applied to the analysis of fiber mixtures and is then usually called a dot-count because the tally of fibers is kept as the preparation is moved past a point or dot in the ocular.

A variety of methods can be used to simplify recognition of the different components. These include chemical stains or dyes and enhancement of optical differences such as refractive indices, dispersion or color. Often, however, one relies on the differences in morphology, *e.g.*, counting the percent of rayon fibers in the sample of "silk".

Example: A dot-count of a mixture of fiberglass and nylon shows:

nylon	262
fiberglass	168

The percentage, by number, of nylon is, therefore:

$$\%\ \text{nylon} = (262/262 + 168) \times 100$$
$$= 60.9\% \text{ by number.}$$

However, although both fibers are smooth cylinders they do have different densities and, usually, different diameters. To correct for diameter, one must measure the average diameter of each type of fiber and calculate the volume of a unit length of each.

	aver. diam., µm	volume of 1-µm slice, µm^3
nylon	18.5	268
fiberglass	13.2	117

The percent by volume is, then:

$$\%\ \text{nylon} = \frac{262 \times 268}{262 \times 268 + 168 \times 117} \times 100$$

$$= 78.1\% \text{ by volume.}$$

Still we must take into account the density of each in order to calculate the weight percent. If the densities are 1.6 for nylon and 2.2 for glass, then the percent by weight is:

---

29. Patterson, H. S., and W. Cawood, Trans. Faraday Soc. 32, 1084 (1936).

30. Delly, J. G., Particle Analyst 89, Ann Arbor Science Publishers, Ann Arbor, MI (1968).

31. Fairs, G. L., Chemistry and Industry 62, 374 (1943). Also, J. Roy. Micros. Soc. 71, 209 (1951).

32. Cadle, R. D., Particle Size: theory and industrial applications, Reinhold, NY, 1965.

$$\% \text{ nylon} = \frac{262 \times 268 \times 1.6}{262 \times 268 \times 1.6 + 168 \times 117 \times 2.2} \times 100$$

$= 72\%$ by weight.

Example: A count of quartz and gypsum shows:

quartz	283
gypsum	467

To calculate the percent by weight, we must take into account the average particle size, the shape and the density of each.

The average particle size with respect to weight, $\bar{d}_4$, must be measured for each and the shape factor must be determined. Since gypsum is more platelike than quartz, each particle of gypsum is thinner. The shape factor can be approximated or can be roughly calculated by measuring the actual thickness of a number of particles. We might find, for example, that gypsum particles average 80% of the volume of the average quartz particle; this is our shape factor. The final equation for the weight percent is:

$$\% \text{ quartz} = \frac{283 \times \pi \bar{d}_4/6 \times Dq}{283 \times \pi\, \bar{d}_4/6 \times Dq + 467 \times \pi\, \bar{d}_4^*/6 \times 0.80 \times Dg} \times 100 \quad (38)$$

where Dq and Dg are the densities of quartz and gypsum respectively; 0.80 is the shape factor, and $\bar{d}_4$ and $\bar{d}_4^*$ are the average particle sizes with respect to weight for quartz and gypsum respectively.

F. AREAL ANALYSIS

The relative ar^as of two or more substances as presented by a polished surface or thin section are directly related to the percentages of each by volume. If the various densities are known the percentages by weight can also be calculated. This technique is usually applied to metals but has been applied to the analysis of cast high explosive mixtures such as amatol (TNT plus ammonium nitrate), the percentage of filler in plastics, pore size in filters etc. The relative amounts of particles can also be accurately measured if the particulate mixture is first cast in plastic, then cut and polished to give a plane section. It must be possible to differentiate between the different phases in the thin section by the differences in optical properties or in the polished surface, when necessary, by use of an appropriate etching agent.

The relative areas may be measured directly in the microscope if a cross-ruled ocular reticle is used. The areas may also be measured on a photomicrograph or a drawing made with a camera lucida (drawing camera) and a planimeter.

Many types of reticles and graticules have been designed for various methods of areal, counting, and particle size analysis. Delly[30] reviews this subject succinctly.

G. MICROSTEREOLOGY

Microstereology is a technique for deducing the internal structure of a three-dimensional body from the microscopical study of two-dimensional sections through it[33]. By applying statistical methods, line intersection or point counts can be used to determine complicated structures quantitatively. The technique is very tedious, but advances in micro-stereology are now being spurred by the application of automated counting and measuring methods.

---

33. Watts, J. T., Microscope 18, 35 (1970).

H. AUTOMATED SIZE ANALYSIS

Several authors[34,35] have made an effort to simplify the job of recording data from microscopical particle size determination. These procedures are based usually on setting movable cross lines in the ocular to the particle diameter and then pressing a button to record that setting. These procedures seem to have little advantage over a tally counter for each size frequency range.

A slightly different system has been promoted by Zeiss which is applied to photomicrographs or electron micrographs[36]. An iris diaphram is oriented over each particle in turn and set to an equivalent area. Pressing a button records the diaphram setting, hence the equivalent diameter. This system works quite rapidly and many hundreds of particles per hour can be measured by a patient operator.

Integrating stages are mechanical stages with several identical spindles, any of which can be used to traverse a sample. The distance moved using any one spindle is shown by a scale on that spindle. Such stages are used for linear analysis of mixtures: by traversing each component with its own spindle the accumulated distances for each are measured separately and recorded automatically on the spindle.

Finally, however, there are completely automatic systems* involving highly sophisticated electronics to take all of the drudgery out of the microscopical procedure[37,38,39]. The microscope is still used, but the image itself is analyzed by electronic means. Particles can be counted and measured at very high rates of speed. These instruments can be programmed to count particles, measure projected areas or measure any of the standard diameters such as Martin's, Feret's, or projected area diameter. They certainly take the drudgery out of the microscopical procedure and eliminate the problem of representative sample size. Thousands of particles can be counted and sized in seconds.

Automatic image analyzers still have the problem of accurate imagery of the particle before measurement and are especially prone to error at high magnification when some particles are usually out of focus. The microscopist normally focuses up and down for best focus on each particle whereas the automatic image analyzer measures all particles in the field of view with one focus setting. Nevertheless, one can, with these precautions in mind, obtain very accurate particle size distribution data. They also, of course, quickly calculate any of the averages or percentages needed, thus encouraging mathematical analysis of the data which would be too tedious without automatic computation.

---

34. Lark, P. D., _Microscope_ **15**, 1 (1965).

35. Humphries, D. W., _Microscope_ **15**, 267 (1966).

36. Endter, F., and H. Gebauer, _Optik_ 13, 97 (1956).

37. Jesse, A., _Microscope_ **19**, 21 (1971).

38. Manalen, D. A., and J. S. Glass, _Am. Lab._, 45 (October 1970).

39. Fisher, C., _Microscope_ **19**, 1 (1971).

* _e.g._, Texture Analyzer (Leitz), πMC (Millipore), Quantimet 720 (IMANCO, Metals Research, Micro-Videomat (Zeiss), Telecounter (Schaefer), Digiscan (Kontron) and QMS (Bausch & Lomb).

I. ANGULAR MEASUREMENT

The rotating stage of the polarizing microscope is graduated in degrees and can be read to some fraction of a degree (usually to tenths) with a vernier. To measure an angle in a specimen the stage is rotated to bring, in turn, each side of the angle parallel to the crossline. Depending upon how the specimen was rotated and which crosslines were used, 90° may need to be added to, or subtracted from the difference of the readings to compute the angle. Several sets of readings should be taken and averaged.

If rotating the specimen is undesirable or impossible, goniometer oculars are available in which the crosslines are rotated through angles read on a graduated scale at the periphery of the ocular.

GENERAL REFERENCES ON MICROMETRY

Allen, T., Particle Size Measurement, Chapman and Hall, London, 1968.

Cadle, R. D., Particle Size, Theory and Industrial Applications, Reinholt, New York, 1965.

Chamot, E. M., and C. W. Mason, Handbook of Chemical Microscopy, vol. 1, third ed., John Wiley and Sons, New York, 1958, pp. 419-488.

Draftz, R. G., The Particle Analyst, pp. 145-152, Ann Arbor Science Publishers, Ann Arbor, Michigan, 1968.

Draftz, R. G., "Optical and electron microscopy," chapter in Handbook of Powder Technology, ed. M. R. Jackson, Marcel Dekker, Inc., New York, 1972.

VIII. CRYSTAL MORPHOLOGY

A. INTRODUCTION

The chemical microscopist must learn some crystallography since all particles except glasses and some polymers are crystalline. It is almost as important to be able to detect the absence of crystallographic properties as to be able to use them in identifying individual particles. Although most dust particles can be identified from simple morphological properties, the use of crystallography greatly extends the number of particles that can be identified. In any case, crystallography greatly aids particle analysis by making identifications more definite and rapid. Crystallographic properties are among those physical properties which may affect the performance of a substance used in a process or a product. Thus, they are particularly important to the industrial microscopist.

Crystals are characterized by repetition of their constituent atoms in a three-dimensional array, quite unlike the randomly placed atoms in a liquid or other noncrystalline material (Figure 73). The orderly arrangement may vary, but only according to definite rules. Many metals, all of the alkali halides and many other compounds have a very simple arrangement of atoms, called cubic; all such materials are said to be in the cubic system. There are six systems in all, differing in atom arrangement. Figure 74 shows top, front and side views of the lattice for each system.

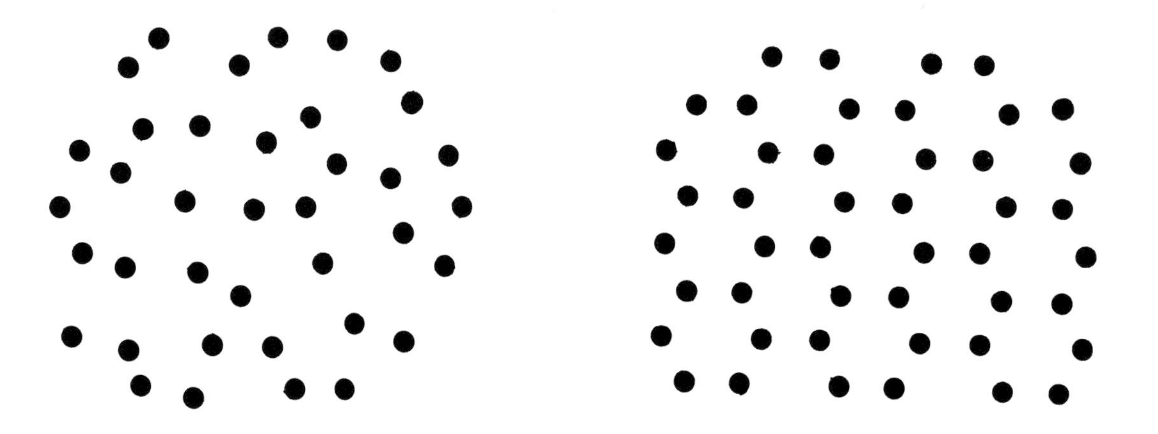

Figure 73. Noncrystalline vs crystalline materials

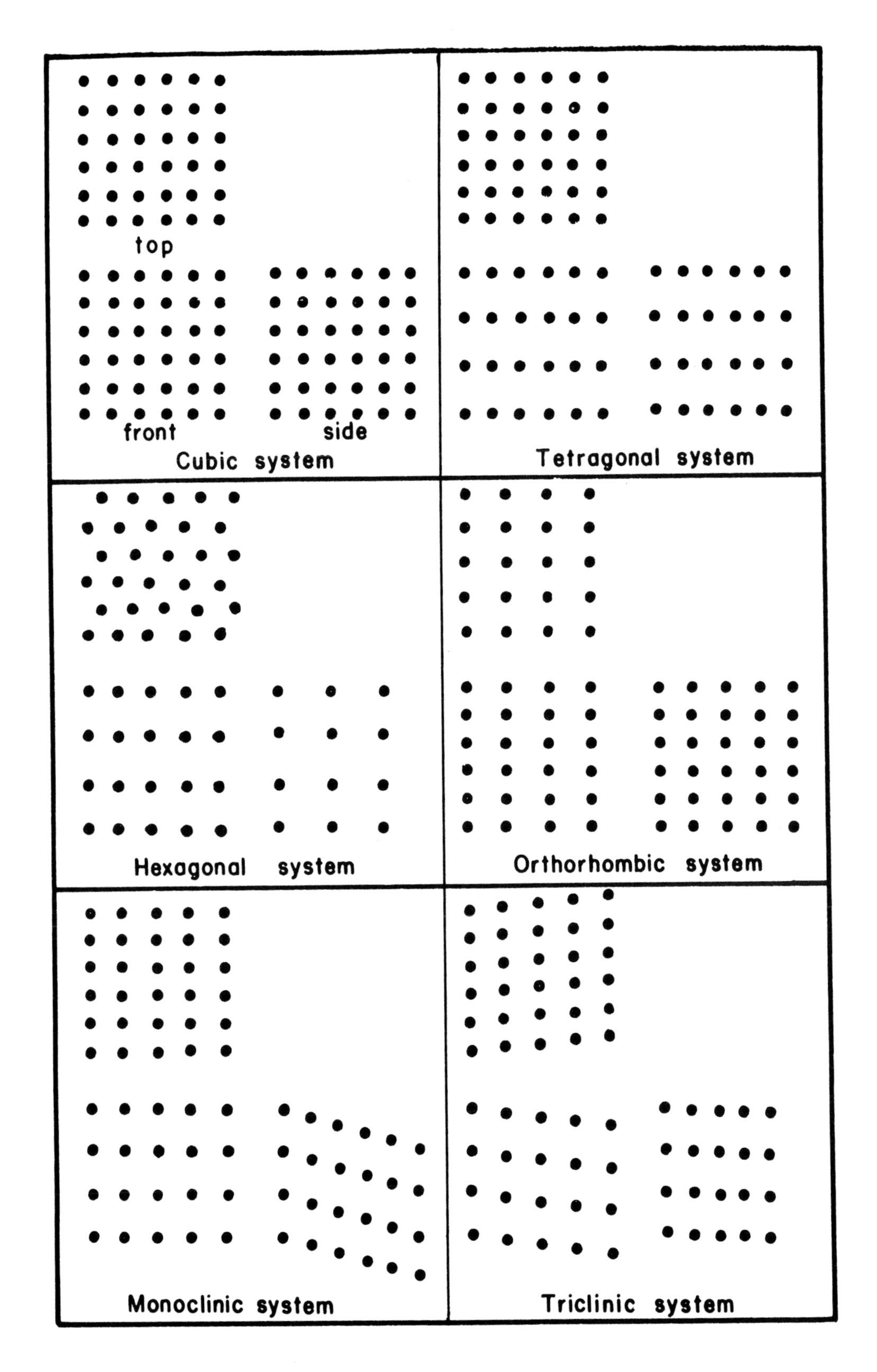

Figure 74. Orthoscopic views for crystals in each of the six crystal systems

B. THE SIX CRYSTAL SYSTEMS

Atomic arrangements in the different crystal systems are verbally described in Table XVII.

Table XVII. Orientation of the crystallographic axes

System	Spacing of atoms and orientation of axes
Cubic	equal spacing of atoms along three mutually perpendicular axes
Tetragonal	equal spacing of atoms along two mutually perpendicular axes; normal to the plane thus formed, the spacing is different
Hexagonal	equal spacing of atoms along three axes 120° apart in the same plane, normal to the plane thus formed, the spacing is different
Orthorhombic	unequal spacing of atoms along three mutually perpendicular axes
Monoclinic	unequal spacing of atoms along two mutually oblique axes; normal to the plane thus formed, the spacing is different
Triclinic	unequal spacing of atoms along three mutually oblique axes

The atomic arrangement in solid materials is obviously very important since it explains, first, why some solids are crystalline and, second, why different crystalline materials have different shapes (Figure 75).

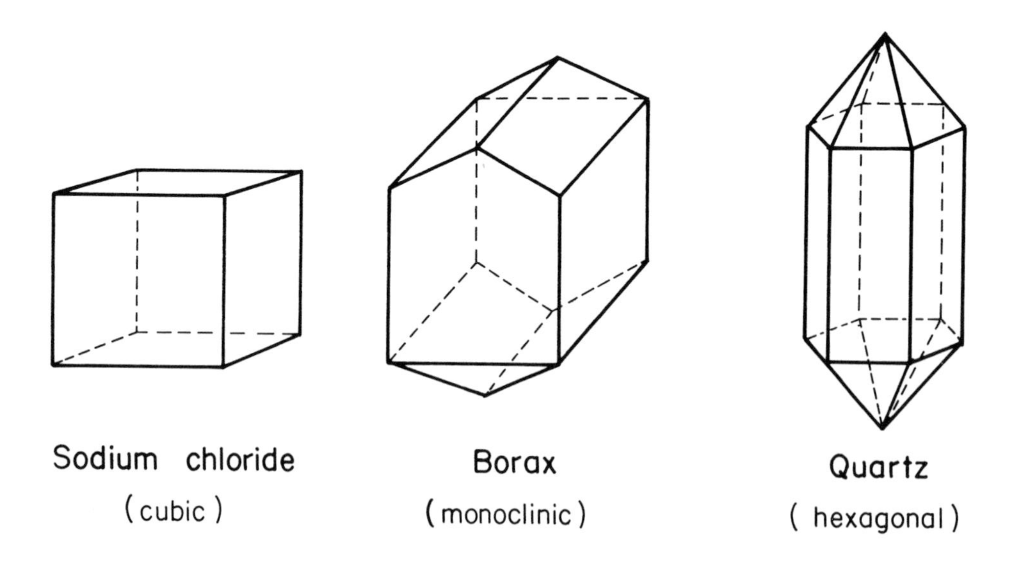

Figure 75. Typical crystals

1. The Cubic System

Crystal faces are directly related to the internal arrangement of atoms, the crystal lattice. Figure 76 shows possible positions of crystal faces on a crystal lattice in two and three dimensions.

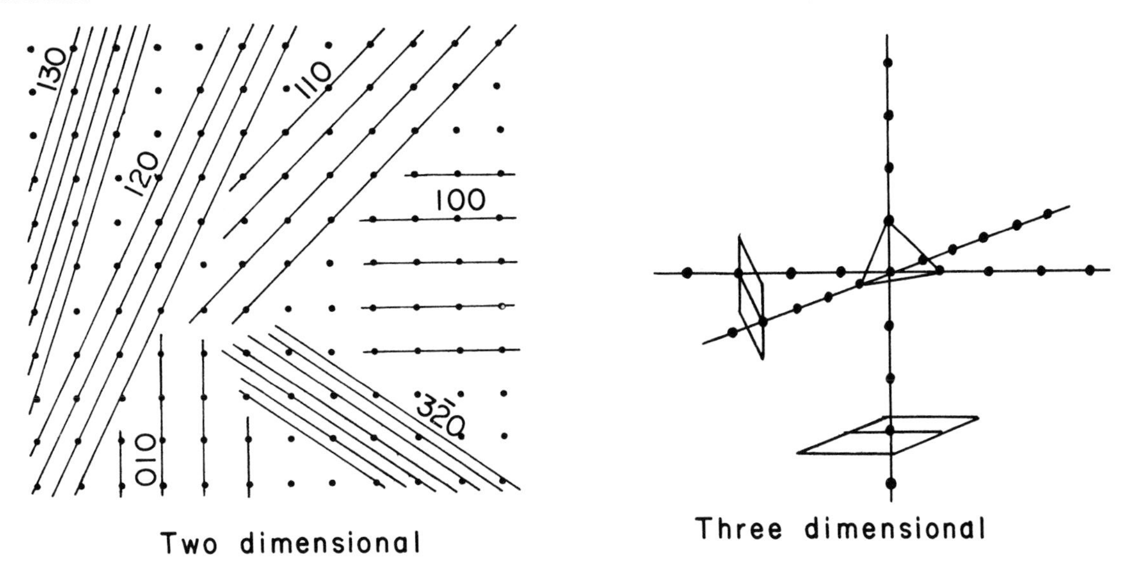

Figure 76. Possible crystal faces on a crystal lattice

In the cubic system, there are three common crystal forms: cube, octahedron and rhombic dodecahedron. The cube (Figure 77) results when every crystal face parallels two of the three axes and cuts only one axis. The octahedron (Figure 78) has eight faces, all of which intersect all three crystal axes. The rhombic dodecahedron (Figure 79) has twelve faces, all of which intersect two of the three crystallographic axes.

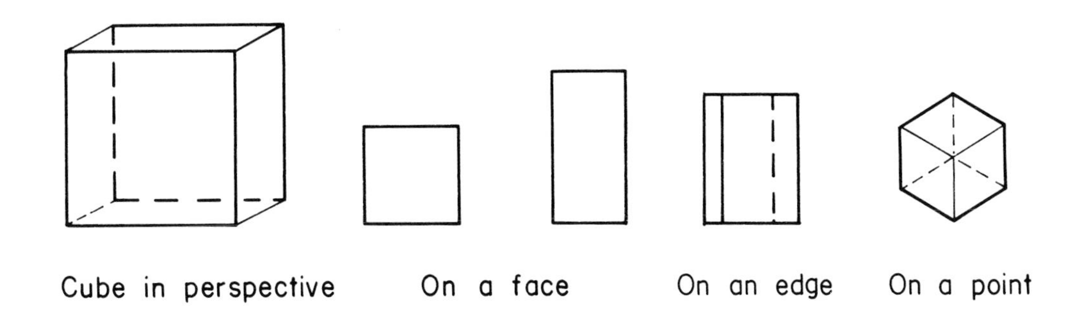

Figure 77. Ideal cube and typical distortions

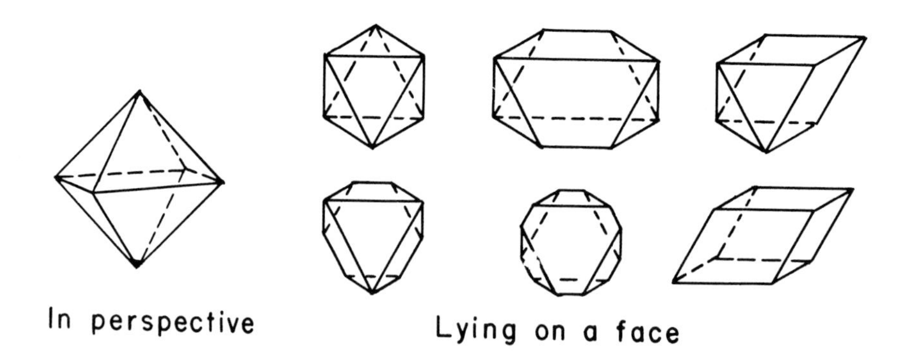

Figure 78. Ideal octahedron and typical distortions

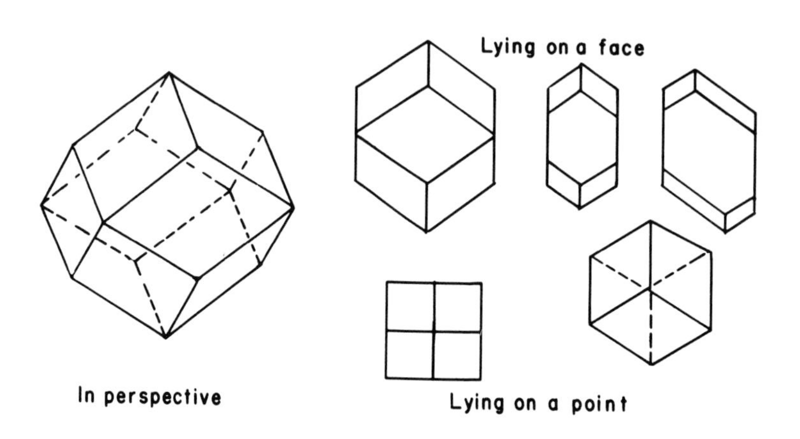

Figure 79. Ideal rhombic dodecahedron and typical distortions

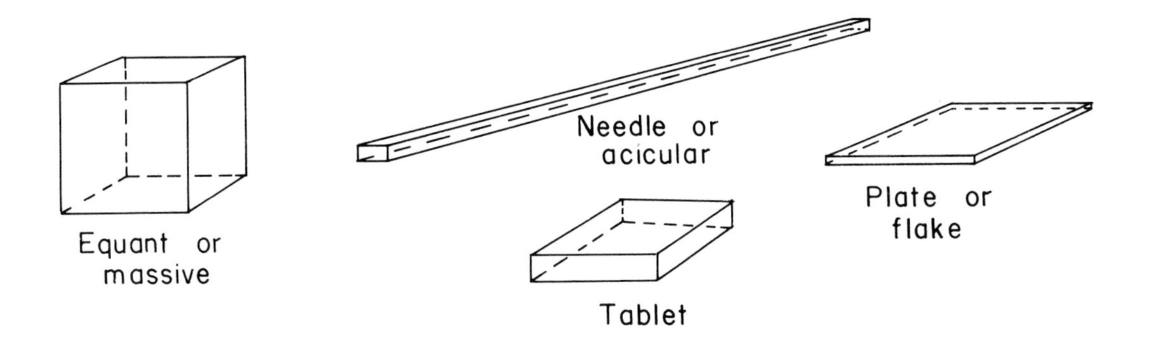

Figure 80. Common crystal habits

2. Distortions

Figures 77-79 illustrate the growth of distortions. The sketches on the left show the ideal crystals on which all of the faces are equally developed. The sketches of crystals lying on a face in each figure show distortions in which at least one face has developed more than the others. At this point, it is important to notice that all of these crystals, both ideal and distorted, have identical interfacial angles. In any distortion, the crystal faces move only parallel to themselves, thus changing the general shape but not the angles. This is an expression of the law of constancy of interfacial angles.

3. Habit

A crystal's general shape is usually referred to as its habit. Figure 80 shows some common habits. The same descriptive terms are used for crystals in any of the crystal systems. Words like cube, hexagon, prism and rhomb are best not used unless the crystal system is known. Rods, plates, tablets, needles etc. can occur in any of the six crystal systems.

4. Cleavage

Another important feature of many crystalline substances is cleavage. Some minerals, e.g. limestone (calcite), have excellent cleavage, so that the powdered mineral breaks up into quite well-formed small crystals called rhombohedra (Figure 81). Few other minerals show as perfect cleavage as calcite and many show no regular cleavage. In fact, quartz, a crystalline material, and glass, a noncrystalline material, show almost identical fracture, usually called conchoidal. These two materials when finely ground cannot be differentiated microscopically except by optical tests.

Crystalline materials are usually described morphologically in terms of habit, form, interfacial angles and, when possible, system. The angles most often used are the so-called profile angles shown when a crystal is observed lying on a face. A cube, for example, shows 90°; an octahedron, 60° or 120°; a rhombic dodecahedron, 70° 32' and its supplement, 109° 28'; and a calcite rhombohedron, 101° 54' and its supplement, 78° 6'.

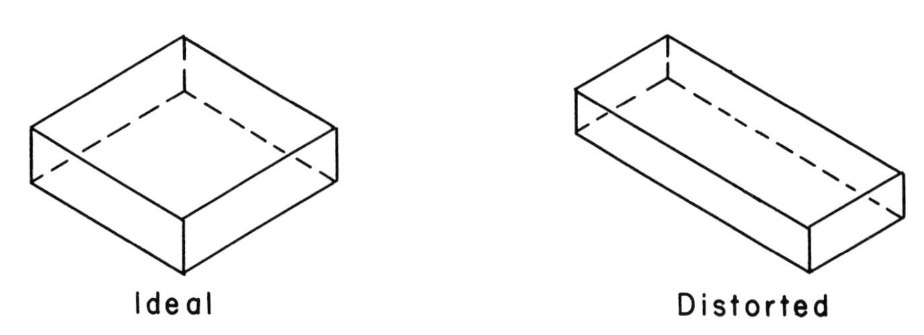

Figure 81. Rhombohedra

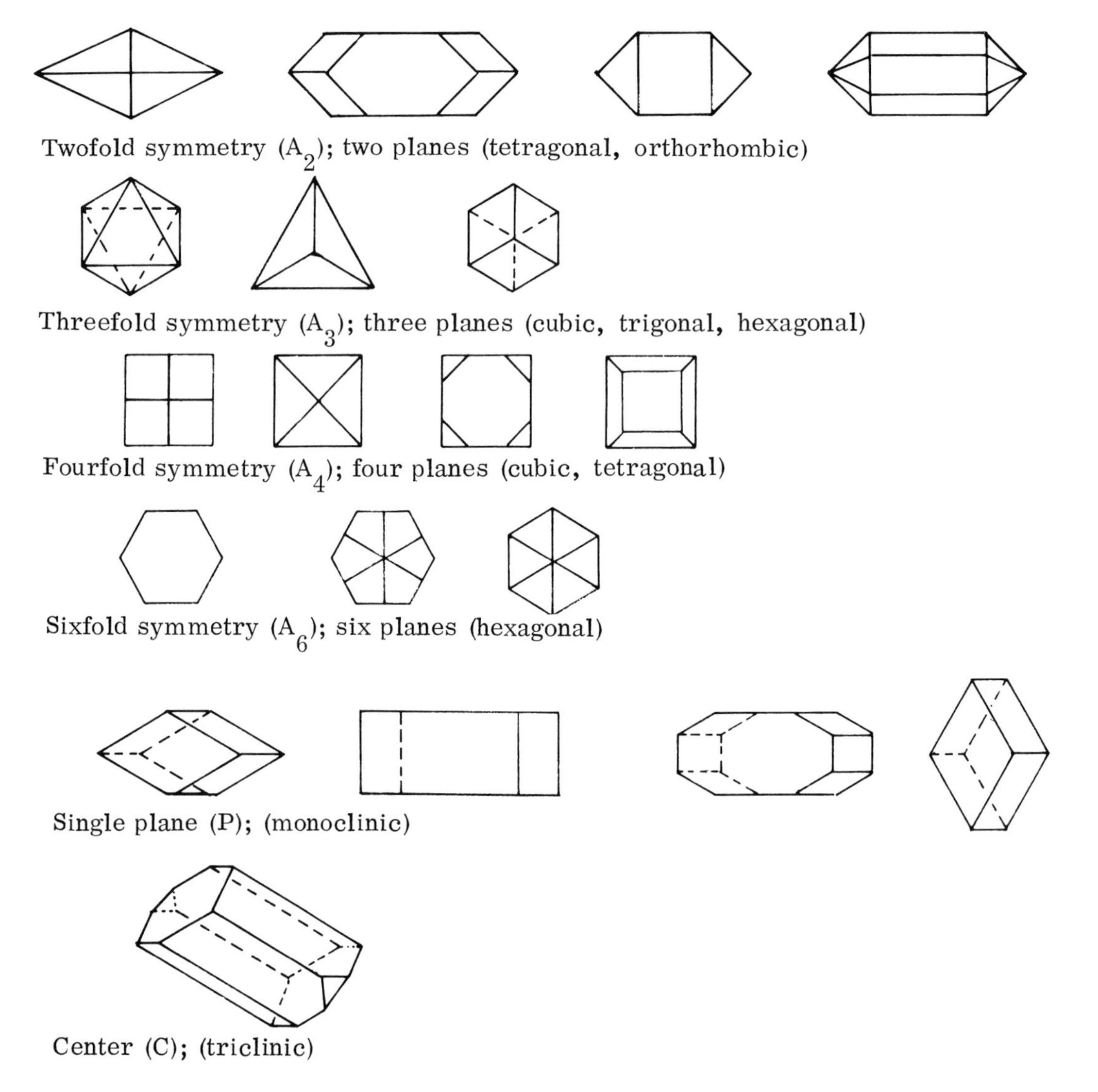

Figure 82. Elements of symmetry

5. Laws of Crystallography

Crystal forms and habits are governed by two laws of crystallography. The first is the law of CONSTANCY OF INTERFACIAL ANGLES: In all crystals of the same substance, angles between corresponding faces are constant. The second is the law of RATIONAL INDICES: The lengths of intercepts of different crystal faces on a crystallographic axis are in ratios of small integers.

6. Symmetry

External crystal symmetry provides one of the easiest ways to differentiate between crystal systems. The elements of external symmetry include the plane (P), the axis (A) and the center (C). The axis of symmetry may be twofold (A_2), threefold (A_3), fourfold (A_4) or sixfold (A_6). Each crystal system has different symmetry requirements (Table XVIII and Figure 82).

Table XVIII. Symmetry elements in the six crystal systems*

	Crystal System						
			hexagonal				
Symmetry element	cubic	tetragonal	trigonal	hexagonal	orthorhombic	monoclinic	triclinic
A_2	6	4	3	6	3	1	0
A_3	4	0	1	0	0	0	0
A_4	3	1	0	0	0	0	0
A_6	0	0	0	1	0	0	0
P	9	5	3	7	3	1	0
C	1	1	1	1	1	1	1

* In every crystal syst em there are forms having less symmetry than that indicated here.

7. Assignment of Axes

The assignment of crystallographic axes is based on the positions of external symmetry elements and on internal lattice spacings. To make the assignment using the following rules requires consideration of an undistorted crystal showing the full symmetry of that system (holosymmetric). Crystals showing less symmetry (hemisymmetric or tetartosymmetric and hemimorphic; Figure 83 must be visualized as the holosymmetric equivalent for the purpose of axis assignment.

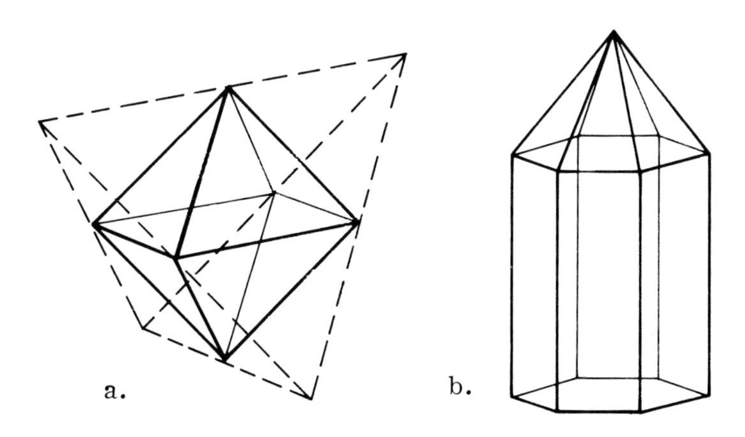

Figure 83. Examples of hemisymmetric (a) and hemimorphic forms (b)

Cubic	Three identical, mutually perpendicular directions are chosen as the three *a* axes so that the required four A_3 axes form body diagonals of a cube defined by the three *a* axes.
Tetragonal	The required A_4 axis is chosen as *c* and two mutually perpendicular *a* axes are chosen in the plane perpendicular to *c*.
Hexagonal or Trigonal	The A_3 or A_6 axis is chosen as *c* and three *a* axes are chosen 120° apart in the plane perpendicular to *c*.
Orthorhombic	The intersection of two planes of symmetry becomes one axis; the other two axes lie in the two planes and perpendicular to the intersection. If the crystal has three planes of symmetry, the three intersections become the three axes. The three axes are chosen as *a*, *b* and *c* with $c < a < b$ as determined by diffraction or, if necessary, by interfacial angles.
Monoclinic	The unique axis perpendicular to the plane of symmetry is *b*. Axes *a* and *c* are chosen from directions in the plane of symmetry so that $c < a$ and so that simple forms result.
Triclinic	Three prominent directions are chosen as *a*, *b* and *c* so that $c < a < b$ and so that simple forms result.

8. Crystal Forms

Similar external crystal faces are classified as forms, *e.g.* the six faces of a cube are similar and constitute the cube form. In general, a single face having no equivalent is called a *pedion*; two parallel faces having no equivalents comprise the form called a *pinacoid*. A form made up of two nonparallel faces is called a *dome* if it is symmetrical about a plane of symmetry, a sphenoid if it is not (except in the tetragonal system where all such faces are called *sphenoids*). *Prisms*, made up of three, four, six, eight or twelve similar faces, are called *trigonal* with three similar faces, *tetragonal* or *orthorhombic* with four, *hexagonal* with six, *ditetragonal* with eight and *dihexagonal* with twelve. A group of three, four, six, eight or twelve similar faces intersecting in a point is called a *pyramid* and has the same modifying adjective as the corresponding prism.

9. Miller Indices

The notations usually used for naming crystal faces are called *Miller indices* and have the form $\{hkl\}$. These notations are based on the above-described assignment of axes and on an expression of the intercepts of the face on the three (hexagonal has four) axes. The axes are taken in order: a_1, a_2, a_3 (cubic); a_1, a_2, c (tetragonal); a_1, a_2, a_3, c (hexagonal); a, b, c (orthorhombic), monoclinic and triclinic). Crystals are always oriented so that a_1 or a lies front (positive) to back (negative), a_2 or b lies right (positive) to left (negative) and a_3 or c lies top (positive) to bottom (negative) (Figure 84).

If we take the simplest face that intersects all of the axes (Figure 85), we can express the intercepts as OA:OB:OC (these may be calculated from appropriate angles or, preferably, the actual lattice spacings may be found by x-ray diffraction). In any case, the Miller index, hkl, of another face on the same crystal, having intercepts OH, OK, OL (Figure 85), is:

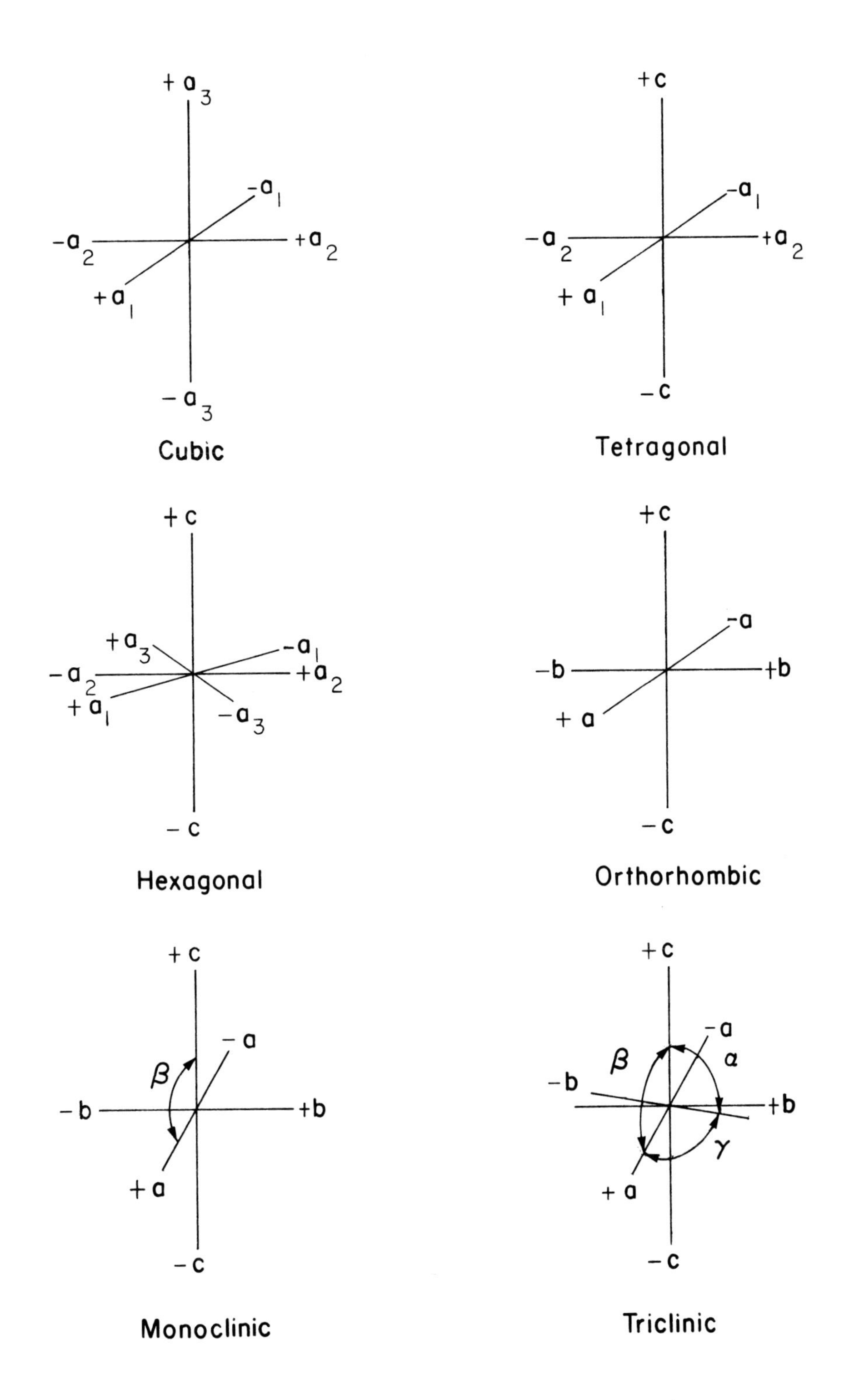

Figure 84. Crystallographic axes in the six crystal systems

$$hkl = \frac{OA}{OH} : \frac{OB}{OK} : \frac{OC}{OL} = 1/4 : 1/3 : 1/2 = \frac{12}{4} : \frac{12}{3} : \frac{12}{2}$$

$$= 3 : 4 : 6 = 346$$

where hkl is expressed as the simplest whole numbers. OA, OB and OC are usually referred to as $\underline{a}_o$, $\underline{b}_o$ and $\underline{c}_o$, the lattice parameters. When h, k or l is 0, that face is parallel to the corresponding axis. Pedions and pinacoids are usually 100, 010 or 001, domes and prisms are usually hk0, h0l or 0kl and bipyramids and sphenoids are usually hkl. A single symbol, hkl, is a face symbol; {hkl}, a form symbol, includes all of the faces in that form; [hkl] is a zone symbol which includes all faces having parallel intersection edges.

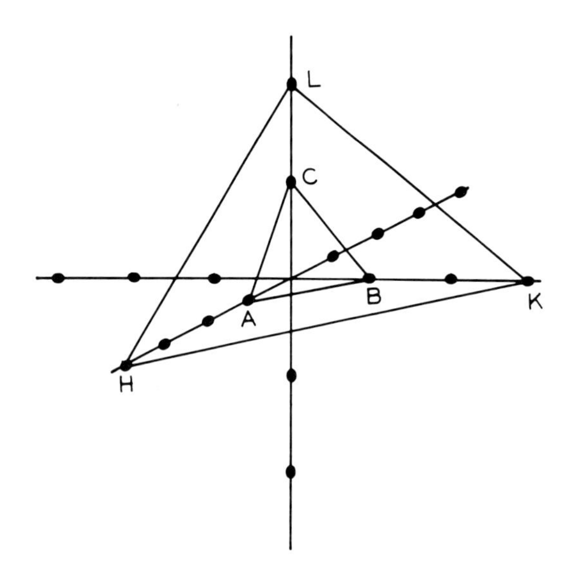

Figure 85. Miller indices: Face ABC = 111, HKL = 346

Each crystal system includes many forms. The most important to the chemical microscopist are tabulated below. In each case the form-symbol, {hkl}, is the generalized Miller index of that face in the form which most closely fits the description — "in front, on top, to the right" (in that order) — when the crystal is "set up" properly. When properly "set up", the crystal is visualized with the $\underline{c}$ axis vertical ($\underline{a}_3$ vertical in the cubic system), the $\underline{a}$ axis front to back ($\underline{a}_1$ in the cubic and tetragonal systems) and the $\underline{b}$ axis right to left ($\underline{a}_2$ in the cubic and tetragonal systems). The positive ends of these axes are, respectively, top, front and right. Crystals in the hexagonal system are set up with $\underline{a}_1$ and $-\underline{a}_3$ symmetrically arranged pointing to the front, $\underline{a}_1$ 30° to the left and $-\underline{a}_3$ 30° to the right; $\underline{a}_2$ then lies right to left with $-\underline{a}_2$ to the left.

Table XIX lists the forms and faces most commonly observed in microscopical work.

10. Twinning

Crystals are often twinned. Twinning usually results in formation of re-entrant angles as in the examples shown in Figure 86. Generally, but not always, different portions of the twinned crystal can be differentiated by optical tests. It is most important to reserve the use of the words twins and twinning to those situations where the two or more interpenetrating crystals bear a definite, unique and characteristic angular relationship. The common composition plane, or twin-plane, always coincides with a possible crystallographic face. This face nearly always has simple Miller indices and is usually a pinacoid face. Randomly intergrown crystals are not twins.

Table XIX. Forms and faces in the six crystal systems

Cubic system

	Form-symbol	Miller indices of all faces
	Holosymmetric forms	
Cube	$\{100\}$	$100, \bar{1}00, 010, 0\bar{1}0, 001, 00\bar{1}$
Rhombic dodecahedron	$\{110\}$	$110, \bar{1}\bar{1}0, \bar{1}10, 1\bar{1}0, 101, \bar{1}0\bar{1}$ $\bar{1}01, 10\bar{1}, 011, 0\bar{1}\bar{1}, 0\bar{1}1, 01\bar{1}$
Octahedron	$\{111\}$	$111, \bar{1}\bar{1}\bar{1}, \bar{1}11, 1\bar{1}\bar{1}, 1\bar{1}1, \bar{1}1\bar{1},$ $11\bar{1}, \bar{1}\bar{1}1$
Trapezohedron	$\{hkl\}$	$211, 121, 112, \bar{2}\bar{1}\bar{1}, \bar{1}\bar{2}\bar{1}, \bar{1}\bar{1}\bar{2},$ $\bar{2}11, \bar{1}21, \bar{1}12, 2\bar{1}\bar{1}, 1\bar{2}\bar{1}, 1\bar{1}\bar{2},$ $2\bar{1}1, 1\bar{2}1, 1\bar{1}2, \bar{2}1\bar{1}, \bar{1}2\bar{1}, \bar{1}1\bar{2},$ $21\bar{1}, 12\bar{1}, 11\bar{2}, \bar{2}\bar{1}1, \bar{1}\bar{2}1, \bar{1}\bar{1}2$
	Hemisymmetric forms (one half of faces in corresponding holosymmetric form are missing)	
Pyritohedron		
positive	$\{210\}$	$210, 2\bar{1}0, 102, 10\bar{2}, 021, 02\bar{1},$ $\bar{2}\bar{1}0, \bar{2}10, \bar{1}0\bar{2}, \bar{1}02, 0\bar{2}\bar{1}, 0\bar{2}1$
negative	$\{120\}$	$120, \bar{1}20, 201, 20\bar{1}, 012, 0\bar{1}2,$ $\bar{1}\bar{2}0, 1\bar{2}0, \bar{2}0\bar{1}, \bar{2}01, 0\bar{1}\bar{2}, 01\bar{2}$
Tetrahedron		
positive	$\{111\}$	$111, \bar{1}\bar{1}1, 1\bar{1}\bar{1}, \bar{1}1\bar{1}$
negative	$\{1\bar{1}1\}$	$\bar{1}\bar{1}\bar{1}, 11\bar{1}, \bar{1}11, 1\bar{1}1$

Note: Pyritohedra and trapezohedra are rarely observed except with minerals; these and other more complex forms do not crystallize quickly enough for recrystallizations on a microscope slide.

Tetragonal system

	Form-symbol	Miller indices of all faces
	Holosymmetric forms	
Prism		
first order	$\{110\}$	$110, \bar{1}\bar{1}0, \bar{1}10, 1\bar{1}0$
second order	$\{100\}$	$100, \bar{1}00, 010, 0\bar{1}0$

Dipyramid		
first order	$\{111\}$	$111, \bar{1}\bar{1}\bar{1}, \bar{1}11, 1\bar{1}\bar{1}, 1\bar{1}1, \bar{1}1\bar{1}, 11\bar{1}, \bar{1}\bar{1}1$
second order	$\{101\}$	$101, \bar{1}0\bar{1}, \bar{1}01, 10\bar{1}, 011, 0\bar{1}\bar{1}, 0\bar{1}1, 01\bar{1}$
Pinacoid	$\{001\}$	$001, 00\bar{1}$
	Hemisymmetric forms	
Disphenoid	$\{hkl\}$	$111, \bar{1}\bar{1}1, 1\bar{1}\bar{1}, \bar{1}1\bar{1}$
	$\{h\bar{k}l\}$	$1\bar{1}1, 11\bar{1}, \bar{1}11, \bar{1}\bar{1}\bar{1}$
	$\{h0l\}$	$101, \bar{1}01, 01\bar{1}, 0\bar{1}\bar{1}$
	$\{0kl\}$	$011, 0\bar{1}1, \bar{1}0\bar{1}, 10\bar{1}$

Note: More complex forms, e.g. scalenohedrons and disphenoids, are extremely rare in microscopy.

Hexagonal system

	Form-symbol	Miller indices of all faces
	Holosymmetric forms	
Prism		
first order	$\{10\bar{1}0\}$	$10\bar{1}0, \bar{1}010, 0\bar{1}10, 01\bar{1}0, \bar{1}100, 1\bar{1}00$
second order	$\{11\bar{2}0\}$	$11\bar{2}0, \bar{1}\bar{1}20, 2\bar{1}\bar{1}0, \bar{2}110, \bar{1}2\bar{1}0, 1\bar{2}10$
Dipyramid		
first order	$\{h0\bar{h}l\}$	$h0\bar{h}l, \bar{h}0hl, 0\bar{h}hl, 0h\bar{h}l, \bar{h}h01, h\bar{h}01$ etc.
second order	$\{hh\cdot\overline{2h}\cdot 1\}$	$hh\cdot\overline{2h}\cdot 1, \bar{h}\bar{h}\cdot 2h\cdot 1, 2h\cdot\bar{h}\bar{h}l, \overline{2h}\cdot hhl$ etc.
Pinacoid	$\{0001\}$	$0001, 000\bar{1}$
	Hemisymmetric forms	
Pyramid		
first order, positive	$\{h0\bar{h}l\}$	$h0\bar{h}l, \bar{h}0hl, 0\bar{h}hl, 0h\bar{h}l, \bar{h}h01, h\bar{h}01$
first order, negative	$\{h0\bar{h}\bar{l}\}$	$h0\bar{h}\bar{l}, \bar{h}0h\bar{l}, 0\bar{h}h\bar{l}, 0h\bar{h}\bar{l}, \bar{h}h0\bar{1}, h\bar{h}0\bar{1}$

second order, positive	$\{hh\cdot\bar{2}\bar{h}\cdot l\}$	$hh\cdot\bar{2}\bar{h}\cdot l$, $\bar{h}\bar{h}\cdot 2h\cdot l$, $2h\cdot\bar{h}\bar{h}l$, $\bar{2}\bar{h}\cdot hhl$ etc.
second order, positive	$\{hh\cdot\bar{2}\bar{h}\cdot\bar{l}\}$	$hh\cdot\bar{2}\bar{h}\cdot\bar{l}$, $\bar{h}\bar{h}\cdot 2h\cdot\bar{l}$, $2h\cdot\bar{h}\bar{h}\bar{l}$, $\bar{2}\bar{h}\cdot hh\bar{l}$ etc.
Dipyramid		
first order	$\{h0\bar{h}l\}$	$h0\bar{h}l$, $\bar{h}h0l$, $0\bar{h}hl$, $h0\bar{h}\bar{l}$, $\bar{h}h0\bar{l}$, $0\bar{h}h\bar{l}$
second order	$\{hh\cdot\bar{2}\bar{h}\cdot l\}$	$hh\cdot\bar{2}\bar{h}\cdot l$, $\bar{2}\bar{h}\cdot hhl$, $h\cdot\bar{2}\bar{h}\cdot hl$, $hh\cdot\bar{2}\bar{h}\cdot\bar{l}$, $\bar{2}\bar{h}\cdot hh\bar{l}$, $h\cdot\bar{2}\bar{h}\cdot h\bar{l}$

Rhombohedral subsystem (trigonal)

Hemisymmetric forms

Rhombohedron		
positive	$\{h0\bar{h}l\}$	$h0\bar{h}l$, $\bar{h}h01$, $0\bar{h}hl$, $\bar{h}0h\bar{l}$, $h\bar{h}0\bar{l}$, $0h\bar{h}\bar{l}$
negative	$\{0h\bar{h}l\}$	$0h\bar{h}l$, $\bar{h}0hl$, $0h\bar{h}l$, $0\bar{h}h\bar{l}$, $h0\bar{h}\bar{l}$, $0\bar{h}h\bar{l}$
Pedion	$\{0001\}$	0001
	$\{000\bar{1}\}$	$000\bar{1}$

Note: Again, the hemisymmetric forms possible in the hexagonal system are rarely observed microscopically.

Orthorhombic system

	Form-symbol	Miller indices of all faces
	Holosymmetric forms	
Prism	$\{hk0\}$	$hk0$, $\bar{h}\bar{k}0$, $\bar{h}k0$, $h\bar{k}0$
	$\{h0l\}$	$h0l$, $\bar{h}0\bar{l}$, $\bar{h}0l$, $h0\bar{l}$
	$\{0kl\}$	$0kl$, $0\bar{k}\bar{l}$, $0\bar{k}l$, $0k\bar{l}$
Dipyramid	$\{hkl\}$	hkl, $\bar{h}\bar{k}\bar{l}$, $\bar{h}kl$, $h\bar{k}\bar{l}$, $h\bar{k}l$, $\bar{h}k\bar{l}$, $hk\bar{l}$, $\bar{h}\bar{k}l$
Pinacoid	$\{100\}$	100, $\bar{1}00$
	$\{010\}$	010, $0\bar{1}0$
	$\{001\}$	001, $00\bar{1}$
	Hemisymmetric forms	
Pyramid		
positive	$\{hkl\}$	hkl, $\bar{h}\bar{k}l$, $\bar{h}kl$, $h\bar{k}l$
negative	$\{hk\bar{l}\}$	$hk\bar{l}$, $\bar{h}\bar{k}\bar{l}$, $\bar{h}k\bar{l}$, $h\bar{k}\bar{l}$
Disphenoid		
right	$\{hkl\}$	hkl, $\bar{h}k\bar{l}$, $\bar{h}\bar{k}l$, $h\bar{k}\bar{l}$
left	$\{h\bar{k}l\}$	$h\bar{k}l$, $\bar{h}\bar{k}\bar{l}$, $hk\bar{l}$, $\bar{h}kl$

Dome	$\{0kl\}$	$0kl$, $0\bar{k}l$
	$\{0k\bar{l}\}$	$0k\bar{l}$, $0\bar{k}\bar{l}$
	$\{h0l\}$	$h0l$, $\bar{h}0l$
	$\{h0\bar{l}\}$	$h0\bar{l}$, $\bar{h}0\bar{l}$
Pedion	$\{001\}$	001
	$\{00\bar{1}\}$	$00\bar{1}$

Monoclinic system

	Form-symbol	Miller indices of all faces
	Holosymmetric forms	
Prism	$\{hk0\}$	$hk0$, $\bar{h}\bar{k}0$, $\bar{h}k0$, $h\bar{k}0$
	$\{0kl\}$	$0kl$, $0\bar{k}\bar{l}$, $0\bar{k}l$, $0k\bar{l}$
	$\{hkl\}$	hkl, $\bar{h}\bar{k}\bar{l}$, $\bar{h}k\bar{l}$, $h\bar{k}l$
	$\{\bar{h}kl\}$	$\bar{h}kl$, $h\bar{k}\bar{l}$, $hk\bar{l}$, $\bar{h}\bar{k}l$
Pinacoid	$\{001\}$	001, $00\bar{1}$
	$\{010\}$	010, $0\bar{1}0$
	$\{100\}$	100, $\bar{1}00$
	$\{h0l\}$	$h0l$, $\bar{h}0\bar{l}$
	$\{\bar{h}0l\}$	$\bar{h}0l$, $h0\bar{l}$

Hemisymmetric forms

	Form-symbol	Miller indices of all faces
Sphenoid		
first order, right	$\{0kl\}$	$0kl$, $0k\bar{l}$
first order, left	$\{0\bar{k}l\}$	$0\bar{k}l$, $0\bar{k}\bar{l}$
third order, right	$\{hk0\}$	$hk0$, $\bar{h}k0$
third order, left	$\{h\bar{k}0\}$	$h\bar{k}0$, $\bar{h}\bar{k}0$
fifth order, right negative	$\{hkl\}$	hkl, $\bar{h}k\bar{l}$
fifth order, right positive	$\{\bar{h}kl\}$	$\bar{h}kl$, $hk\bar{l}$
fifth order, left negative	$\{h\bar{k}l\}$	$h\bar{k}l$, $\bar{h}\bar{k}\bar{l}$
fifth order, left positive	$\{\bar{h}\bar{k}l\}$	$\bar{h}\bar{k}l$, $h\bar{k}\bar{l}$
Dome		
first order, upper positive	$\{0kl\}$	$0kl$, $0\bar{k}l$
first order, lower negative	$\{0k\bar{l}\}$	$0k\bar{l}$, $0\bar{k}\bar{l}$
third order, front	$\{hk0\}$	$hk0$, $h\bar{k}0$
third order, back	$\{\bar{h}k0\}$	$\bar{h}k0$, $\bar{h}\bar{k}0$
fourth order, upper negative	$\{hkl\}$	hkl, $h\bar{k}l$

fourth order, upper positive	$\{\bar{h}kl\}$	$\bar{h}kl$, $\bar{h}\bar{k}l$
fourth order, lower negative	$\{\bar{h}k\bar{l}\}$	$\bar{h}k\bar{l}$, $\bar{h}\bar{k}\bar{l}$
fourth order, lower positive	$\{hk\bar{l}\}$	$hk\bar{l}$, $h\bar{k}\bar{l}$

Pedion 010, 100, 001, $0\bar{1}0$, $\bar{1}00$, $00\bar{1}$, $h01$, $\bar{h}0\bar{1}$, $\bar{h}01$, $h0\bar{1}$

<u>Triclinic system</u>

Form-symbol

Holosymmetric forms

Pinacoid $\{hkl\}$, $\{\bar{h}kl\}$, $\{h\bar{k}l\}$, $\{hk\bar{l}\}$, $\{h01\}$, $\{\bar{h}01\}$
$\{0kl\}$, $\{0\bar{k}l\}$, $\{hk0\}$, $\{h\bar{k}0\}$, $\{100\}$, $\{010\}$, $\{001\}$

Hemisymmetric forms

Pedion (Each of the individual pinacoid faces may appear alone as a pedion.)

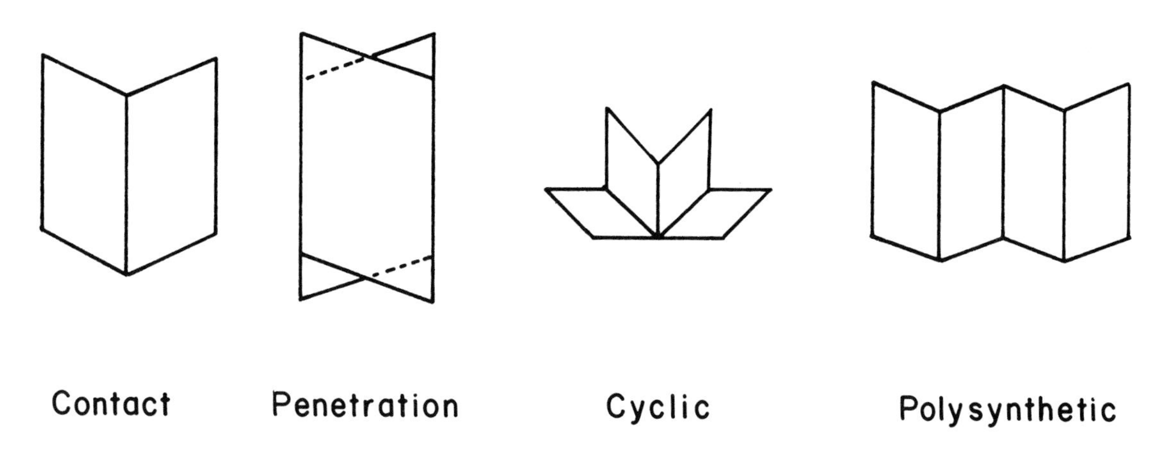

Figure 86. Twinned crystals

11. Polymorphism

Polymorphism is a phenomenon observable with most, and probably all, elements and compounds. A compound showing polymorphism can crystallize with different internal lattices, thereby giving corresponding different external crystal morphology and internal physical properties. The crystal systems of the two (or more) modifications of the compound are usually, but not always, different. A few of the most common examples of polymorphism are carbon, crystallizing as diamond (cubic) and graphite (hexagonal); calcium carbonate, crystallizing as calcite (rhombohedral), aragonite (orthorhombic) and vaterite (hexagonal); titanium dioxide, crystallizing as rutile, brookite and anatase; and the various forms of silica: quartz, tridymite and cristobalite.

Table XX. Examples of isomorphous series of closely related compounds

Series			
1	2	3	4
$KClO_4$	$NH_4H_2PO_4$	$La_2(C_2O_4)_3 \cdot 10H_2O$	p-dichlorobenzene
$KMnO_4$	KH_2PO_4	$Ce_2(C_2O_4)_3 \cdot 10H_2O$	p-dibromobenzene
NH_4ClO_4	$NH_4H_2AsO_4$	$Pr_2(C_2O_4)_3 \cdot 10H_2O$	
NH_4BF_4	KH_2AsO_4	etc.	

12. Isomorphism

Isomorphism is, in a sense, the opposite of polymorphism; instead of different crystalline modifications of the same compound, isomorphism involves closely similar lattices as well as closely similar forms and habits for different compounds. It is usually observed when molecules of several different substances are nearly the same size and shape. The groups shown in Table XX would be expected to be isomorphous on the basis of known atomic radii and this is actually observed to be the case.

In each of these groups the external crystal forms of the pure compounds appear to be identical, but careful measurement shows that the corresponding interfacial angles are slightly different. Isomorphous compounds also form mixed crystals, usually in all proportions and with intermediate crystal angles and optical properties.

IX. CRYSTAL OPTICS

A. INTRODUCTION

The spacing of atoms along crystallographic axes determines not only the crystal system, shape and angles but also the optical properties. Each crystal in the cubic system, for example, has identical spacing of atoms along its three mutually perpendicular crystallographic axes; hence it should not be surprising that the optical properties of all cubic crystals are identical in all directions within the crystal.

B. COLOR

Solids may show absorption colors or reflection colors. Substances which absorb transmitted light very strongly, or which are opaque, may show a characteristic color or luster in reflected light. Some substances show one color by transmission and another by reflection, e.g., fuchsin which is red by transmission, green by reflection.

Some substances are pleochroic, that is, they show different colors in different directions. This phenomenon is characteristic of colored anisotropic substances. The colors shown change with orientation and can be seen only with polarized light. Colored fibers, when they are pleochroic, show different light absorption parallel to the length than crosswise. Tourmaline is pleochroic as are azurite and malachite. Sphalerite, garnet and other cubic crystals are not pleochroic, though they are colored, because they are isotropic. The strongest absorption almost invariably occurs parallel to the vibration direction showing the highest refractive index. The difference in absorption is also proportional to the birefringence of the substance.

C. REFRACTIVE INDEX

The optical property of most importance to the microscopist is the refractive index, **n**. The refractive index of any substance is a measure of the extent to which light is slowed down as it passes through that substance. Light travels most rapidly in a vacuum, almost as fast in air (see Table I, Chapter II. In water it is slowed by about 25% ($n = 1.33$) and, in general,

$$n_s = c_o / c_s \qquad (39)$$

where, c_o is the velocity of light in a vacuum

c_s is the velocity of light in a substance whose refractive index is n_s.

Thus glass, with a refractive index slightly higher than 1.50, must slow light by about 33%. Higher atomic weight elements are more effective in slowing the light; therefore compounds containing high atomic weight elements usually have higher refractive indices (Table XXI).

The refractive index of any solid may be determined microscopically by successively immersing fine particles of the solid in liquids of known refractive index until, in one such liquid, the particles become invisible. When this occurs the refractive index of the invisible particles must be identical with that of the liquid medium in which they are immersed. Similarly, the index of the solid may be determined by varying the index of the mounting liquid until it matches that of the solid and then measuring the index of the liquid. The index may be varied by mixing liquids, by varying the temperature and/or the wavelength of light and by dissolving varying amounts of solid in a liquid.

In microscopical immersion methods for determining the refractive indices of a solid, the medium which matches a given solid particle must be found by trial and error. There are, however, two tests which speed this process considerably; the Becke test and the oblique illumination test. Both tests enable the microscopist to tell whether the index of a given medium is greater or less than that of a particle immersed therein.

Table XXI. Refractive indices of some compounds arranged in order of increasing molecular weights

Compound	Molecular weight	Refractive index
Sodium fluoride	42.00	1.336
Sodium chloride	58.45	1.5442
Sodium chloride	102.91	1.6412
Sodium iodide	149.92	1.7745

1. Becke Test

The Becke line is a bright halo, near the boundary of a transparent particle, that moves with respect to that boundary as the microscope is focused up and down. The halo will always move up to the higher refractive index medium as the position of focus is raised. The halo crosses the boundary to the lower refractive index medium when the microscope is focused down. The particle must be illuminated with a narrow cone of axial light, which may be obtained by closing the iris diaphram of the condenser until only a small aperture remains.

Several explanations of the Becke line have been proposed. All of them depend on reflection and refraction of light rays at an interface between two materials of different refractive index, so as to concentrate the light as a bright halo. The rays are always concentrated towards the medium of higher index as illustrated in Figure 87.

Figure 87 illustrates a prep which might be a crystal or particle mounted in a liquid. When the microscope is focused exactly on the crystal, no Becke line is visible. Raising the focus shifts the image in focus to the region above the crystal; where the light is concentrated within the boundary of the higher index medium. Lowering the tube (or raising the stage) to the region below the best focus for the crystal causes this Becke line to cross the interface to the lower index medium. The movement of the Becke line across the interface is explained by Figure 88, after Wahlstrom[40]. Further treatment of this aspect of the subject is found in Wahlstrom, pages 106-14 and Johannsen[41], pages 271-5.

A second bright line which moves in the direction opposite to the Becke line may sometimes be observed especially with thick particles or when the index difference between particles and mountant is large. This is called the false Becke line and results from the concentration of reflected light at the low index side of an interface. The false Becke line can be weakened or eliminated by reducing the angular aperture of the microscope, i.e., by closing the substage iris and by using low power, low-AA objectives. This also enhances the true Becke line formed by the refracted and internally reflected rays on the opposite side of the interface. Usually, however, the true Becke line becomes easier to see upon changing to a higher power objective.

---

40. Wahlstrom, E. E., Optical Crystallography, 4th ed., John Wiley and Sons, New York, 1969.

41. Johannsen, A., Manual of Petrographic Methods, 2nd ed., McGraw-Hill, New York, 1918; reprinted in facsimile by Hafner Publishing Company, New York, 1968.

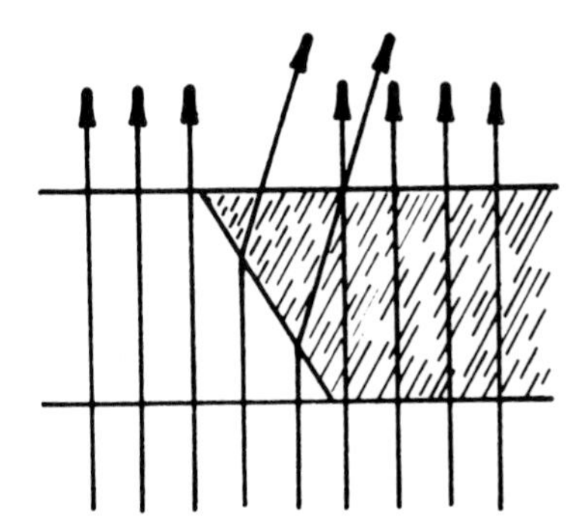

a. Rays passing from lower to higher index medium are bent towards the normal to the interface.

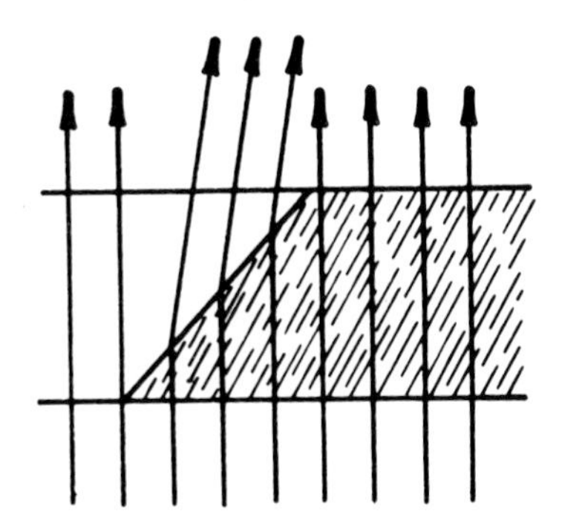

b. Rays passing from higher to lower index medium are bent away from the normal to the interface.

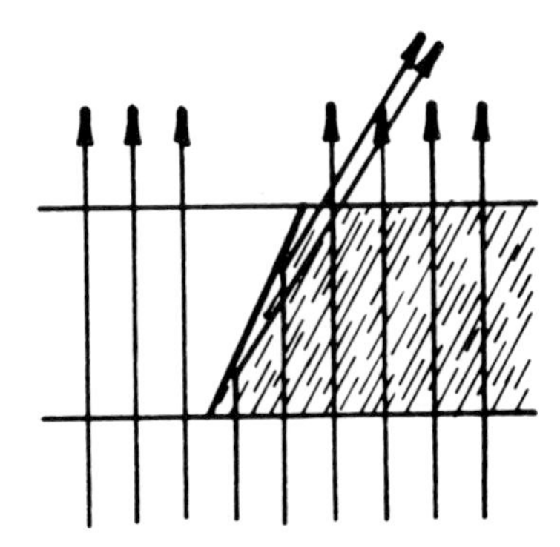

c. Rays passing through the higher index medium strike the interface at more than the critical angle and are reflected back into the denser medium.

Figure 87. Parallel rays (from axial illumination) crossing inclined interfaces. The shaded substance has the higher refractive index.

Another way to avoid the false Becke line is to use the oblique illumination method (see 3. below).

2. Central Illumination Method

Most particles and crystals are somewhat lens-shaped — thin at the edges and thicker at the center. Immersed in a medium they act as biconvex lenses, refracting light toward a spot above the particle center when the medium has a lower index, and away when the medium is of higher index (Figure 89). Thus, as the focus of the microscope is raised, the center brightens if the particle has the higher index and darkens if it has the lower. The Becke line also moves toward or away from the center of the particle under the same circumstances — therefore the two effects enhance and reinforce each other, and in practice are not separable. Central illumination is a separate phenomenon, however, not identical with the Becke line. Figure 90 illustrates a particle showing both.

When the focus is lowered the central illumination effects are reversed, just as is the behavior of the Becke line.

It is useful to remember that the closer the refractive indices of the particle and the medium, the less will be the contrast shown by the particle boundary. Hence, from the appearance of a particle in a refractive index medium, the Becke test tells whether particle or medium has the higher index and the degree of contrast tells how much higher.

The accuracy of the Becke test is ordinarily about ± 0.001 but monochromatic light and temperature control can bring it to ± 0.0004. Phase contrast illumination permits improvement of accuracy to ± 0.0001 or better for isotropic materials.

3. Oblique Illumination Method

The sensitivity of the Becke test is reduced if the edges of the particles are steeply inclined, or when a false Becke line is present. In such cases the oblique illumination test may be preferred.

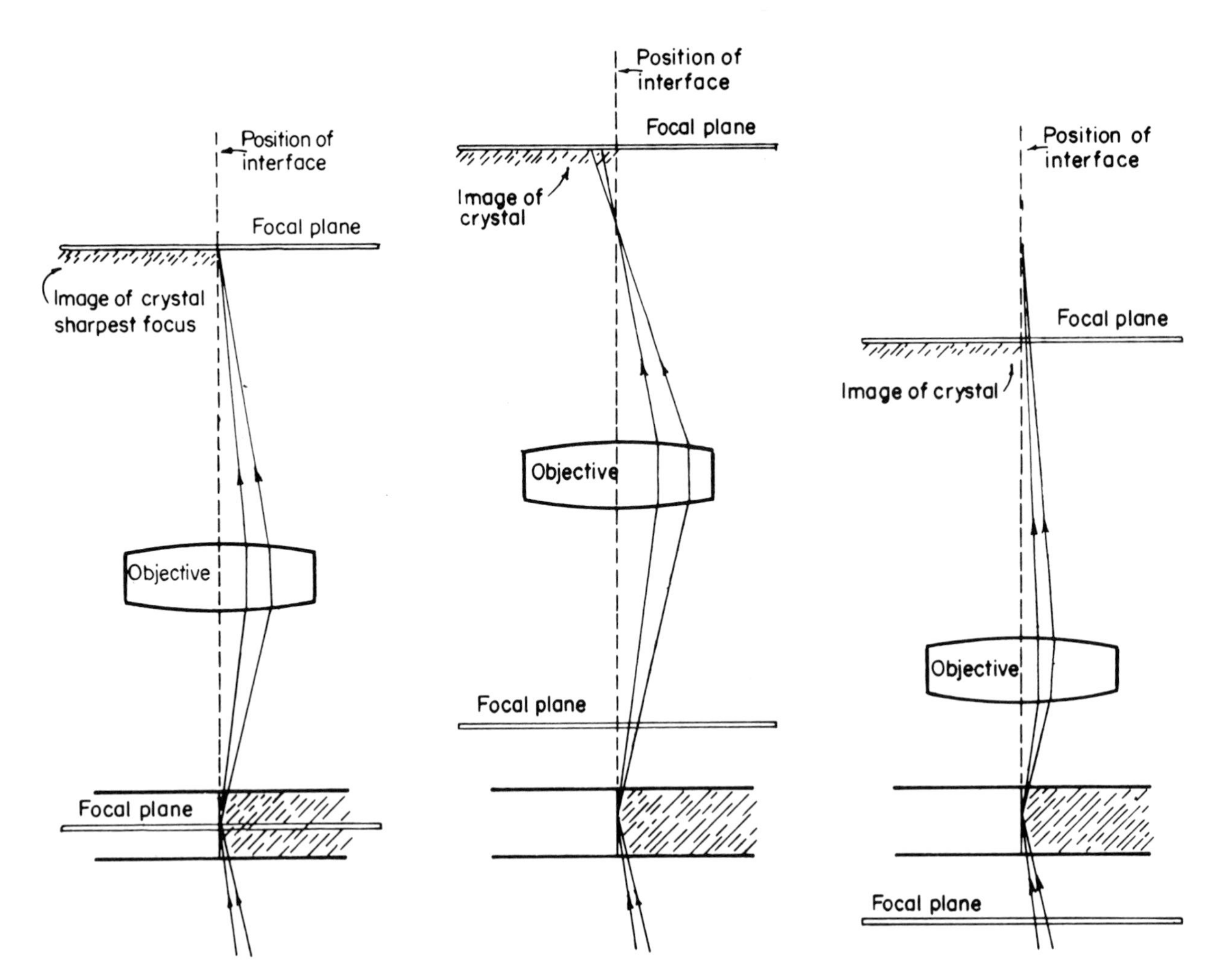

a. The objective is focused on the crystal (shaded), which has a higher index than the mounting medium. No Becke line is observed.

b. The objective is focused above the crystal. The Becke line is observed within the crystal boundary. (The microscope reverses the image from right to left.)

c. The objective is focused below the crystal. The Becke line has crossed the boundary and appears on the low index side.

Figure 88. Diagram illustrating lateral movement of the Becke line as focus is changed. Two rays reflected from the high-index side of a vertical interface are shown to represent the Becke line. (Other rays contributing to the Becke line are omitted for simplicity.)

If particles immersed in a medium of different refractive index are illuminated by unilateral oblique transmitted light, they will appear unsymmetrically shaded. The side of the particle on which the shading is seen depends upon the index of the particle relative to the medium.

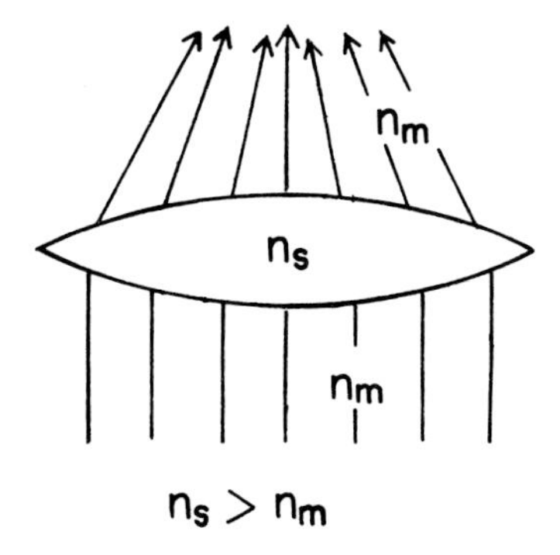

a. The particle has a higher index than the medium and refracts light to form a bright spot above the particle center.

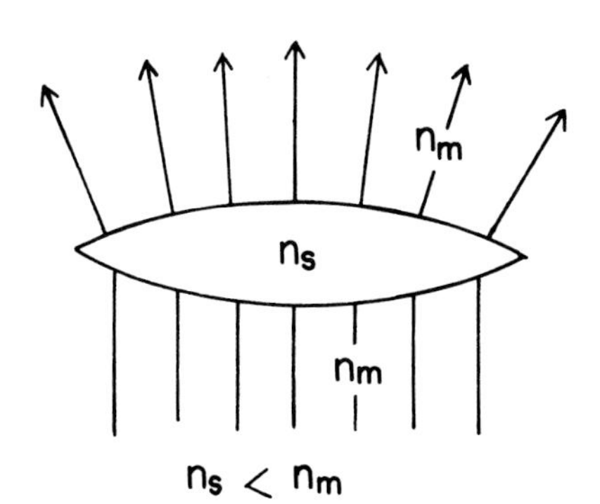

b. The particle has a lower index than the medium and refracts light away from the particle center.

Figure 89. Central illumination; refraction of light by a lens-shaped particle

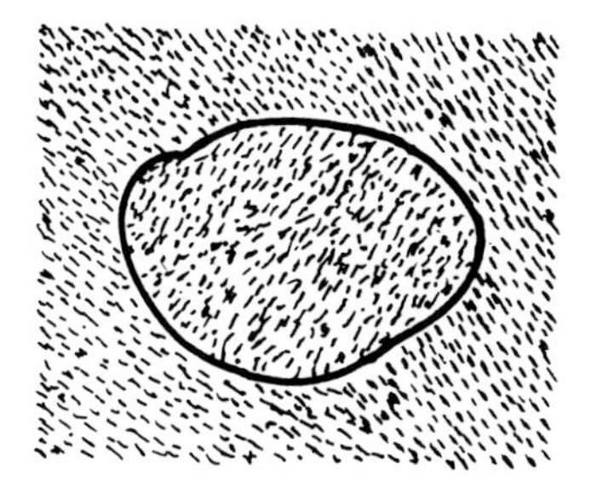

a. Particle in focus

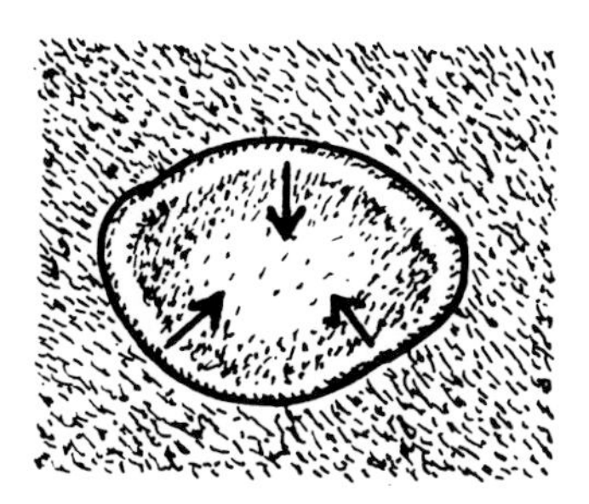

b. Focus slightly raised, index of particle higher than of medium. Becke line moves toward bright center due to central illumination.

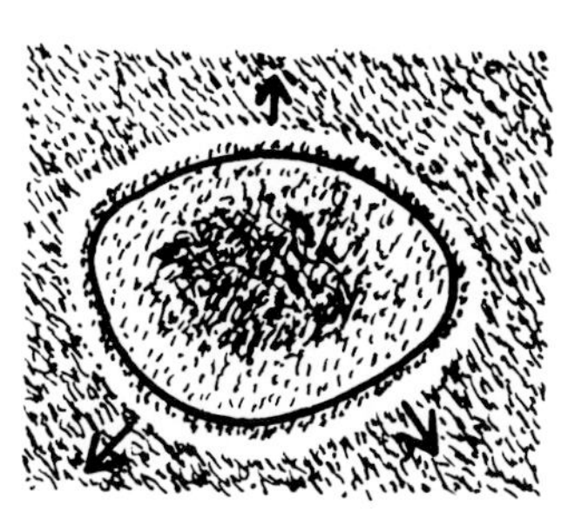

c. Focus slightly raised, index of particle lower than of medium. Becke line moves away from dark center due to central illumination.

Figure 90. Central illumination and the Becke line (after Wahlstrom)

Unilateral oblique illumination is obtained by covering one side of the fully opened iris diaphram of the substage condenser with a finger or piece of cardboard. The condenser should be racked up close to the stage to focus on the prep (if the condenser is lowered so as to focus much below the level of the prep the shadowing effect is reversed). When the condenser and shade are properly positioned about half the field appears dark. A particle of lower index will be shaded on the side opposite to the dark side of the field (Figure 91). It is a good idea to check the condenser adjustment with a solid immersed in a liquid, both of known index.

If the index of particle and liquid are the same, no shadow will appear. The particle shading is due to convergence or divergence of the oblique light by the particle, which behaves as a rough lens. One side of the particle tends to transmit light in such directions that it can enter the microscope objective, while the light from the other side of the particle is lost. The side transmitting to the objective will, of course, appear bright. Figure 92 indicates the paths of the light rays greatly exaggerated. Since the microscope gives an inverted image the effects, as observed, are reversed.

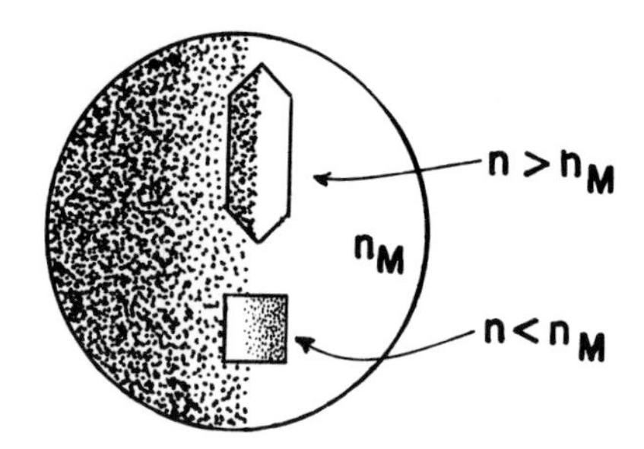

Figure 91. Oblique illumination test for refractive index determination

Most liquids used in refractive index work have a greater dispersive power than solids. When as in Figure 92c, a particle and mounting liquid have the same index for a wavelength of light in the middle of the visible spectrum (e.g. sodium light), the particle will have the higher index for red light and lower index for blue light. In white light the particle will tend to converge red light and diverge blue, as in Figure 92c, and in oblique light opposite edges of the particle will be tinted with these colors. In axial light the particle will show a reddish fringe on the inside and a bluish fringe on the outside. Occurrence of such color fringes at the edges of the image of the particle indicates that the particle and the liquid have the same index for light in the middle of the spectrum. If more accurate results are necessary, monochromatic light must be used. The phenomenon is known as the Christiansen effect and is the basis for dispersion staining methods (see section 4.b. below).

Sensitivity of the oblique illumination test may be improved by Saylor's double-diaphram method[42]. An additional diaphram is inserted at the back focal plane of the objective, on the same side as the diaphram already placed below the condenser (Figure 93). These two aperture diaphrams can be adjusted to give what approaches unilateral dark field illumination and extend the accuracy of the oblique illumination test to ± 0.0001.

4. Index Variation Immersion Methods

These methods are applied to the basic Becke or oblique illumination tests to reduce the number of liquids needed and for other special purposes.

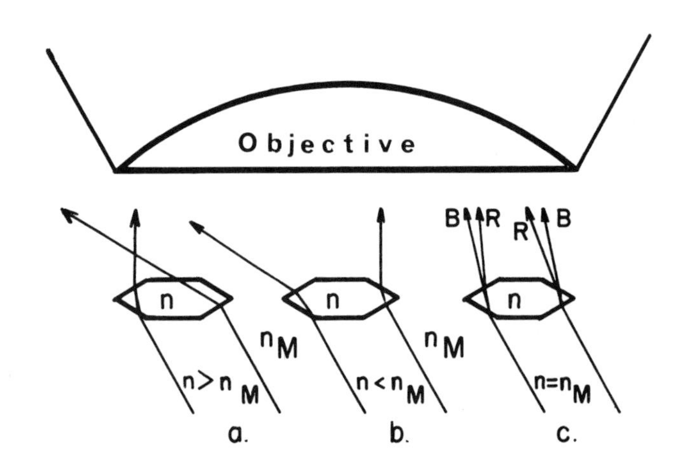

Figure 92. Light paths in oblique illumination test

42. Saylor, C. P., U. S. Bur. Stds. J. Res. 15, 277 (1935).

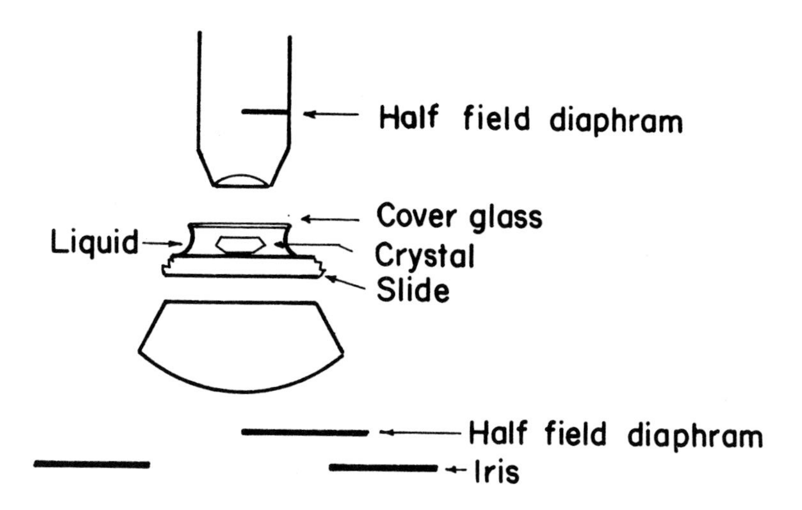

Figure 93. Saylor's double diaphram method

a. Mixing two liquids

Instead of mounting particles of a substance successively in single liquids of known refractive index, the particles may be mounted in a liquid of somewhat higher index and small increments of a lower index liquid may be added until a match in index is observed (by the Becke line or oblique illumination test). The index of the mixed liquid is then determined with a microrefractometer.

This procedure has the advantage that, after one or two initial tests to determine the range of index involved, only two liquids need be used. The procedure uses a minimum of sample and permits "back-titration" if the end point is overreached.

b. Variation of wavelength

Variation of the index of a substance with the wavelength of light is called dispersion of refractive index. All refractive indices of solids vary with wavelength and this variation can be used to assist in the identification of unknown particles (Table XXII).

Variation of the wavelength may be achieved by using a succession of filters, spectrally limited lamps (sodium, hydrogen, mercury etc.) or a monochromator. By varying the wavelength, a greater range of refractive index values is obtained from a single immersion liquid. Normally, of course, a sodium lamp, or at least a yellow filter, is used for the measurement of refractive index, hence the indices for sodium light are of the most interest. Consider what would be observed, however, when a particle is mounted in a refractive index medium having different dispersion. Even though the refractive indices may be identical at one wavelength, they will be different at all other wavelengths. In particular, consider quartz particles mounted in Cargille index liquid $n_D^{25} = 1.560$, which has a refractive index with red light very near the two indices of quartz. With blue light, the indices of quartz and liquid 1.560 are different. The result observed with darkfield illumination is a blue outline around each quartz crystal because one sees only the light which is scattered at the crystal-liquid interface. An identification procedure has been based on a very old observation by Christiansen[43]

43. Christiansen, C. (Experiments on determination of the refractive index of white powders by the liquid immersion method), Ann. der Physik und Chemie (Wiedemann's Annalen) 23, 298-306 (1884).

Table XXII. Dispersion of refractive index for several common compounds

Compound	Index	Wavelength (nm)		
		486.1[1]	589.3[2]	656.3[1]
Sodium chloride	n	1.5534	1.5443	1.5407
Mercuric iodide	ω	-	2.748	2.600
Magnesium silicofluoride	ε	1.3634	1.3602	1.3587
Water	ω	1.3133	1.3090	1.3071
Periclase	n	1.7475	1.7376	1.7335
Zincite	ε	2.081	2.020	2.000
Quartz	ω	1.5497	1.5442	1.5419
Potassium bicarbonate	α	1.383	1.380	1.379
Potassium perchlorate	γ	1.4812	1.4769	1.4750
Barium sulfate	β	1.641	1.636	1.634

(1) Hydrogen lines; (2) Sodium doublet; average wavelength 589.3

who found he could make up emulsions of colorless components and, by proper choice of the components, vary the refractive index and dispersion to obtain colored gels. These chromatic emulsions, as he called them, would also change color as the temperature was changed since this also changed the wavelength at which the two components had identical indices. This same phenomenon, without the temperature change, is called dispersion staining by Crossmon[44], who has applied it to the analysis of particulate substances important in industrial hygiene (for full details, see Chapter X).

c. Variation of temperature

Refractive indices of both liquids and particles decrease as the temperature increases. The change per degree centigrade, dn/dT, is, however, much greater for liquids. Hence, the index change during heating or cooling of immersed particles can be neglected while the change in index of the immersion liquid is varied thermally to provide a larger index range from a single liquid. The method is not generally so well suited for use with organic compounds including polymers since, for these, dn/dT may be relatively large.

The expression dn/dT is called the temperature coefficient of refractive index. It remains essentially constant for each liquid over the normal range of temperatures and, once determined, can be used to calculate the refractive index of the liquid at any other temperature. The commercially available Cargille liquids have their dn/dT marked on each bottle.

---

44. Crossmon, G. C., Anal. Chem. 20, 976 (1948).

The temperature coefficient of refractive index for a liquid can be measured conveniently on a microscope hot stage. Particles are chosen of known refractive index sufficiently below that of the liquid so that they are visible when immersed at room temperature. Then they are heated until the liquid index matches that of the particle. Repeating this determination with a solid of slightly lower refractive index (about 0.02-0.03) gives a second, higher, temperature at which a match is obtained. From the two temperatures and the two refractive indices, the coefficient can be calculated.

Because dn/dT for mounting liquids is appreciable, the room temperature during an ordinary refractive index determination (not by a variation method) should always be noted and used, if necessary, to correct the observed refractive index. Example: a liquid has a refractive index of 1.532 at 25° C. The temperature coefficient,

-dn/dT, is 0.0004. What is the index at 20° C?

$$n_{20^\circ C} = 1.532 + 25-20\ (0.0004)$$
$$= 1.534$$

d. Double variation method

Before the Cargille sets of refractive index liquids became available, it was advantageous to vary the indices of the few available liquids. By combining the techniques of changing both wavelength and temperature, the greatest range of index values can be obtained from each liquid. This procedure, first applied by Emmons[45], became known as the double variation method. This method has more recently been applied to the forensic comparison of glass by McCrone[46].

5. Liquids for Measuring Refractive Indices

Lists of standard immersion liquids for refractive index determinations are found in Chamot and Mason's Handbook of Chemical Microscopy[47] and other texts. Needham[48] gives directions for preparing sets of mixed liquids.

Sets of calibrated mixed liquids can be purchased from R. P. Cargille Laboratories[49]. These liquids, available with indices from 1.30 to 2.11, come in 1/4 ounce bottles marked with the temperature coefficient and dispersion. A special set of liquids with high dispersion is available with indices from 1.500 to 1.645.

Organic compounds often dissolve in the Cargille liquids. If the particle is not very soluble, the index can sometimes be determined by working quickly or by saturating the Cargille liquid with the compound and using the solution as the index medium. In this case the refractive index of the matching saturated Cargille liquid must be measured with a refractometer. Otherwise a liquid in which the compound is less soluble must be found. Solutions of potassium iodide and excess mercuric iodide in water cover the refractive index range up to 1.73 and solutions of the same salt in glycerin extend the range up to 1.79. If a more viscous medium is desired, as when crystals may need to be rolled

---

45. Emmons, R. C., Am. Mineral. 11, 115 (1926); 13, 504 (1928); 14, 414, 441, 482 (1929).

46. McCrone, W. C., JAOAC 55 834-39 (1972); 56 1223-6 (1973); 57 668-70 (1974).

47. Chamot, E. M., and C. W. Mason, Handbook of Chemical Microscopy, vol. 1, 3rd ed., 331, John Wiley and Sons, New York, 1959.

48. Needham, G. H., The Practical Use of the Microscope, 201-03, Charles C. Thomas, Springfield, IL, 1959.

49. R. P. Cargille Laboratories, Inc., 55 Commerce Road, Cedar Grove, NJ 07009.

into better positions, these salts can be dissolved in Aquaresin®, a borate of ethylene glycol sold by Glyco Products[50]. Such mixtures give indices up to about 1.77. It is usually possible to dissolve considerable mercuric iodide in these liquids to raise the index further and, indeed, this may be necessary to prevent separation of solid potassium iodide.

6. Isotropic and Anisotropic Materials

Only glasses, some plastics and crystals in the cubic system show a single refractive index (Table XXIII), they are called isotropic. Other transparent particles, called anisotropic, have different refractive indices depending on the vibration direction of light through the particle. Anisotropic materials show either two (Table XXIV) or three (Table XXV) "principal" refractive indices in particular directions but any refractive index between these principal values may be observed in intermediate directions.

Table XXIII. Refractive indices of a few isotropic crystals, n_D^{25}

NaF	1.326	RbI	1.647
K_2SiF_6	1.339	SrCl	1.650
KF	1.352	K_2SnCl_6	1.657
KCN	1.410	CsI	1.661
Na alum	1.439	LiCl	1.662
K alum	1.456	KI	1.667
NH_4 alum	1.459	CsBr	1.698
KCr alum	1.481	NH_4I	1.703
KCl	1.490	NH_4Br	1.711
$NaClO_3$	1.515	MgO	1.736
$CaCl_2$	1.52	As_2O_3	1.755
CsCl	1.534	NaI	1.775
NaCl	1.544	$ZnO \cdot Al_2O_3$	1.780
RbBr	1.553	$Pb(NO_3)_2$	1.782
KBr	1.559	LiBr	1.784
CdF_2	1.56	CsI	1.788
$Ba(NO_3)_2$	1.572	K_2PtCl_6	1.825
CsF	1.578	CaO	1.837
$SiBr_4$	1.579	SrO	1.870
CsBr	1.582	LiI	1.955
$Sr(NO_3)_2$	1.588	BaO	1.98
Na_2CaSiO_4	1.600	AgCl	2.07
$NaBrO_3$	1.617	Sb_2O_3	2.087
NH_4Cl	1.640	CaS	2.137
NaBr	1.641	AgBr	2.25
Li_2O	1.644	Diamond	2.42

7. Polarized Light

Polarized light must be used to study anisotropic crystals. Ordinary light can be polarized with a Polaroid® filter and this method is used on contemporary microscopes. Polarized light differs from ordinary light in that it has a single vibration direction perpendicular to the direction of propagation. Polarization of ordinary light by a Polaroid® filter is depicted in Figure 94. If the filter is rotated through any angle, the vibration plane of the polarized light rotates correspondingly.

50. Glyco Chemicals, Division of Charles L. Huisking and Company, Inc., Williamsport, PA 17701

Table XXIV. Refractive indices and signs of a few common tetragonal and hexagonal crystals, n_D^{25}

Compound	Omega	Epsilon	Sign
Ice, H_2O	1.544	1.553	+
$NH_4H_2PO_4$	1.525	1.479	-
Quartz, SiO_2	1.544	1.553	+
Beryl, $3BeO \cdot Al_2O_3 \cdot 6SiO_2$	1.581	1.575	-
Chile saltpeter, $NaNO_3$	1.587	1.336	-
Potassium periodate, KIO_4	1.617	1.652	
$KAg(CN)_2$	1.623	1.625	+
Apatite, $CaF_2 \cdot 3Ca_3P_2O_8$	1.634	1.632	-
Beryllium silicate, $BeSiO_4$	1.654	1.668	+
Calcite, $CaCO_3$	1.658	1.486	-
Magnesium chloride, $MgCl_2$	1.670	1.658	-
Dolomite, $CaCO_3 \cdot MgCO_3$	1.682	1.503	-
Zinc chloride	1.687	1.713	+
Willemite, Zn_2SiO_4	1.694	1.723	+
Magnesite, $MgCO_3$	1.700	1.509	-
Beryllium oxide, BeO	1.719	1.733	+
Corundum, Al_2O_3	1.768	1.760	-
Smithsonite, $ZnCO_3$	1.818	1.618	-
Cobalt carbonate, $CoCO_3$	1.855	1.60	-
Siderite, $FeCO_3$	1.875	1.633	-
Calcium tungstate, $CaWO_4$	1.9200	1.937	+
Zircon, $ZrSiO_4$	1.924	1.968	+
Cassiterite, SnO_2	1.997	2.093	+
Zincite, ZnO	2.008	2.029	+
Silver iodide, AgI	2.218	2.229	-
Cadmium sulfide, CdS	2.506	2.529	+
Anatase, TiO_2	2.554	2.493	-
Rutile, TiO_2	2.616	2.903	+
Carborundum, SiC	2.654	2.697	+
Mercuric iodide	2.748	2.455	-
Cinnabar, HgS	2.854	3.201	+
Hematite, Fe_2O_3	3.22	2.94	-

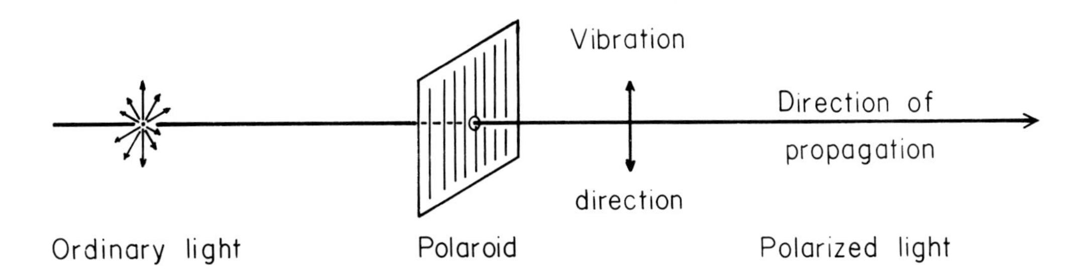

Figure 94. Polarization of light

Table XXV. Refractive indices and signs of a few common orthorhombic, monoclinic and triclinic crystals, n_D^{25}

Compound	Alpha	Beta	Gamma	Sign
Borax, $Na_2B_4O_7 \cdot 10H_2O$	1.447	1.469	1.472	(-)
Gypsum, $CaSO_4 \cdot 2H_2O$	1.521	1.523	1.530	(+)
Orthoclase, $K_2O \cdot Al_2O_3 \cdot 6SiO_2$	1.518	1.524	1.526	(-)
Albite, $Na_2O \cdot Al_2O_3 \cdot 6SiO_2$	1.525	1.529	1.536	(+)
Biotite, $(K,H)_2(Mg,Fe)_2(Al,Fe)_2(SiO_4)_3$	1.541	1.574	1.574	(-)
Anhydrite, $CaSO_4$	1.569	1.575	1.613	(+)
Anorthite, $CaO \cdot Al_2O_3 \cdot 2SiO_2$	1.576	1.583	1.589	(-)
Pyrophyllite, $Al_2O_3 \cdot 4SiO_2 \cdot H_2O$	1.552	1.588	1.600	(-)
Talc, $3MgO \cdot 4SiO_2 \cdot H_2O$	1.539	1.589	1.589	(-)
Muscovite, $K_2O \cdot 3Al_2O_3 \cdot 6SiO_2 \cdot 2H_2O$	1.561	1.590	1.594	(-)
Topaz, $(Al,F)_2SiO_4$	1.619	1.620	1.627	(+)
Wollastonite, $CaSiO_3$	1.616	1.629	1.631	(-)
Barite, $BaSO_4$	1.637	1.638	1.649	(+)
Diopside, $CaMg(SiO_3)_2$	1.664	1.671	1.694	(+)
Olivine, $(Mg,Fe)_2SiO_4$	1.662	1.680	1.699	(+)
Aragonite, $CaCO_3$	1.530	1.681	1.685	(-)
Malachite, $CuCO_3 \cdot Cu(OH)_2$	1.655	1.875	1.909	(-)
Sulfur, S	1.950	2.038	2.241	(+)
Cerussite, $PbCO_3$	1.804	2.076	2.078	(-)
Realgar, As_2S_2(Li light)	2.46	2.59	2.61	(-)
Stibnite, Sb_2S_3	3.194	4.046	4.303	(-)

8. Uniaxial Crystals

Anisotropic crystals in the tetragonal and hexagonal systems have one unique crystallographic direction and either two (tetragonal) or three (hexagonal) directions which are alike and perpendicular to the unique direction (Figure 95). The unique direction corresponds to the c axis; the others correspond to the two or three a axes. The arrangement of structural units (atoms, ions) along the a axes is identical. Hence, light travels along the c axis (propagation direction) with a single velocity because the vibration directions lie in an isotropic plane. The refractive index corresponding to the c direction is ϵ, the index perpendicular to it is ω.

If now a crystal like one shown in Figure 96 is observed with polarized light, the refractive index observed will be ϵ, ω or some intermediate index, ϵ', depending on the crystal orientation relative to the vibration direction of the polarized light. The refractive index actually observed can, therefore be controlled by rotating the crystal to the proper position relative to the polarizer vibration direction.

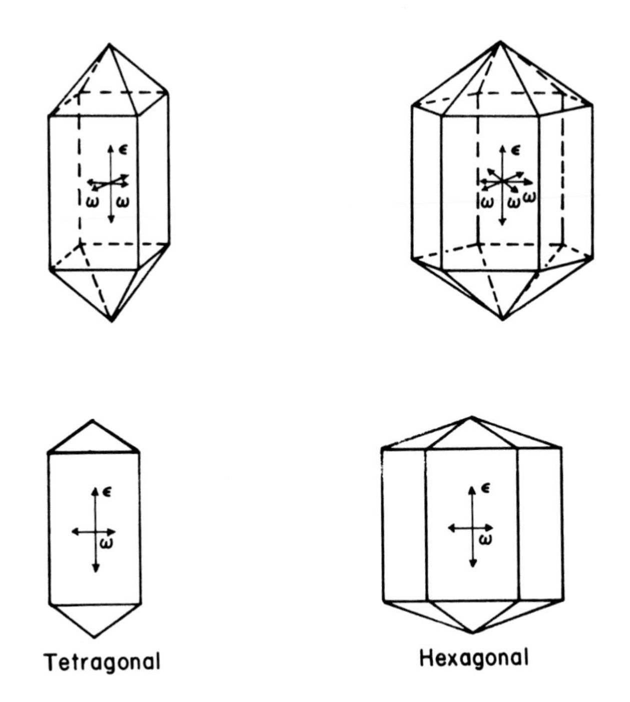

Figure 95. Refractive indices of tetragonal and hexagonal crystals

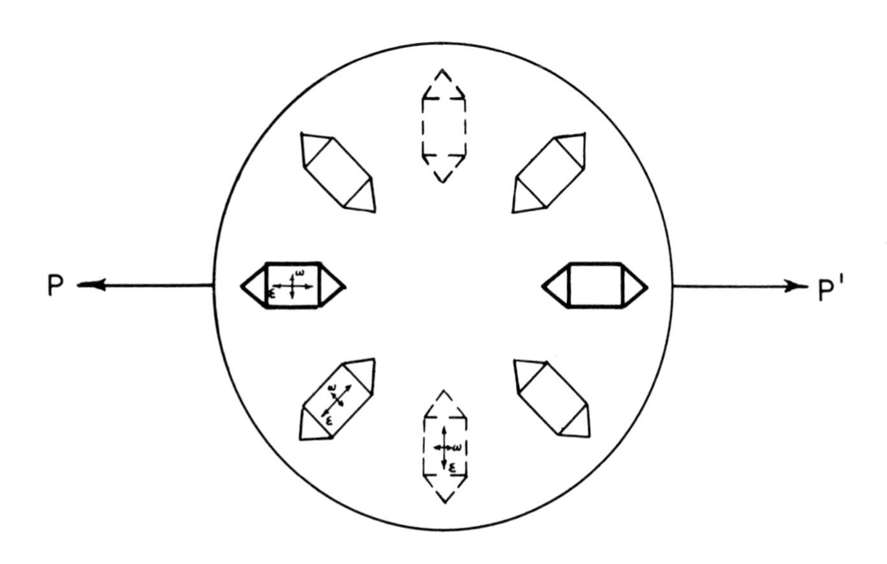

Figure 96. Variation in contrast of uniaxial crystal in liquid of refractive index near omega

If a uniaxial crystal is oriented on the slide with its c axis (ϵ vibration direction) perpendicular to the optical axis of the microscope, and if it is mounted in a liquid medium having a refractive index equal to ω, the crystal will disappear twice at two positions 180° apart during each complete rotation of the stage. At the two positions 90° from those two positions in which the omega index is observed, the crystal boundaries will appear most strongly (greatest contrast) since here the epsilon refractive index is being observed (Figure 96).

All of the crystals shown in Table XXIV will appear and disappear as the stage is rotated if they are mounted in a medium with an index equal to either one of their two principal refractive indices and if they are properly oriented. In contrast to this behavior, of course, each isotropic compound listed in Table III has a single refractive index and disappears in all positions when it is mounted in a liquid of the same refractive index.

In other orientations of the two crystals in Figure 95 different refractive indices are observed (Figure 97). It is important to notice, however, that every view of the tetragonal crystal shows the omega refractive index itself at two positions 180° apart during rotation of the stage even though most of the views show only a component of epsilon. The view parallel to the c axis, in fact, shows only the omega index in all positions of rotation of the stage. If a given particle is suspected of being calcite, it should be mounted in a liquid having the omega refractive index for calcite (1.658). All calcite crystals in 1.658, no matter how they are tipped, will disappear completely every 180° during rotation of the stage. Each tetragonal and hexagonal crystal has one such refractive index, although this is not true for anisotropic crystals in any other crystal system.

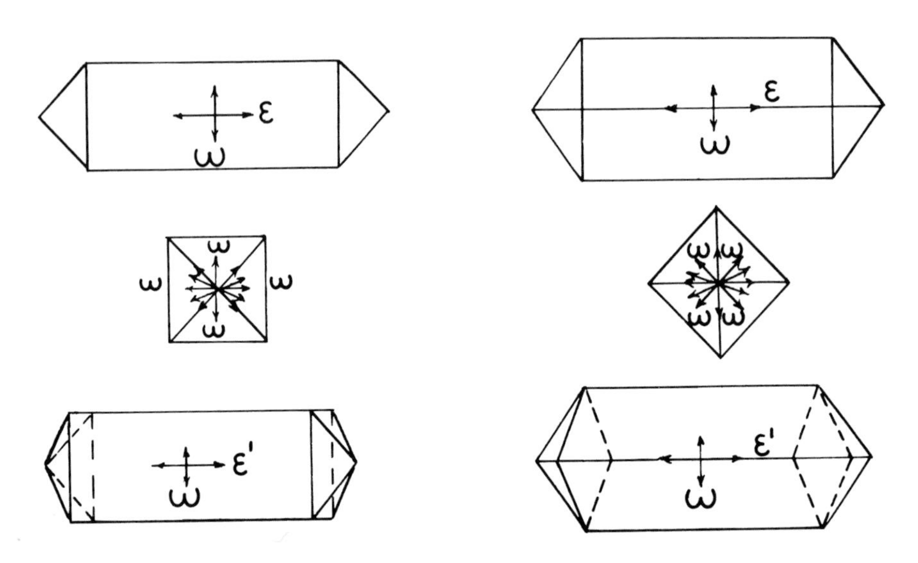

Figure 97. Refractive indices shown by a tetragonal crystal in different orientations

By this test the ω index can be found even for randomly oriented crystal fragments showing no recognizable morphology. The particles are immersed successively in different index liquids until one liquid is found in which every particle disappears as the stage is rotated at two positions 180° apart.

To find the ϵ index successive mounts again are made until another liquid is found, beyond which no fragment shows a more extreme refractive index. Now, however, the only particles that will disappear during stage rotation are those oriented to show ϵ itself. All the others will show prime (') values of ϵ and remain visible throughout 360°.

The ϵ refractive index can be observed only when the microscope axis is exactly perpendicular to the optic axis of a uniaxial crystal, i. e., when the crystal lies on a face or edge parallel to c. As Figure 96 illustrates, both ϵ and ω may be determined for a crystal in this position.

Uniaxial crystals are described as optically positive or negative, depending upon whether or ω is the greater (Table XXIV).

If $\epsilon > \omega$, the crystal is positive (+).

If $\omega > \epsilon$, the crystal is negative (-).

If the position of the optic axis is known (e. g., from morphology), the optic sign, or sign of double refraction, can be determined in one of two ways:

1) By inspection — if the particle is mounted in a known index liquid, then the Becke line and degree of contrast indicate the orientation of the higher index. If the higher index is ϵ, the crystal is positive; if ω, negative.

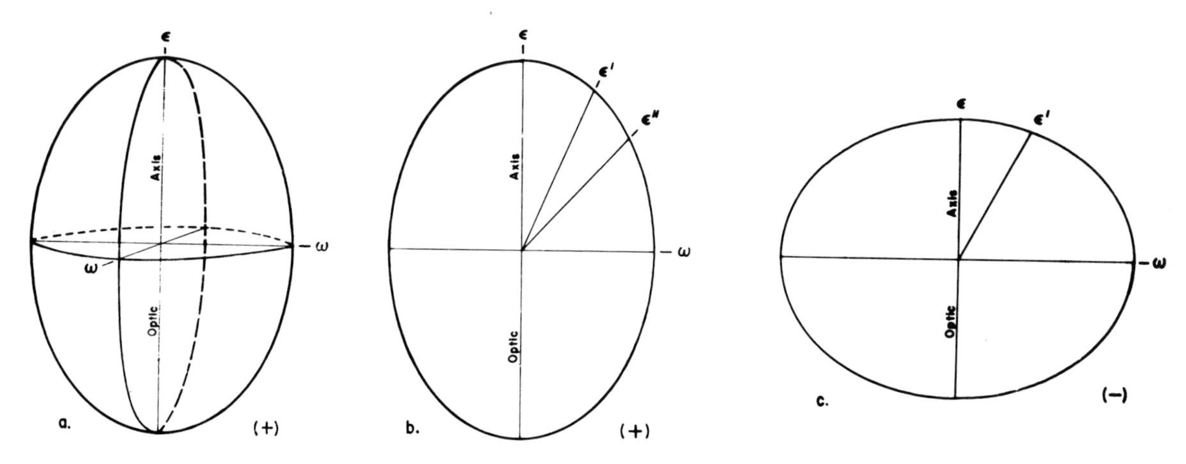

Figure 98. Uniaxial indicatrix

2) By compensators — these crystal plates of known sign can be superimposed over the oriented particle to give the high index direction of that particle (see section D. 3 below).

If the position of the optic axis is not obvious from morphology, the sign can again be determined in two ways.

1) By interference figure — superimposing a compensator over the interference figure gives the sign of birefringence directly (see section D. 4 below).

2) By successive mounting in different index liquids — two liquids can be found which match the highest and lowest indices of any particle in any orientation. If all of the particles show the lowest index at some point during rotation, that index is ω and the sign is positive. If all show the highest index, that index is ω and the sign is negative. Conversely, if only some of the particles show the lowest (or highest) index during rotation, that index is ϵ.

Figure 97 shows the effect of tilting a uniaxial crystal on the observation of refractive indices. No matter what position the crystal assumes, ω will always be visible because the ω vector component has the same velocity in all propagation directions through the crystal. The refractive index of the ϵ component, however, varies for different directions and is least (negative crystals) or greatest (positive crystals) for the vibration direction parallel to the optic axis. The observed value, ϵ', varies from ϵ, when the optic axis is parallel to the stage, to ω, when the optic axis is normal to the stage.

The relationships can be conveniently represented with a three-dimensional construction called the indicatrix. Radii proportional in length to refractive index values may be used to represent indices observed in all directions of light propagation through a crystal. Such radii, then, describe solid geometrical forms whose surfaces define refractive index values observed for any orientation of a crystal. The radius for ω is constant and describes a circle. The radii for all other indices vary from ω through an infinite number of ϵ' values to ϵ, describing an ellipse. When the ellipse and the circle are placed at right angles to each other and rotated about ϵ, the resulting form is an ellipsoid of revolution whose maximum (optically negative crystals) or minimum (optically positive crystals) dimension is the same as the diameter of the circle resulting from ω. Figure 98a shows the uniaxial indicatrix thus produced; b and c are two-dimensional sections of the indicatrices for positive and negative crystals and show the relationship of ϵ, ϵ' and ω for each.

The optic sign of a crystal is not the same property as the sign of elongation, although frequently the two agree. Many uniaxial substances crystallize with one dimension much longer than the other two. When the crystal length is also the high index direction, the sign of elongation is positive. The optic sign of such a crystal would be positive if the refractive index parallel to the length were ϵ, negative if it were ω. When the crystal length corresponds to the low index direction, the sign of elongation is negative but the optic sign would be negative only if that low refractive index were ϵ.

The sign of elongation is often applied to fibers, most of which behave as though optically uniaxial.

9. Biaxial Crystals

A third group of crystals, also anisotropic, has three principal refractive indices instead of one (cubic; isotropic) or two (tetragonal or hexagonal; uniaxial). These crystals, in the orthorhombic, monoclinic or triclinic systems, are termed biaxial. These three systems differ optically as well as morphologically from other systems and reflect the fact that biaxial crystals have unequal spacing of atoms along the three different crystallographic axes (Figure 99). The optical differences are represented in this figure.

Most crystalline compounds (about 75%) fall in the third group, those with three principal refractive indices (Table XXV) and designated α, β and γ. The lowest refractive index is always α, γ is always the highest and β always the intermediate.

The indicatrix for such crystals is a triaxial ellipsoid rather than an ellipsoid of revolution as for uniaxial crystals. There are, however, two circular sections through the triaxial ellipsoid; light traveling perpendicular to these sections splits into two components of equal velocity corresponding to the refractive index β. The two directions of travel are the two optic axes (OA) of biaxial crystals (Figure 100b). The three axes of the ellipsoid are α, β and γ (Figure 100a).

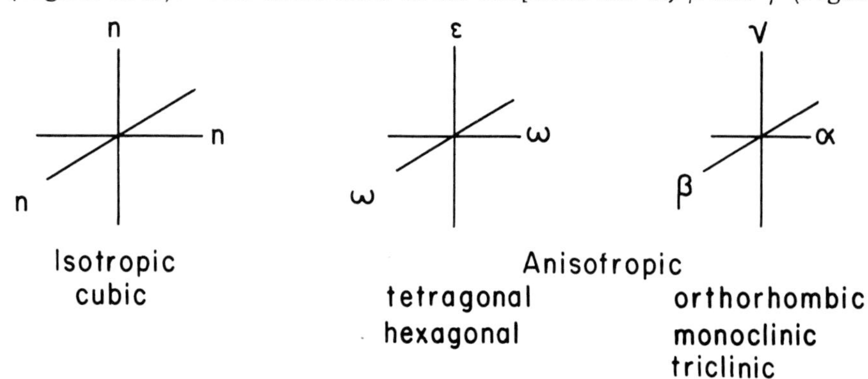

Figure 99. Vibration directions for the principal refractive indices of cubic, uniaxial and biaxial crystals

The two optic axes form an acute angle, designated as the optic axial angle (2V), in the plane of α and γ. The optic axial angle is a constant for any particular substance and can be measured and used to characterize that substance. Figures 100c and d are sections of the biaxial indicatrix showing the $\alpha - \gamma$ plane and 2V. The bisector of the optic axial angle is called the acute bisectrix (Bx_a); in 100c it is γ; in 100d it is α. It is also apparent in 100c that the angle supplementary to 2V is obtuse and is bisected by α. Therefore α is called the obtuse bisectrix (Bx_o) in 100c and 100d, Bx_o is γ. The plane containing the optic axes as well as α and γ is called the optic axial plane (OAP).

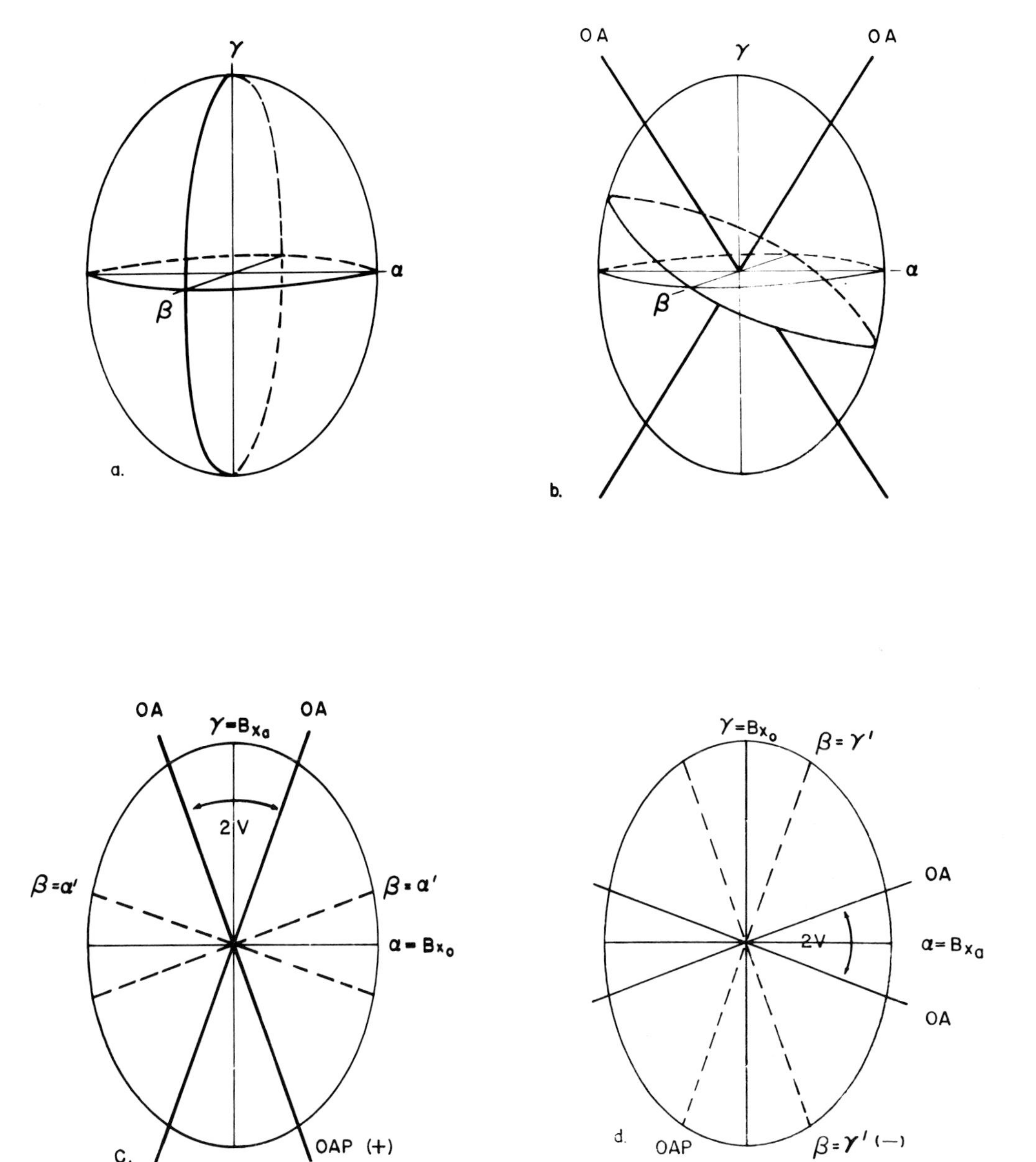

Figure 100. Biaxial indicatrix

Biaxial crystals have two isotropic directions, the optic axes. Prime values of all three refractive indices may be observed on properly oriented crystals. If the crystal is oriented on the microscope stage so that one is looking directly parallel to β (i.e., looking down β), the principal values of α and γ are observed. Since the principal value of β is between those of α and γ, an index equal to β can also be designated on the section of the biaxial indicatrix showing the $\alpha - \gamma$ plane. In Figure 100c, β is closer in numerical value to α than γ and is equal to α'. In Figure 100d, the numerical value of β is closer to γ and therefore equal to γ'. The dividing line between α' and γ' is the numerical midpoint between the values of α and γ. Figure 100 also shows that the lines designating the value of α' and γ' equal to β are perpendicular to the optic axes.

Biaxial crystals are described as optically positive or negative:

If $\gamma - \beta > \beta - \alpha$, the crystal is positive (+).
If $\gamma - \beta < \beta - \alpha$, the crystal is negative (-).

The refractive indices actually observed for a compound depend on the orientation of the crystal, and each different view presents a different but specific maximum or minimum refractive index (Figure 101). Since, by convention, all crystals with three principal refractive indices have alpha as their lowest index and gamma as their highest, no crystal can show, in any orientation, a refractive index lower than alpha or higher than gamma. This is the basis for a procedure mineralogists use to identify unknown minerals. The mineralogist assumes that the unknown could be in any of the six crystal systems. He mounts a few crystals of his sample in any refractive index medium in the neighborhood of 1.60. After checking the Becke line and the degree of contrast in all positions of rotation of several crystals, he can choose a second refractive index medium. Then, by successive approximation, he finds the refractive index media which have the two extreme indices for that compound.

If all of the crystals disappear in a single medium in all positions of rotation of the stage, the compound is isotropic and therefore cubic (if not a glass). If all of the crystals disappear in a given medium every 180° during rotation of the stage but are visible in other positions, they are tetragonal or hexagonal and have two indices — omega and epsilon. Finally, if only a very few crystals in the preparation show the highest or lowest refractive index on rotation of the stage, the compound is biaxial and the two indices measured are alpha, the lowest, and gamma, the highest. To locate β, the intermediate index, the interference figure can be used as discussed below in section D.4.b.7). Obviously, the sample must contain crystals of a single solid state species.

The relationship between refractive index and crystal morphology for all crystal systems is discussed below in section E.

D. CROSSED POLARS

Other optical tests can be used to assist in identifying an unknown; some of these are based on the use of crossed polars. Thus far we have used only a single polar with plane polarized light to pick out and measure a particular refractive index for an anisotropic crystal. If, now, we insert a second polar, called an analyzer, above the preparation, we see that the field itself becomes black (the vibration directions for light passing the two polars must be perpendicular). If we place a mixture of crystalline materials on the stage between the two polars, some of the crystals appear colored, some appear white and some are invisible against the black background. The latter are isotropic, and thus must be in the glassy state or in the cubic system. The other crystals which appear white or colored are anisotropic and must have at least two principal refractive indices.

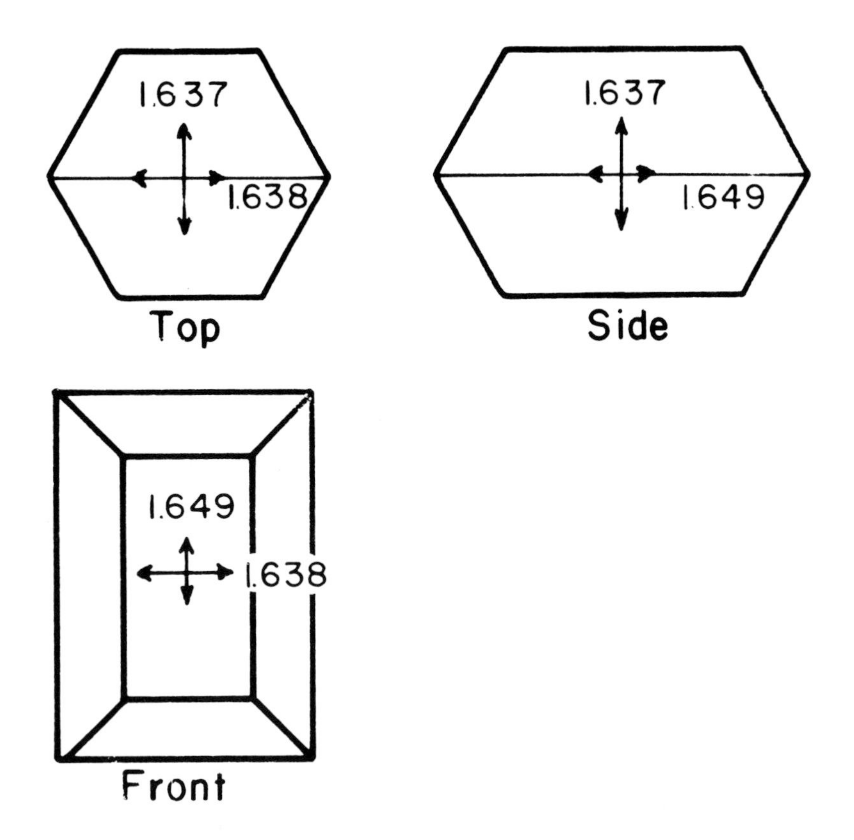

Figure 101. Refractive indices shown by different views of $BaSO_4$

1. Extinction

If the orientation of the crystals that appear white or colored between crossed polars is changed by rotating the stage, all of the crystals will be observed to disappear (become black) four times during complete rotation of the stage. These positions, 90° apart, are called extinction positions and reveal the vibration directions of each crystal. These directions will parallel the vibration direcions of the two polars when the crystal is extinct and are represented on drawings of the crystal by two arrows crossing at right angles (Figures 97, 101 and 102).

Uniaxial and biaxial crystals may exhibit any of three types of extinction: parallel, symmetrical or oblique. These are observed with the ocular cross lines parallel to the vibration directions of the polarizer and analyzer (so that they represent these directions). The stage is rotated until the crystal appears dark, the analyzer is removed and the position of the cross lines with respect to the long side or prominent angles of the crystal is noted. As Figure 102 illustrates, extinction is parallel if a cross line parallels the long direction of the crystal, symmetrical if a cross line bisects a prominent angle and oblique if the cross line is oblique to the long direction of the crystal. The angle between the nearer cross line and a prominent direction of the crystal is called the extinction angle; it must never exceed 45°.

The extinction position is useful in locating a particular refractive index of a crystal. Take calcite as an example (Figure 103). Calcite crystals show extinction between crossed polars when the bisectors of the rhomb angles are parallel to the vibration directions of the polars.

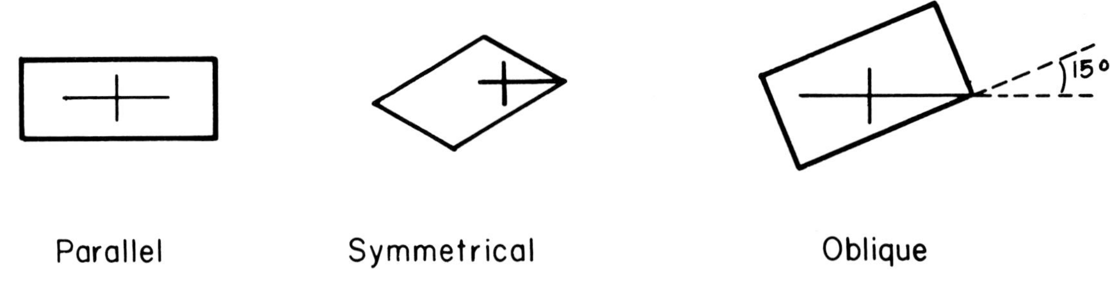

Figure 102. Extinction in anisotropic crystals

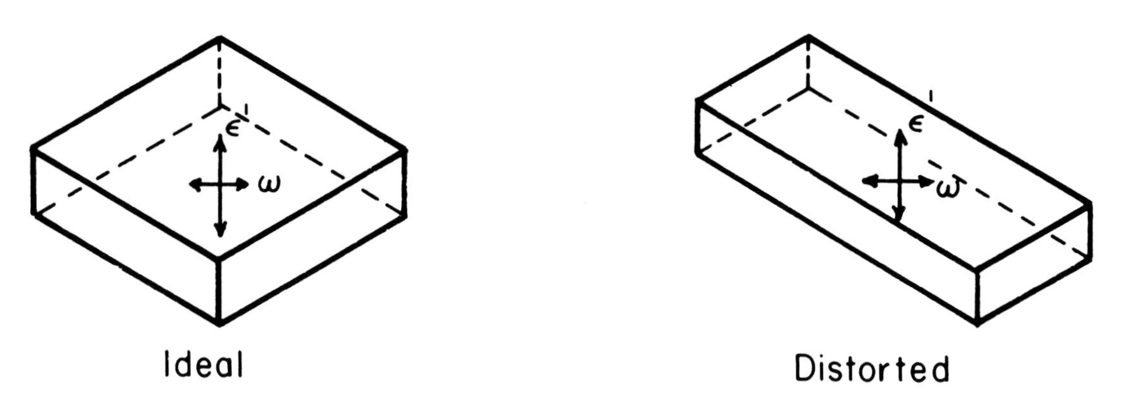

Figure 103. Extinction in rhombs of calcite

The vibration directions in the calcite crystal are therefore parallel to the bisectors of the rhomb angles. Each of the crystals in Figure 103 will show extinction when the two vibration directions are parallel to the cross lines in the ocular and will show interference colors in all intermediate positions. To observe the omega index, rotate the crystal to an extinction position such that the vibration direction of the polarizer bisects the acute rhomb angle, then remove the analyser. This step illuminates the microscope field with light whose vibration direction is parallel to omega.

2. Birefringence

Differences in atomic, ionic or molecular arrangement in the principal crystallographic directions of anisotropic crystals give rise to differences in optical properties, such as refractive index, in these same directions. An anisotropic crystal may be pictured as presenting different amounts of interaction with light in different directions, along which the arrangement of atoms will be different. This interaction slows down the light (increases refractive index) as it travels through the crystal.

Polarized light is limited to a single vibration direction for any position of the polarizer (Figure 94). Therefore a crystal can be oriented so that a particular crystallographic direction is parallel to the vibration direction of the polarized light. Observation of refractive indices when a crystal is thus aligned has been described in section C.8, above.

When polarized light enters an optically anisotropic crystal, the light is resolved into components vibrating in two perpendicular planes. Splitting plane polarized light into two vector components is called double refraction. Since the components follow two principal vibration directions having different refractive indices, they move through the crystal at different rates; they emerge with one retarded by a definite amount which depends on the difference in the two refractive indices, n_2-n_1, and the thickness. The actual distance of one behind the other is called the retardation.

If the crystal is oriented so that one of its principal refractive indices is parallel to the vibration direction of the polarizer, the second vector component becomes zero. All light emerging from the crystal has the same vibration direction as the polarizer; it is cut off (absorbed) by the analyzer, whose vibration direction is perpendicular to that of the polarizer. Crystal and field, therefore, both appear dark and the crystal is said to be at extinction.

If the crystal is not so oriented, the emerging vector components will recombine in the vibration plane of the analyzer. Since one component is retarded, interference on recombination of the two components by the analyzer will cause the image to appear colored. Interference will destroy some wavelengths of light and reinforce others. The latter will constitute the interference color for the crystal. Substances exhibiting these interference (or polarization) colors are termed birefringent, anisotropic or doubly refractive. The actual colors depend upon the retardation which, in turn, depends upon the thickness and the differences in refractive indices (birefringence). If the crystal thickness varies, several colors may be observed. The colors are brightest when the crystal is rotated farthest from an extinction position, i.e., 45° away.

The interference colors, directly related to the retardation, are in turn related to birefringence and thickness according to:

$$\text{retardation} = 1000 \times t(n_2 - n_1) \tag{40}$$

where n_1 and n_2 are the two refractive indices for any particular view of the crystal, t = thickness and $n_2 - n_1$ is the birefringence.

Since the retardation is normally measured in nanometers and thickness in micrometers, a factor of 1000 must be added.

If a wedge of regularly increasing thickness of any anisotropic material, such as quartz. is turned to the 45° position and observed between crossed polars, a definite sequence of interference colors is seen. The sequence is known as Newton's series and is divided into "orders" by the red bands which occur periodically as the thickness increases. The first-order colors are black, gray white, yellow, orange and red, as thickness increases. The higher orders include blues and greens in place of gray and white; the colors become paler until they approach "high-order white at about the tenth order. The color series is shown in the Michel Lévy chart[51] (see Figure 105) which includes scales relating thickness, birefringence and interference color so that any one of these variables can be determined when the other two are known.

With some experience, the microscopist becomes adept at estimating birefringence from the interference color and can use it for quick checks of the difference between the refractive indices of a substance and for distinguishing substances in a mixed preparation of similar sized particles. Birefringence can be determined as a numerical constant from the measured thickness and retardation of a given crystal.

Interference colors are also observed with parallel polars, i.e., with the vibration directions of polarizer and analyzer parallel to each other. The colors are the complements of those seen with crossed polars and appear on a bright field.

3. Compensators

When two anisotropic substances are superimposed, addition or subtraction of their individual retardations may occur (Figure 104). If both substances are placed in a position of brightness (45° away from the vibration directions of polarizer and analyzer), and their slower components are

51. Bloss, F. Donald, An Introduction to the Methods of Optical Crystallography, 144 et seq., Holt Rinehart and Winston, 1961.

parallel, the two retardations are added. The interference color resulting from addition is of a higher order than that of either substance alone and is equivalent to the numerical sum of the retardations. The slower component, corresponding to the higher refractive index, is indicated by the shorter arrow.

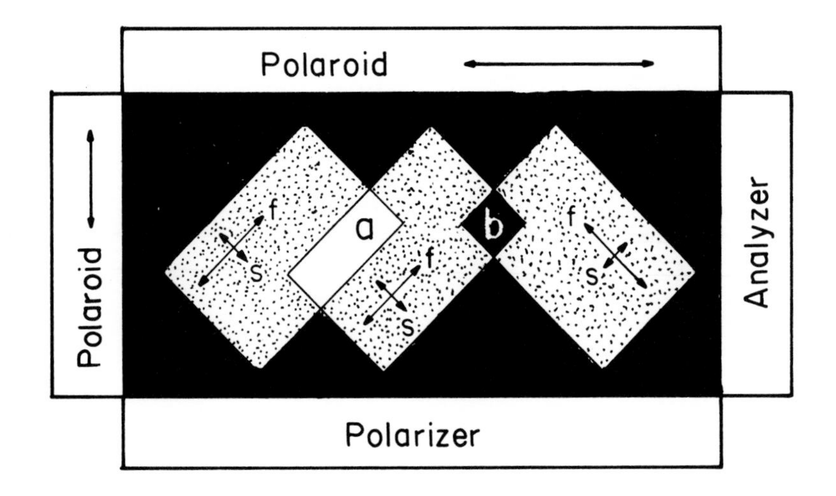

s = slow component
f = fast component
a = addition of retardations resulting in higher order polarization color than either overlapping plate alone.
b = subtraction of retardations resulting in lower order color than at least one of overlapping plate. If the plates cause equal retardation the result is zero (complete compensation), as illustrated.

Figure 104. Addition and subtraction of retardation.

If the slow components are at right angles to each other, subtraction of retardations occurs. The resultant interference color, of lower order than that of one of the substances, corresponds to a net retardation numerically equal to the difference in retardations of the two substances.

The fast and slow vibration directions of a crystal, respectively the lower and higher refractive indices, may thus be located with a compensator. Compensators are of several types but basically they can be represented by one of the anisotropic substances in Figure 104, and a crystal on the microscope stage is represented by the other substance. The slow vibration direction of the compensator is usually marked on the holder and fixed by inserting the compensator in a slot in the body tube at an angle 45° from the vibration directions of the polars. The stage can be rotated to align the crystal so that its slow vibration direction is successively parallel to and perpendicular to the slow direction of the compensator and the interference colors produced by addition or subtraction of retardations can be observed.

The "first-order red" plate compensator is a layer of selenite or quartz of the proper thickness to produce a retardation equivalent to about 550 nm. It is especially useful with very weakly birefringent materials (gray or white when observed by themselves). For these, addition of retardation is indicated by the appearance of second-order blue; subtraction, by first-order yellow.

The quarter-wave plate may also be used as a compensator. This is usually a thin mica plate of uniform thickness having a retardation of about 125 nm. It is most useful for compensation studies with substances of low birefringence; in other words, when the interference colors are in the first order.

The quartz wedge is also used as a compensator. Its variable retardation extends to several orders of interference colors. The retardation which exactly compensates that of a crystal can be found by pushing in the wedge until it reaches a position at which the interference color of the crystal, after going through successively lower orders, appears dark gray or black. Retardation can be compensated only when the slow components of the crystal and the wedge are perpendicular. If the crystal shows successively higher colors as the retardation of the wedge is increased, the slow directions of crystal and compensator are parallel and the stage must be turned 90° before compensation can be achieved.

Compensators of the Berek type consist of a small quartz or calcite plate which can be tilted by a micrometer screw so that progressively thicker sections of the plate are placed in the light path. The effect is the same as that of the quartz wedge, and the micrometer makes it possible to measure the retardation precisely. The Berek compensator is a standard tool in fiber microscopy.

A number of special purpose compensators (Babinet, Sénarmont etc.) are described in advanced texts[40,41,52].

4. Michel-Lévy Chart

The Michel-Lévy interference color chart is a valuable aid to the light microscopist in that it graphically related the thickness, retardation (optical path difference), and birefringence (numerical difference between the refractive indices) for particular views of transparent, colorless or lightly-colored substances. These characteristics allow unknown materials to be identified, or provide important optical information about those which are known.

Retardation increases linearly with both the thickness of a substance and with the birefringence: the greater the thickness, the greater the retardation; the greater the difference between the refractive indices, the greater the retardation (or higher the interference color). That is,

$$r = t(n_2 - n_1)$$

where r is the retardation interference color) expressed in nanometers (nm); t is the thickness, which must also be expressed in nanometers (multiply the reading in micrometers by 1000); and $(n_2 - n_1)$ is the numerical difference between the refractive indices (birefringence).

Michel-Lévy published the chart that bears his name in Paris in 1888. The chart is based on the above equation which also can be written:

$$t = \frac{r}{(n_2 - n_1)} \quad \text{or} \quad n_2 - n_1 = \frac{r}{t}.$$

Hence, any one of these parameters can be determined from the other two.

Referring to the chart, we see the thickness, t, increases along the ordinate on the left side. The path difference, or retardation, r, increases as we go right along the bottom of the chart; the names of the interference colors are also given. First-order red falls at about 550 nm; second-order red at twice 550, or 1,100 nm; third-order red at 1,650 nm etc. The birefringence $(n_2 - n_1)$ starts on the top of the chart at the left and proceeds all the way to the right and down the right side for increasing values.

The diagonal lines represent the birefringence values listed at the upper end of each diagonal line. This is so because $t = r/(n_2 - n_1)$ is the equation for a straight line through the origin of the coordinates, with a slope of

$$\tan \theta = \frac{1}{(n_2 - n_1)}.$$

52. Hartshorne, N. H., and A. Stuart, Crystals and the Polarizing Microscope, 4th ed., Edward Arnold, London, 1970.

Each line is assigned an angle θ, and thereby a special value $(n_2 - n_1)$. The names of numerous substances appear opposite their characteristic maximum birefringence value on some versions of the chart.

Since the Michel-Lévy chart shows the interrelationships between thickness, birefringence and interference color, the microscopist can, as suggested earlier, determine any one from the chart if he knows the other two. Several examples will illustrate this:

Suppose a cylindrical synthetic fiber 15 μm in diameter shows a maximum interference color corresponding to about 900 nm. This is determined by orienting the length of the fiber 45° to the vibration directions of the crossed polars and comparing the color running down the center of the fiber to the colors in the chart. The order is found by noting the number of reds between the center and either edge of the fiber.

One must be very careful here because the colors are very, very close together at the edge of a cylindrical fiber. It often is better to count orders on a taper-cut end of a fiber. In the present example, we pass through only first-order red, indicating the yellow at the center is second order.

To determine the birefringence, we look for 900 nm on the abscissa and move vertically until we reach a horizontal line corresponding to a thickness of 15 μm on the ordinate. There will be a diagonal line where the two lines cross. We now follow the diagonal line to the upper right to read the birefringence, 0.060, at the top of the chart. Looking up this value in a birefringence table for synthetic fibers we learn that a cylindrical fiber having this birefringence is nylon. We could also have calculated birefringence from $(n_2 - n_1) = r/1000t$:

$$r = 1000 \text{ (0 to 40) x } 0.014 = 0 \text{ to } 560 \text{ nm.}$$

Finally, suppose we have a rock thin-section containing the mineral augite (birefringence 0.024) showing an optic normal interference figure and a first-order red interference color (550 nm). It is desired to know the thickness of the section. At the top of the chart we find the birefringence 0.024 and follow the diagonal line until it intersects the 550-nm line on the abscissa. From the coordinates we go directly left to the thickness of the ordinate and find 23 μm. Once again we can find the solution from the equation:

$$t = \frac{550}{1000\ (0.024)} = \sim 23\ \mu\text{m.}$$

Other orientations of augite would show lower order polarization colors.

Thus, the Michel-Lévy chart enables the microscopist to rapidly determine thickness, birefringence or retardation, knowing the other two quantities.

A similar chart can be used to relate thin film colors to film thickness and refractive index. Newton's series of interference colors is displayed in the same color sequence by thin films of increasing thickness. This makes it possible to quickly estimate thickness of corrosion or contamination films.

In the James S. Martin Laboratory of McCrone Associates in Chicago, interference films are constantly used to tell if electron microscope support films are thin enough. The polymer films, e.g., collodion, used to cement small particles to beryllium plates for microprobe analysis interfere with the probe analysis, hence must be very thin. The interference colors shown by such films help control this critical parameter.

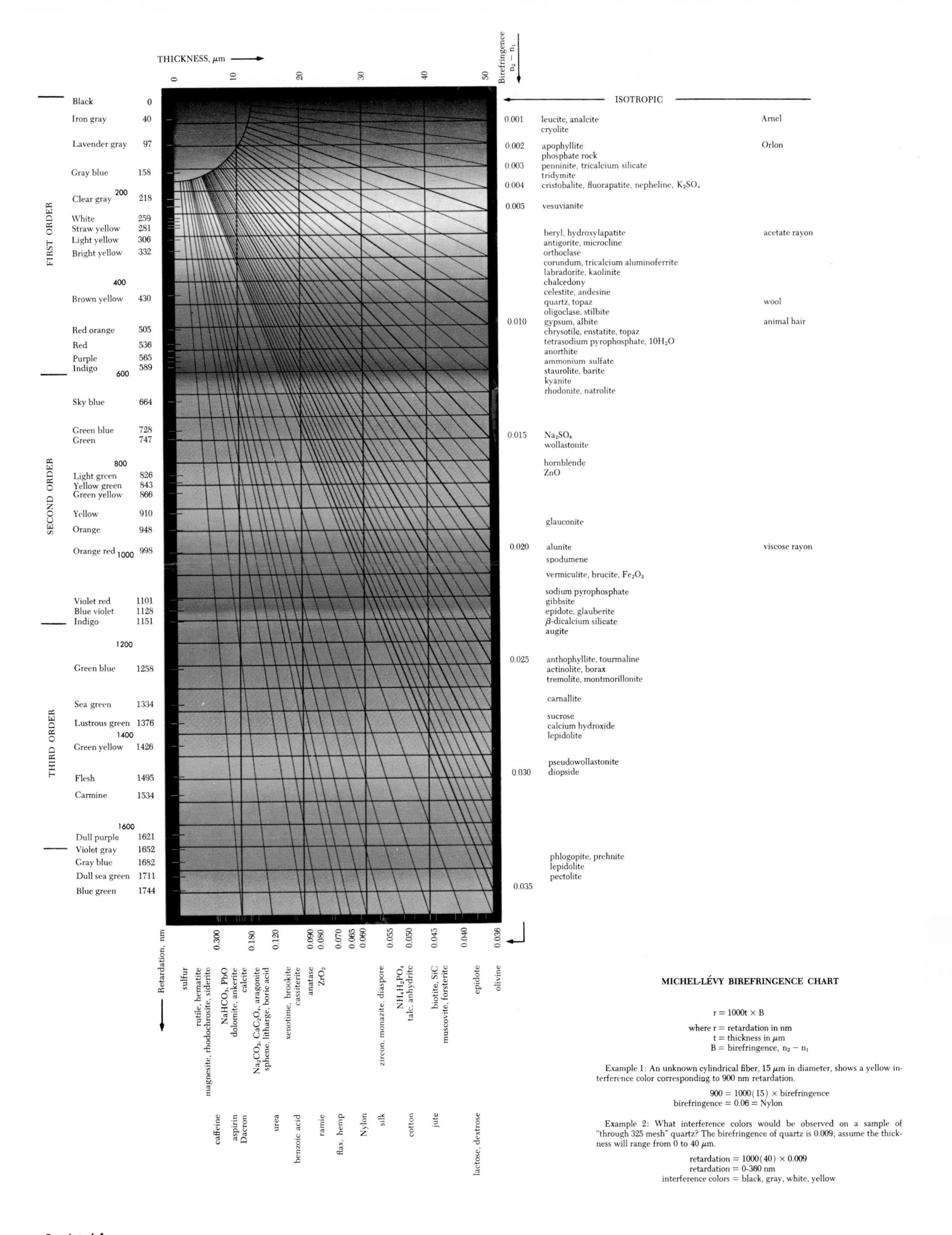

MICHEL-LÉVY BIREFRINGENCE CHART

$r = 1000t \times B$

where r = retardation in nm
t = thickness in μm
B = birefringence, $n_2 - n_1$

Example 1: An unknown cylindrical fiber, 15 μm in diameter, shows a yellow interference color corresponding to 900 nm retardation.

$900 = 1000(15) \times$ birefringence
birefringence = 0.06 = Nylon

Example 2: What interference colors would be observed on a sample of "through 325 mesh" quartz? The birefringence of quartz is 0.009; assume the thickness will range from 0 to 40 μm.

retardation = $1000(40) \times 0.009$
retardation = 0-360 nm
interference colors = black, gray, white, yellow

Reprinted from
THE PARTICLE ATLAS
McCrone and Delly
Ann Arbor Science Publishers, Inc.

5. Interference Figures

Microscopical observation of an object in the usual manner is termed orthoscopic observation. Another way of using the microscope involves study of the back focal plane of the objective by removing the ocular, by using the Bertrand lens, by examining the image at the eyepoint above the ocular with a magnifier or by using a phase telescope. Such observations are called conoscopic, denoting that the observation is associated with a cone of light. It has already been pointed out that observation of the objective back focal plane under ordinary conditions shows the angular aperture of illumination being used. Each point in the back focal plane field is associated with a particular propagation direction of the illuminating radiation. The point in the center of the field corresponds to light traveling parallel to the microscope axis. All points at the edge of the field represent light traveling through the specimen at the maximum angle (highest angular or numerical aperture) relative to the microscope axis. Intermediate angles, of course, are represented by points in the back focal plane of the objective lying between the center and the edge of the field.

Considering, now, the optical indicatrix of a uniaxial crystal, assume that the c axis of the crystal (epsilon vibration direction, or optic axis) is parallel to the microscope axis. In this orientation, observed orthoscopically between crossed polars, the birefringence is zero, i.e., the crystal is isotropic. If the crystal is tipped a few degrees in any direction, the birefringence is no longer zero but is still very small and the interference color is gray. Further tilting of the crystal results in an increase in birefringence and higher-order interference colors. Note that this tilting is equivalent to tilting the propagation direction of light passing through a stationary crystal.

Conoscopic observation of a crystal placed in the center of the microscope field reveals, in the back focal plane of the objective, a pattern of interference colors corresponding to the full cone of directions by which the crystal is illuminated, each direction showing its own interference color. Superimposed on this color pattern is the pattern of extinction positions; the combination of these two patterns is the interference figure.

Figure 105 shows the interference colors (a) and extinction patterns (b) characteristic of a uniaxial crystal observed parallel to c, as well as the corresponding interference figure. Of course, only the final combined figure (c) is observed.

The interference color pattern varies with crystal thickness and with the difference in the two principal refractive indices ($\epsilon - \omega$); an increase in either increases the order of the interference colors at any given point in the figure, i.e., more rings appear in the field. On the other hand, the extinction pattern is unchanged by changes in thickness (except that the arms of the dark cross narrow with increased thickness and/or birefringence).

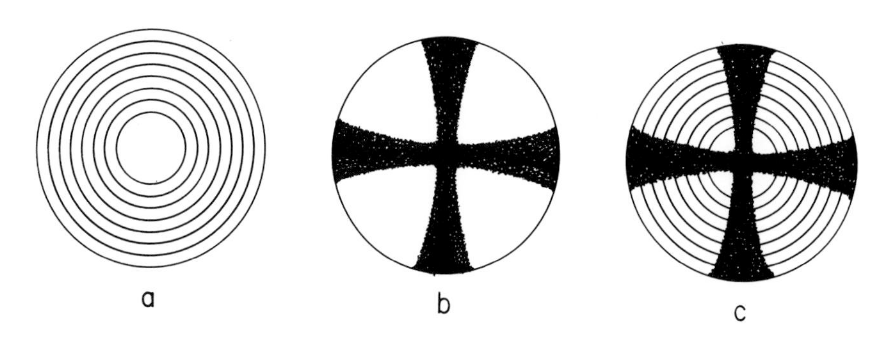

Figure 105. Elements (a. and b.) of a uniaxial interference figure (c.).

All uniaxial crystals (tetragonal and hexagonal) show interference figures of this type when they are viewed parallel to c. Figure 107 shows how the interference figure varies with tilting of a hexagonal crystal.

Biaxial crystals show a different type of interference figure, but one related to the uniaxial figure. The most obvious difference is that, upon rotation, the uniaxial figure does not change whereas the biaxial changes from a simple black cross to a pair of hyperbolic isogyres, or brushes. When the crystal is rotated 45° away from the simple cross position, the isogyres reach their maximum separation, illustrated in Figure 107b and c. (The isochromatic curves have been omitted for simplicity.) Like the black cross of the uniaxial interference figure (Figure 107a, the black brushes are the pattern of extinction positions of the crystal in the biaxial interference figure. At maximum separation, the distance between brushes is related to the optic axial angle, 2V. In 107a, the uniaxial figure 2V may be regarded as zero. In 107b, the extinction brushes are opened up, corresponding to a nearly uniaxial crystal (two of the three principal refractive indices, α and β and/or β and γ, are very close together). In 107c, the brushes are opened more, corresponding to more nearly equal values of $\gamma - \beta$ and $\beta - \alpha$ and, consequently, to a larger optic axial angle.

The two optic axes emerge from the biaxial figure at the midpoints of the brushes in b and c. The acute bisectrix, Bx_a, emerges at the very center of the figure, halfway between the optic axes. This type of interference figure, called a centered acute bisectrix figure, is probably the most useful figure a biaxial crystal can show.

Interference figures are very useful in characterizing and identifying compounds although their effective use is not for novices. They are helpful in determining the orientation of the principal refractive index directions in a crystal, in finding the optic sign, in studying dispersion, in determining the crystal system and, of course, in characterizing and identifying compounds.

a. Determination of optic sign from interference figures

Some orientations of a crystal give interference figures which are diagnostically of little value. When the orientation of the optical indicatrix of the crystal can be recognized from the figure, however, vibration directions for two refractive indices can be determined by noting the extinction positions at various points in the figure. By superimposing compensators on the figure we can observe addition or subtraction of retardations and can thus determine which of these two refractive indices is the higher. From this information we can sometimes deduce the optic sign of the crystal.

1) Uniaxial crystals: $\epsilon > \omega$ (+); $\epsilon < \omega$ (-)

To be sure of being able to determine the optic sign, one must obtain a uniaxial figure in which one can see the optic axis or tell in which quadrant it lies. In such a figure all directions radiating outward from the optic axis correspond to ϵ'. All directions at right angles to these radii correspond to ω (Figure 108, Uniaxial Figure). By superimposing any compensator on the figure, one can determine the direction of the slow (high index) component in any part of the figure. If the slow direction is radial, ϵ' (and therefore ϵ) is the higher index and the crystal is positive. If the slow direction is tangential, ω is the higher index and the crystal is negative.

Figure 108 illustrates the effect of a first-order red plate on the interference figure of positive and negative uniaxial crystals. The most striking color change is observed near the optic axis where retardation due to the crystal approaches zero. The brushes themselves have zero retardation and, therefore, show the first-order red of the compensator. Immediately next to the red brushes the figure turns second-order blue in the two quadrants where retardations add, and first-order yellow in the two where they subtract. (See Figures 119-121, Color Plate.) In Figures 120-122 and 124-126 the slow directions of the compensators are the same as in Figures 108-111.

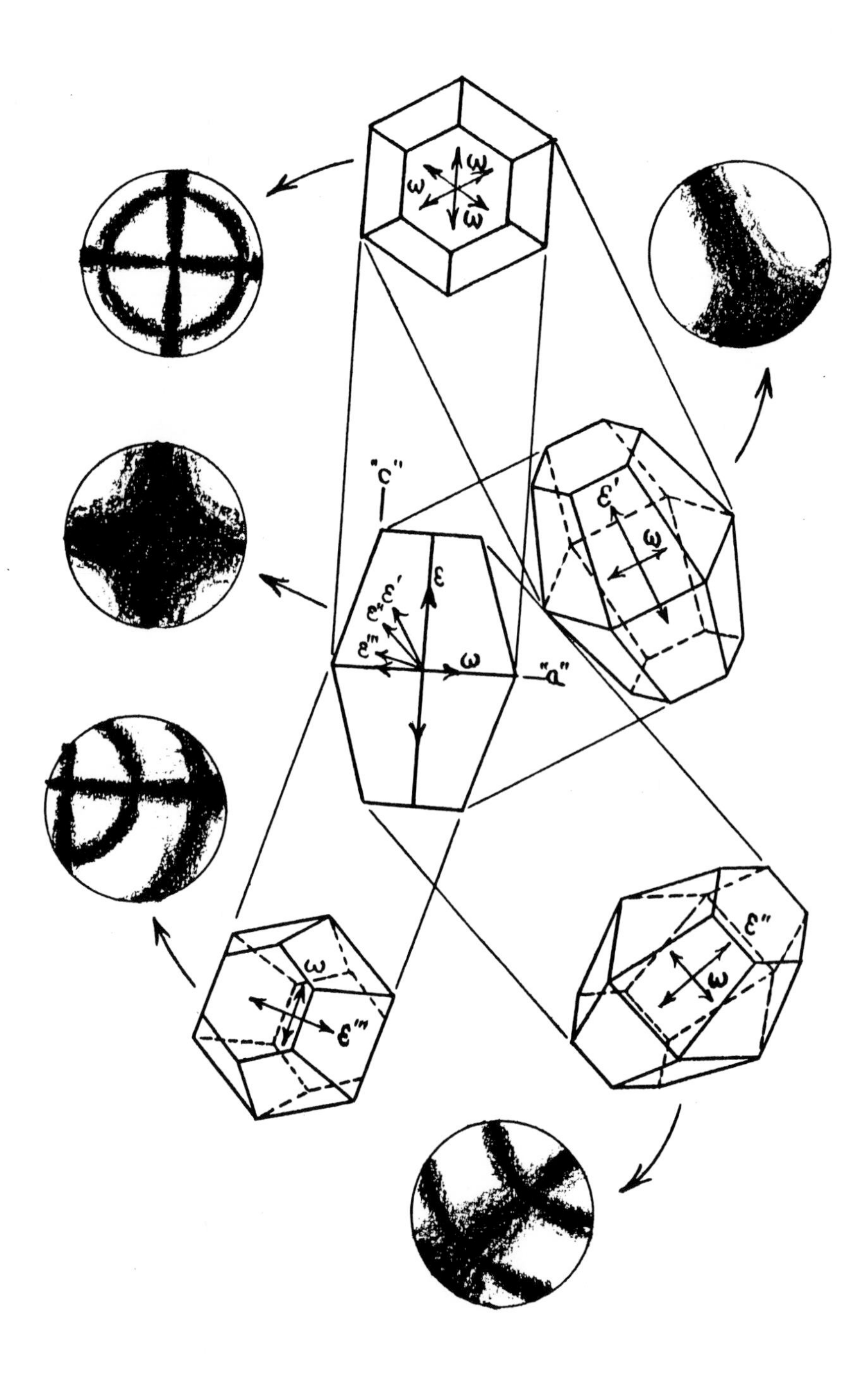

Figure 106. Interference figures seen with various orientations of a hexagonal crystal

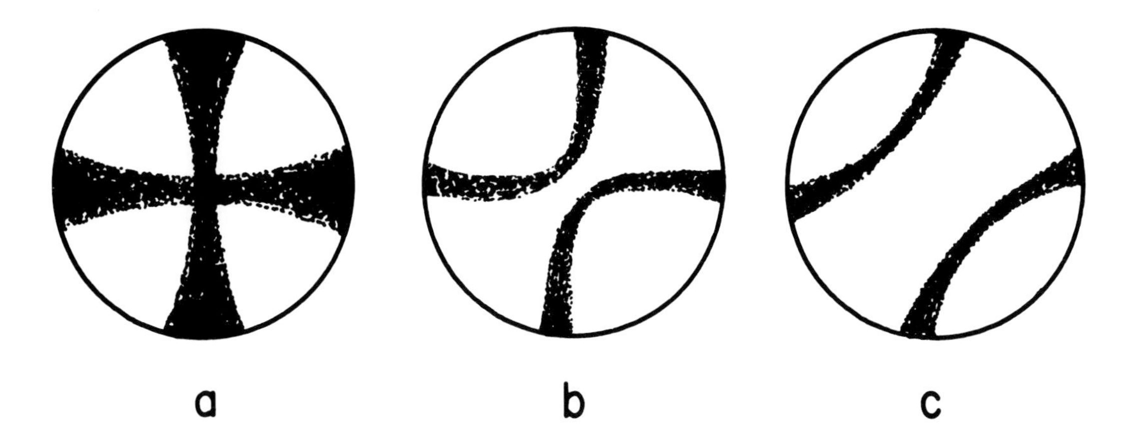

Figure 107. Uniaxial and biaxial interference figures

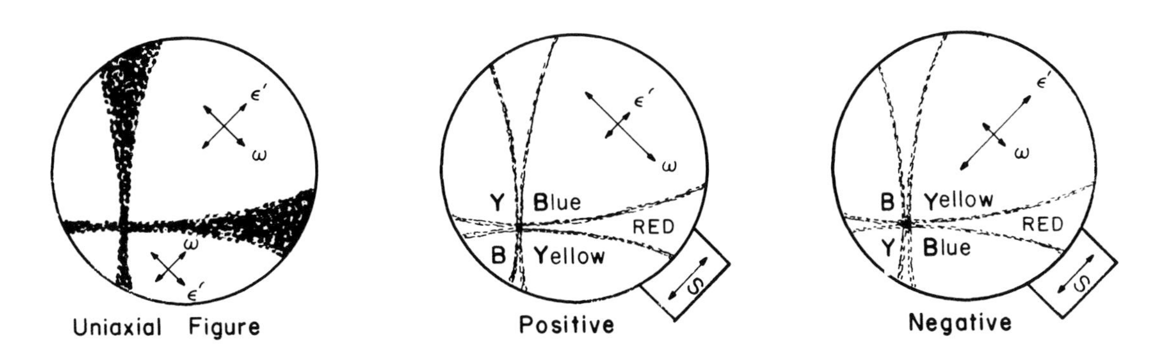

Figure 108. Determination of optic sign of uniaxial crystal with first-order red plate

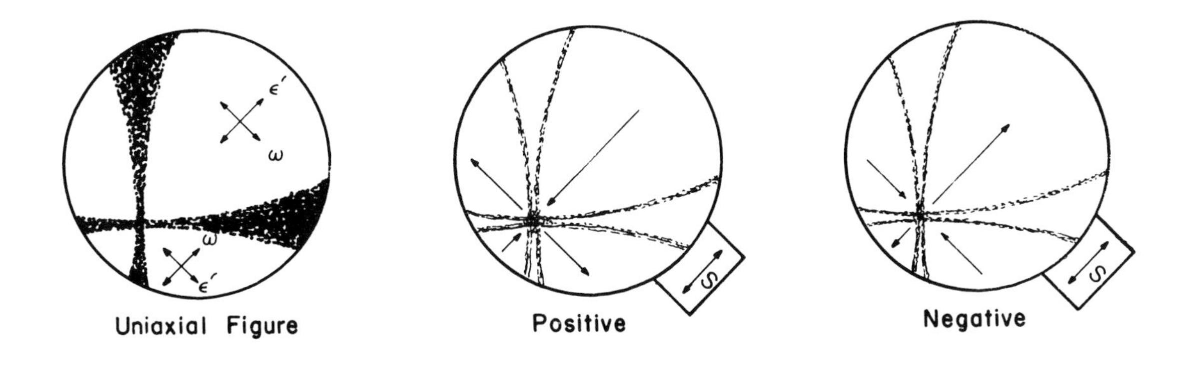

Figure 109. Direction of movement of interference color rings as quartz wedge is inserted; removing the wedge reverses directions of movement.

The quarter-wave compensator, which has only one fourth the retardation of the first-order red plate, turns the black brushes gray. Next to the gray brushes the color goes up to first-order yellow in the two quadrants where additition occurs and down to zero-order black where retardations subtract. (See Figure 122, Color Plate.)

Insertion of a quartz wedge causes the rings of interference colors to move towards the optic axis in quadrants where addition of retardation occurs, and to recede from the optic axis where subtraction occurs, as illustrated in Figure 109.

2) Biaxial crystals: $\gamma - \beta > \beta - \alpha$ (+); $\gamma - \beta < \beta - \alpha$ (-)

To determine the optic sign of a biaxial crystal, one must obtain an interference figure in which the trace of the optic axial plane and Bx_a can be located by seeing at least one optic axis directly or by noting the directions the brushes leave the field as the stage is rotated. For all biaxial crystals β is the refractive index for the vibration direction perpendicular to the optic axial plane. α' and γ' lie in the optic axial plane.

In the Bx_a region of the figure, if β is the slow component, then α or α' must be the refractive index parallel to the trace of the optic axial plane, since α is defined as the lowest index of the crystal. The highest index of the crystal, γ, must be the acute bisectrix of the optic axial angle. Reference to Figure 100c shows that γ is Bx_a, β is closer to α, $\gamma - \beta$ is larger than $\beta - \alpha$, and the crystal is positive. Conversely if β is found to be the low index (fast component) in the Bx_a region of the interference figure, then the high index is parallel to the trace of the optic axial plane and must be γ or γ'; therefore α is Bx_a, $\gamma - \beta$ is less than $\beta - \alpha$, and the crystal is negative (Figure 100d).

Figure 110 illustrates the use of a first-order red plate to determine the sign of a biaxial crystal by means of the interference figure. As in the uniaxial figure, the most striking color change occurs near an optic axis, where retardation in the crystal approaches zero. (See also Figures 123-125, Color Plate). The effect of the quarter-waveplate is shown in Figure 126, Color Plate. Figure 111 illustrates the directions of movement of the polarization color curves upon insertion of a quartz wedge.

b. Determination of optic axial angle

Optic axial angles can serve as analytical data for the identification of unknowns. Also, if two refractive indices and the optic axial angle of a biaxial crystal can be measured, the third refractive index can be calculated (see below).

1) Definitions (see Figure 112)

2V is the true optic axial angle which exists within the crystal.

2E is the apparent optic axial angle when the optical path between the objective and the preparation is occupied by air ($n = 1.0$).

2H is the apparent optic axial angle observed when the optical path between the objective and the top of the preparation is filled with oil ($n = 1.515$).

2E is always larger than 2V. 2H may be either larger or smaller than 2V; larger if the beta refractive index is greater than 1.515 and smaller if β is less than 1.515.

$$\beta \sin V = 1.00 \sin E = 1.515 \sin H \qquad (41)$$

$$\text{or} \qquad \text{Sin } V = \frac{\sin E}{\beta} = \frac{1.515 \sin H}{\beta} \qquad (42)$$

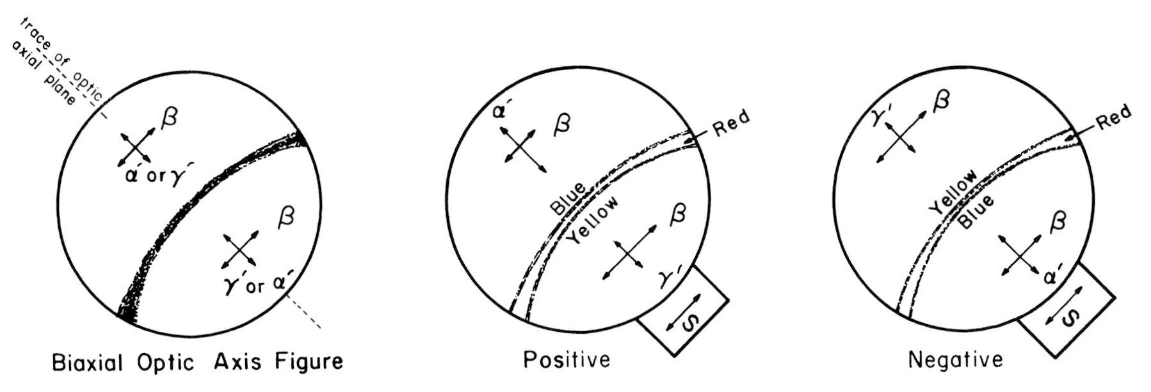

Figure 110. Determination of optic sign of biaxial crystal with first-order red plate

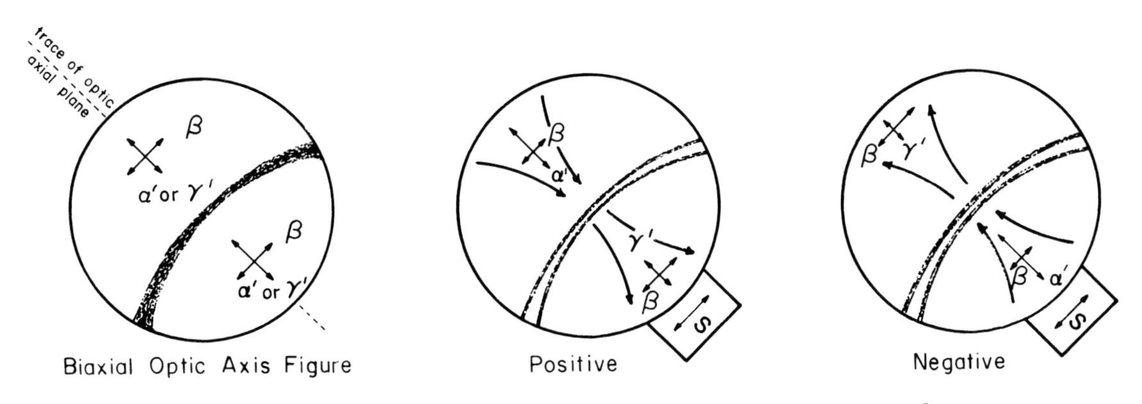

Figure 111. Determination of optic sign of biaxial crystal with quartz wedge

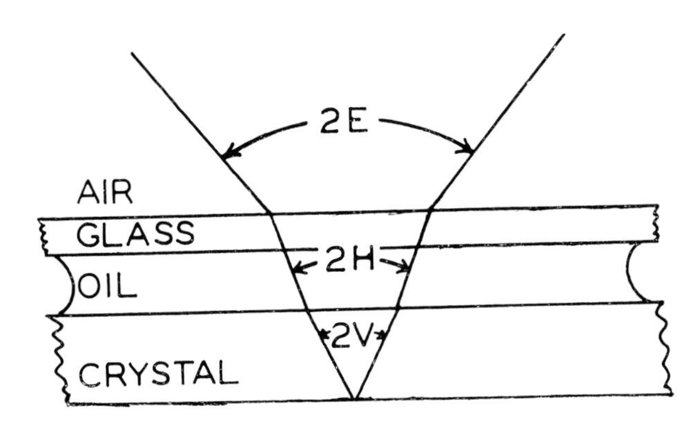

Figure 112. Refraction of light rays passing through media of different refractive index. The two rays are shown passing along the optic axes of a biaxial crystal and into oil, glass and air. The true optic axial angle in the crystal is 2V; the apparent optic axial angle is 2H in the oil, and 2E in air.

2. Determination of 2E

a) Orient a centered acute bisectrix interference figure by rotation of the stage so that the two black brushes are as far apart as possible (the 45° position).

b) Measure the linear distance between the centers of the brushes (d) and the diameter of the field (D) in ocular scale divisions (see Figure 113). These linear distances are proportional, respectively, to the sine of half the optic axial angle, E, and to the sine of half the angular aperture of the objective:

$$\sin E = d\,(\sin \frac{AA}{2})/D \qquad (43)$$

c) Find the NA of the objective (engraved on the side — usually 0.85 for a 40X objective. Since $NA = \sin \frac{AA}{2}$ for a dry objective, sin E can be found by substituting NA in the above equation.

d) Instead of calculating 2E, one may find it by using Figure 113. Calculate the ratio d/D and move from this value upward to the diagonal line corresponding to the NA of the objective and horizontally to the ordinate where 2E is read directly.

For example, in Figure 113, d/D = 0.5. If the NA of the object is 0.85, 2E = 51°.

3) Determination of 2H

a) Measuring d/D and using Figure 113 as described above, the angle read as the ordinate will be 2H, rather than 2E, when an oil immersion objective is used. In Figure 113, for example, if the NA of the objective is 1.25 and d/D = 0.50, then 2H = 48°.

b) When 2H is calculated the equation (43) becomes $\sin H = d\,(\sin \frac{AA}{2})/D$. For an oil immersion objective, however, $NA = n_{oil}\,(\sin \frac{AA}{2}) = 1.515\,(\sin \frac{AA}{2})$.

Thus:

$$\sin H = d\,(\frac{NA}{1.515})/D \qquad (44)$$

c) Alternatively, AA may be read directly from Figure 114 and used in the calculation. For an oil immersion objective, the angular aperture is found by moving up from the NA to the 2H diagonal and across to the ordinate. (For a dry objective the angular aperture is found similarly, using the 2E diagonal.) Use of this procedure emphasizes the relationship between the diameter of the cone of illumination corresponding to the optic axial angle of the crystal and the diameter of the full cone accepted by the objective, but necessitates the use of sine tables. Again: $\sin H = d \sin \frac{AA}{2}/D$.

4. Determination of 2V from 2E or 2H

a) To find 2V, given 2E or 2H, the β refractive index of the crystal must be measured. 2V can then be calculated from the relationship:

$$\sin V = \frac{\sin E}{\beta} = \frac{1.515 \sin H}{\beta}, \qquad (42)$$

or it can be found using Figure 114.

b) For example, if 2E = 51° and β = 1.60, 2V is found by locating 51° on the 2E diagonal, dropping a vertical line to the β = 1.60 diagonal, and reading 2V as the ordinate. In this instance 2V = 31°. Similarly, if for another substance 2H = 48° and β = 1.70, locating 48° on the 2H diagonal and applying the same procedure gives 2V = 43°.

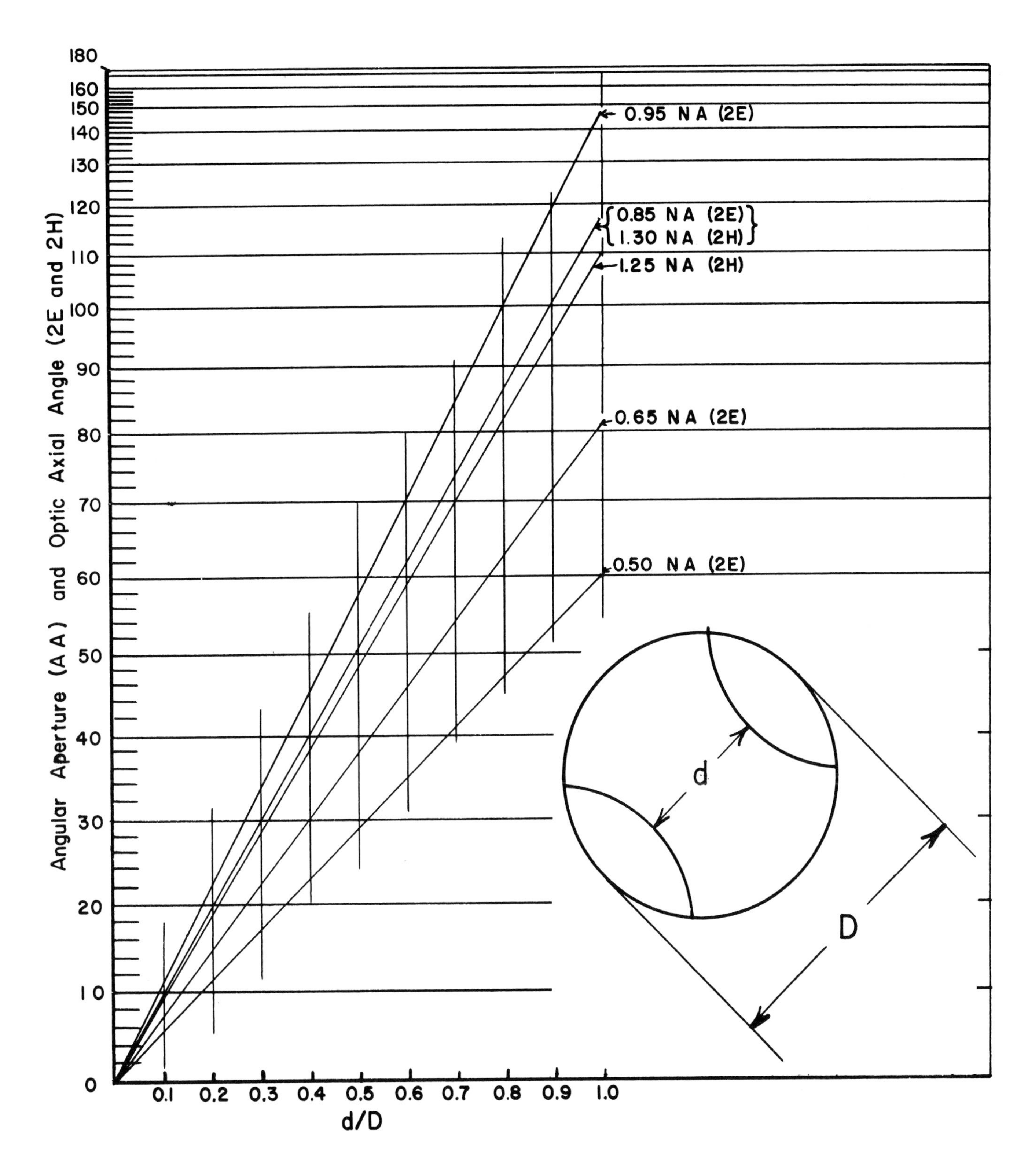

Figure 113. Relationship between angular aperture (AA); optic axial angles 2E, 2H and 2V; and the linear diameters in the back focal plane of the objective.

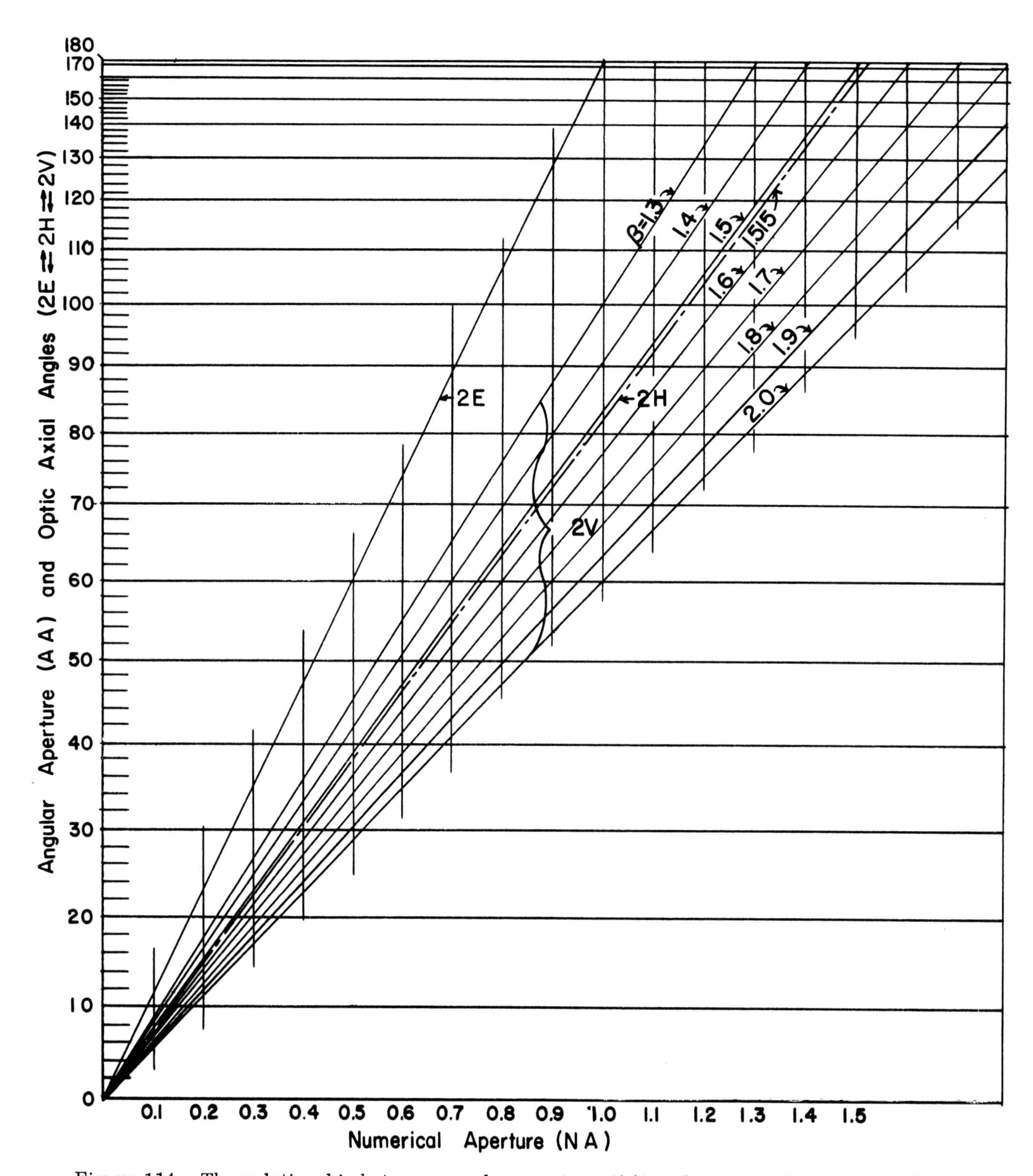

Figure 114. The relationship between angular aperture (AA) and numerical aperture (NA)

5) Determination of 2V by estimating curvature of an isogyre

The value of 2V may be estimated directly by observation of a centered optic axis interference figure in the 45° position. Figure 115 illustrates the appearance of such figures for values of 2V from 0° (uniaxial figure) to 90°. (Only the brushes are shown, not the interference colors.) The estimate is based on the degree of curvature of the black brush centered in the field. If it is sharply curved, 2V is small; if it is nearly straight, 2V approaches 90°. Precision of this method is not high, approximately ±10°.

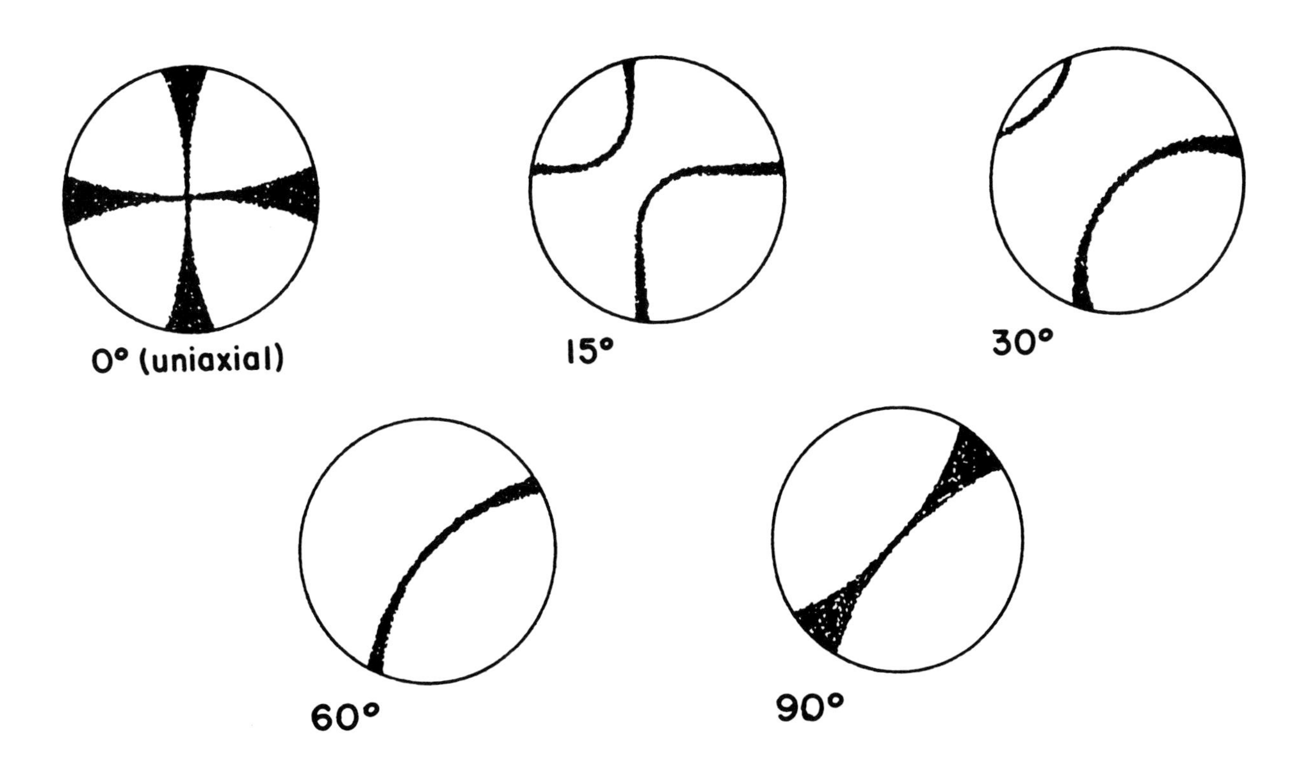

Figure 115. Curvature of isogyres for various optic axial angles

6) Calculation of 2V from α, β and γ.

When the three refractive indices of a crystal are known, 2V can be calculated from the following formulas, V_α being half the optic axial angle of a negative crystal and V_γ that of a positive crystal:

$$\cos^2 V_\alpha = \frac{\gamma^2 (\beta^2 - \alpha^2)}{\beta^2 (\gamma^2 - \alpha^2)}; \quad \cos^2 V_\gamma = \frac{\alpha^2 (\gamma^2 - \beta^2)}{\beta^2 (\gamma^2 - \alpha^2)}. \tag{45}$$

Figure 116 is a nomogram based on these formulas which may be used to find α, β, γ or V when three of these quantities and the optic sign of the crystal are known. As an example assume that $\alpha = 1.480$, $\beta = 1.590$ and $\gamma = 1.652$. 2V may be found as follows:

a) Plot $\alpha = 1.480$ on the left edge of the nomogram (scale of α and β).
b) Plot $\gamma = 1.652$ on the right edge of the nomogram (scale of γ and β).
c) Connect these two points with a straight line.
d) Locate the horizontal line representing $\beta = 1.590$ (read on either scale of β) and plot its intersection with the line joining α and γ.
e) Drop a line from this intersection to the scale of V. In this example, V = 34°, 2V = 68° and the crystal is optically negative.

If 2V and two of the refractive indices are known, the third index may be found provided that the optic sign of the crystal is also known.

7) Measurement of 2V with rotation apparatus

2V can be measured directly on the interference figure of a crystal mounted, in a rotation device, so that it rotates about β or about the plane containing β and Bx_a. If β itself is the rotation axis the crystal can be turned until a centered acute bisectrix interference figure appears in a field of view, as in Figure 117a. Since the optic axes emerge from the crystal at the midpoints of the brushes (melatopes), rotating the crystal to center first one and then the other melatope under an ocular crossline turns the crystal through 2V exactly (Figure 117b and c). The angular positions of the melatopes are read on the scale or drum of the rotation device. The difference between the readings is, of course, 2V and both the acute and obtuse optic axial angles can be found in this manner. The crystal should be immersed in a liquid of refractive index equal, or nearly equal to, β.

If the rotation axis lies in the β-Bx_a plane but is not β, the Bx_a figure observed will be tilted, but 2V can still be measured if the melatopes are in the field of view. Figure 117d illustrates the practical maximum — any further tilting would tip the melatopes out of the field and the angle between them could not be measured, as it can be in Figure 117e and f.

The Hartshorne rotation apparatus, discussed below in section E.2, is well suited for this method.

E. INTEGRATION OF OPTICAL AND MORPHOLOGICAL PROPERTIES OF SOLIDS

Table XXVI summarizes the relationships between optical properties and lattice structure of solid particles. Optical properties are nearly always more diagnostic than morphological, especially in dealing with small particles but, as the table indicates, they can be used to the full only when correlated with morphology. Unless a particle is isotropic its orientation on the microscope stage is all important in controlling which refractive indices can be observed. There are three practical methods of obtaining proper crystal orientation on the microscope stage.

1. Surveying the Preparation

When a sample consists of many particles, one can expect a preparation to exhibit at random all possible crystal orientations. By surveying the field of particles, one can usually find all the views needed to observe the principal refractive indices, unless some shape factor causes the particles to lie on the slide in certain orientations only.

2. Use of Rotation Devices

Another approach is to take one particle or crystal and manipulate it into the necessary orientations. One way is to use a rotation device such as Hartshorne's (Figure 118). This is a particularly well designed version of the spindle stage, in which rotating any randomly oriented crystal about the horizontal needle axes will bring successively into the plane of the microscope stage each of

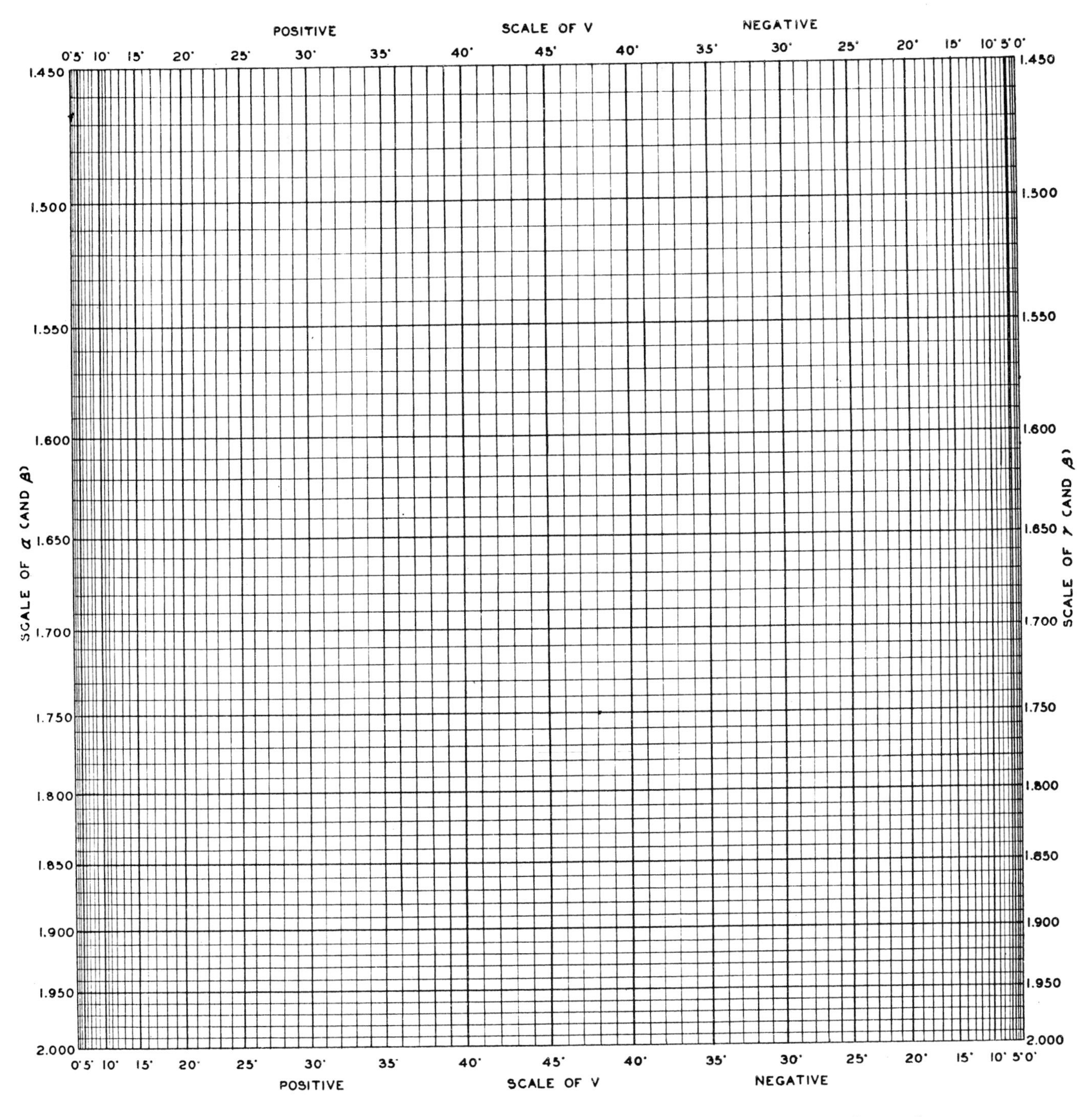

Figure 116. Nomogram of $\sin^2 V_\gamma = \dfrac{\dfrac{1}{\alpha^2} - \dfrac{1}{\beta^2}}{\dfrac{1}{\alpha^2} - \dfrac{1}{\gamma^2}}$ (+) and $\sin^2 V_\alpha = \dfrac{\dfrac{1}{\beta^2} - \dfrac{1}{\gamma^2}}{\dfrac{1}{\alpha^2} - \dfrac{1}{\gamma^2}}$ (-)

JOHN B. MERTIE, JR.
U. S. GEOLOGICAL SURVEY

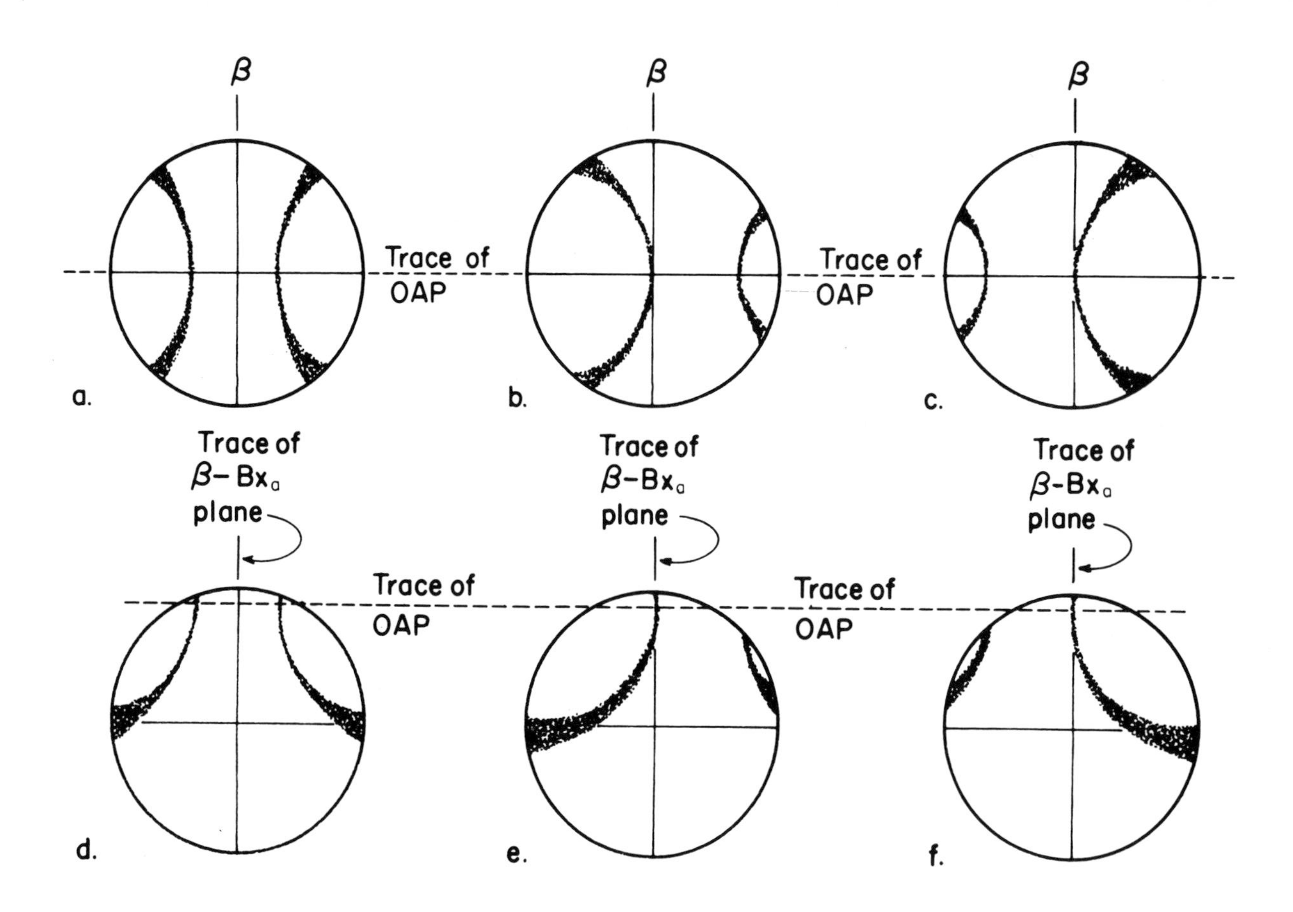

A crystal showing a centered Bx_a figure (a.) is rotated about β to place one melatope on an ocular crossline (b.), and then rotated through 2V to put the other melatope on the crossline (c.). The crystal, now showing a tilted Bx_a figure (d.) is rotated about a direction other than β in the β-Bx_a plane to place one melatope (e.) and then the other (f.) on the crossline. The ocular has been set in the tube with the crosslines 45° from the polar vibration directions.

Figure 117. Measurement of 2V with rotation apparatus

the principal vibration directions for that crystal. Each of these principal refractive indices can therefore be measured by orienting them in this way and then rotating the stage (and the crystal) with relation to the plane of polarized light. The orientation of the mounted crystal is monitored by means of the interference figure. Dr. Saylor[53] has a paper covering this technique which you should read if you really want to do it.

53. Saylor, C. P., "Accurate microscopical determination of optical properties of one small crystal," Advances in Optical and Electron Microscopy, edited by R. Barer and V. E. Cosslett, vol. 1, p. 41-76, Academic Press, New York, 1966.

It simplifies orientation of the crystal if well formed. You can then orient the crystal with the rotation axis parallel to one of the zone axes. Unfortunately, relatively few crystals from a bottle, for example, are sufficiently well formed to make possible orienting and mounting in this fashion. It may not be too difficult, however, to prepare your own crystal for this purpose by recrystallizing it from a drop of solution on a microscope slide. We usually use high boiling liquids such as thymol, nitrobenzene, tetralin, monobromnaphthalene, or ethylene glycol although water works well. One of these liquids should be a reasonably good solvent for any compound you might wish to recrystallize. Furthermore, crystallization should be possible by warming a small droplet containing a few seed crystals. On cooling, the seed crystals usually grow very quickly into well-formed elongated individual crystals.

Let's assume that you manage, in only one or two trials, to obtain a 100 μm long crystal you'd like to mount on the rotation apparatus. This is done by pushing the crystal with a fine tungsten needle through the liquid out through the periphery onto the slide surface. A neck of liquid behind the crystal can be cut with a piece of filter paper and much of the liquid can be soaked up with the same filter paper. To completely dry the crystal it is merely pushed in a snake dance pattern across the surface of the slide, it leaves behind a track of tiny droplets and soon you have a completely dry crystal. To mount this crystal on the rotation apparatus it should be pushed to the edge of the slide and extended over that edge about 27% of its length.

The rotation apparatus needle is then removed from the device and taped to the surface of a second microscope slide with the tip extending past the slide edge a few millimeters. The crystal and the needle are aligned in the microscope field of view with their two axes parallel and the two ends nearly touching. With a second, very fine, tungsten needle a tiny droplet of shellac is placed on the very tip of the rotation apparatus needle. The crystal and/or the needle are then moved into contact and allowed to remain in place until the shellac hardens (about 15 min). You will be able to tell by focusing up and down whether the height of the crystal is the same as the tip of the needle. It is essential that this be so; the slide with the crystal can be shimmed upward by one, two or three coverslips as necessary. The solvent for shellac is alcohol and your crystal should not be soluble, of course, in that solvent. If it is, you could use a tiny droplet of epoxy cement. Another point worth emphasizing is that the tip of the rotation apparatus needle should be ground flat normal to the axis of the needle. This is very important to keep surface tension of the shellac from pulling the crystal around to the side of the needle as the cement dries.

3. Crystal Rolling

We might also describe a technique which is quick and easy and does some of the things, at least, that the rotation apparatus normally does. If the crystal is mounted in a viscous liquid between slide and coverslip, it can be tipped into any desired orientation by sliding the coverslip along the slide with a coarse needle. Using a round coverslip this movement can be made in any direction desired. An ideal liquid for this purpose is Aroclor® 1260. Any high viscosity liquid, however, that does not dissolve your sample should work. Canada, balsam, Karo syrup and Aquaresin have all been used. Some microscopists like to use the very viscous Aroclor 5442. They warm it slightly (to the consistency of Aroclor 1260 at room temperature) and slide the coverslip before the 5442 hardens. The crystal can usually be oriented properly before the 5442 sets and that orientation is then held better than in 1260.

Ideally, perhaps one should use both crystal rolling and the crystal rotation apparatus for quickest results when one is determining crystal morphology and optics. We use crystal rolling to determine crystal system, angles, optic axial angle, optic sign and dispersion, and the relationship between the refractive index directions and the crystallographic axes. This can be done very quickly by rolling the crystal into the two or three orientations necessary to permit viewing along the appropriate zone axes. Once this information is obtained, we mount the crystal on the rotation apparatus to determine the principal refractive indices.

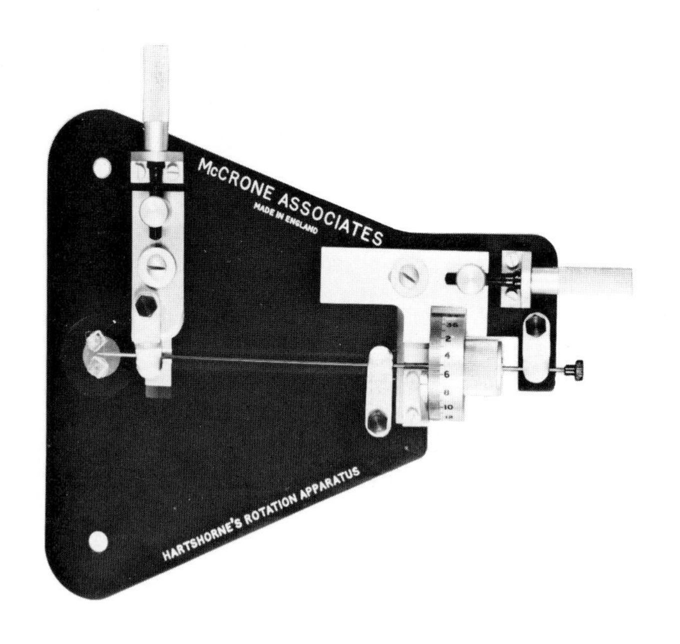

Figure 118. Hartshorne's Rotation Apparatus

GENERAL REFERENCES IN OPTICAL CRYSTALLOGRAPHY

Bloss, F. Donald, An Introduction to the Methods of Optical Crystallography, Holt, Rinehart and Winston, New York, 1961.

Hartshorne, N. H., and A. Stuart, Crystals and the Polarizing Microscope, 4th ed., Edward Arnold, London, 1970.

Hartshorne, N. H., and A. Stuart, Practical Optical Crystallography, 2nd ed., Edward Arnold, London, 1969.

Wahlstrom, Ernest, E., Optical Crystallography, 4th ed., Wiley, New York, 1969.

Winchell, A. N., Optical Properties of Organic Compounds, 2nd ed., Academic, New York, 1954.

Winchell, A. N., and H. Winchell, The Microscopical Characters of Artificial Inorganic Solid Substances: Optical Properties of Artificial Minerals, 3rd ed., Academic, New York, 1964.

Wood, Elizabeth A., Crystals and Light, An Introduction to Optical Crystallography, Van Nostrand, New York, 1964.

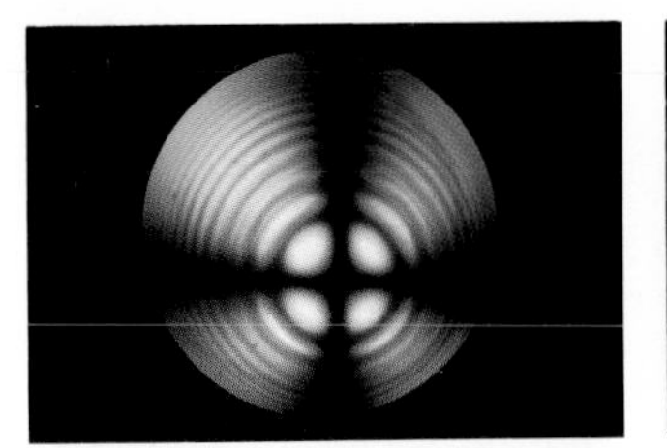

Figure 119. Centered uniaxial optic axis figure.

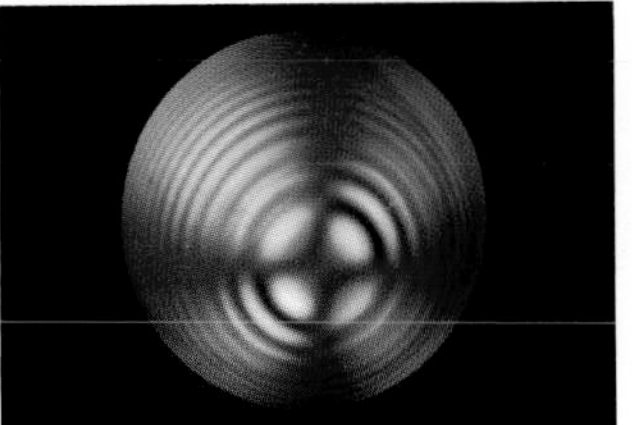

Figure 120. Uniaxial (—) figure with first order red compensator.

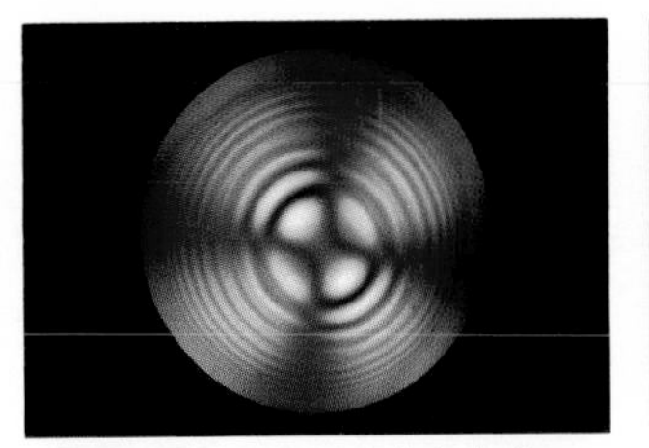

Figure 121. Uniaxial (+) figure with first order red compensator.

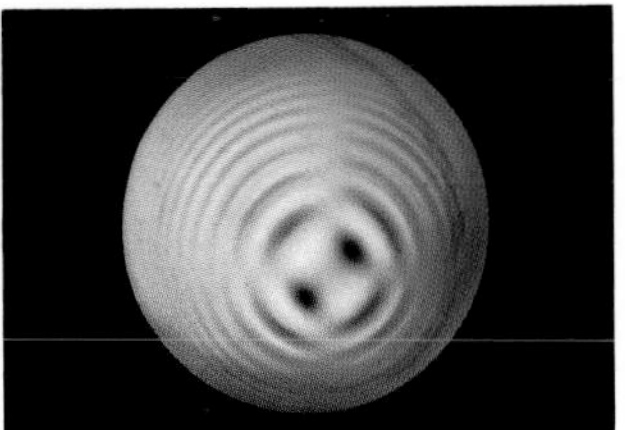

Figure 122. Uniaxial (—) figure with quarter-wave compensator.

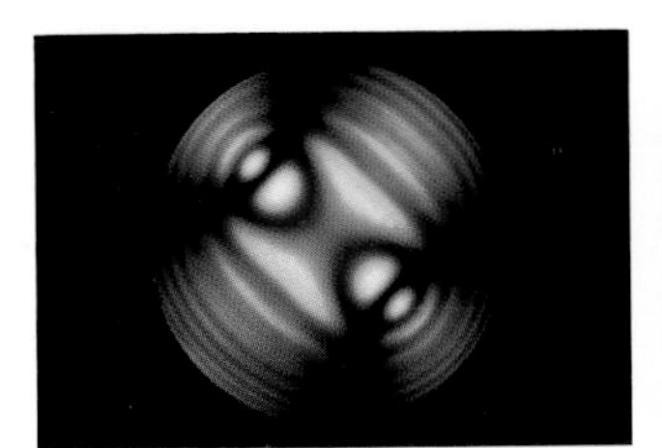

Figure 123. Centered biaxial acute bisectrix figure.

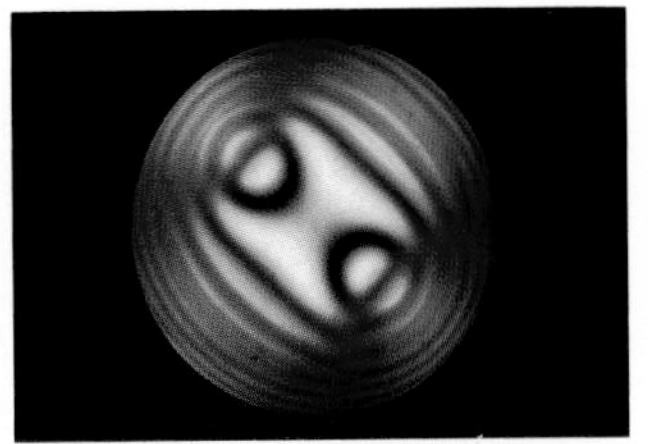

Figure 124. Centered biaxial (—) acute bisectrix figure with first order red compensator.

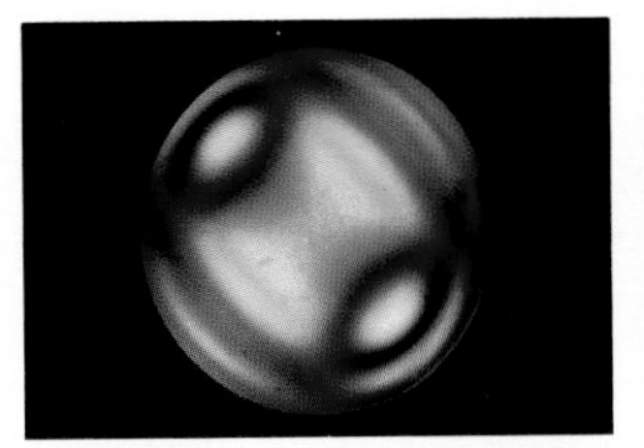

Figure 125. Centered biaxial (+) acute bisectrix figure with first order red compensator.

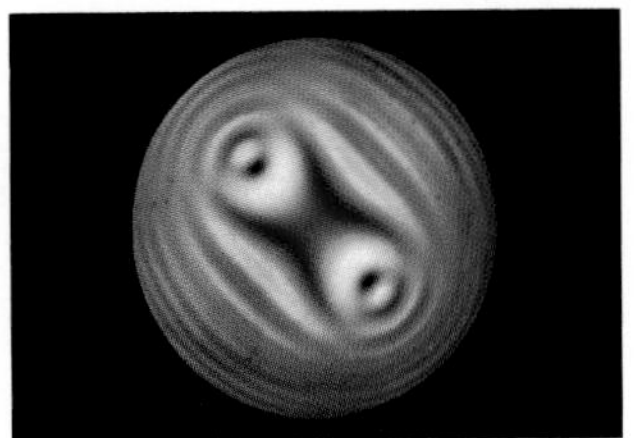

Figure 126. Centered biaxial (—) acute bisectrix figure with quarter-wave compensator.

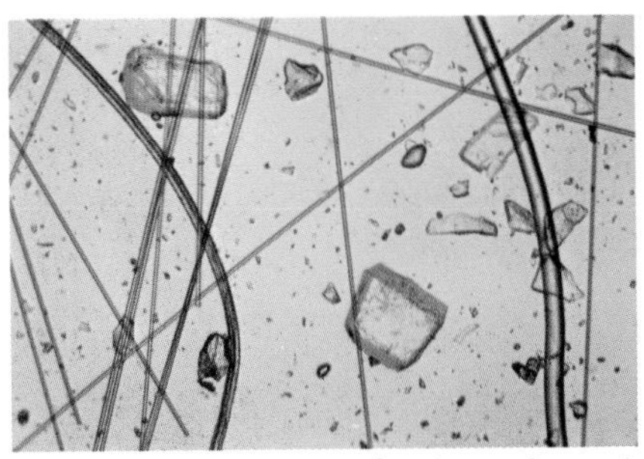

Figure 127. Isotropic and anisotropic particles, unpolarized light.

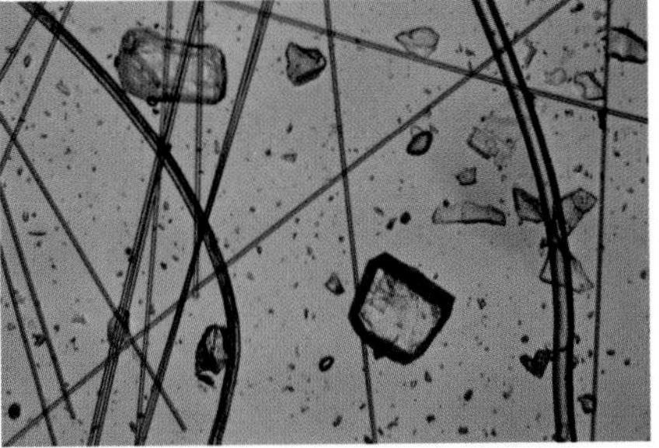

Figure 128. Plane polarized light.

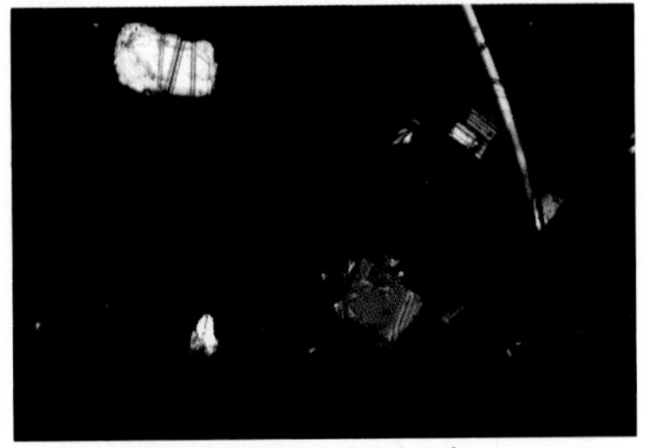

Figure 129. Crossed polars.

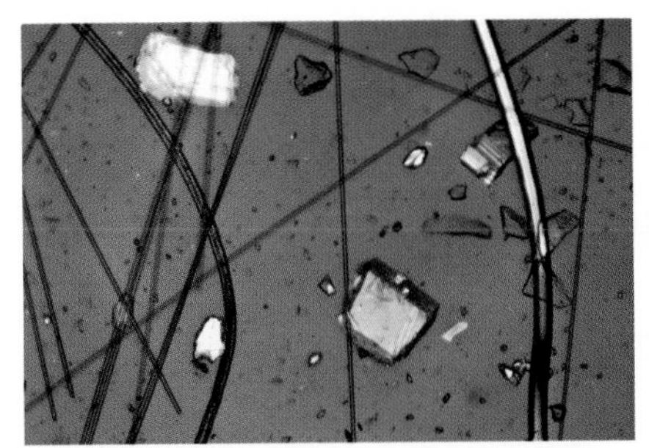

Figure 130. Slightly uncrossed polars.

Figure 131. Parallel polars.

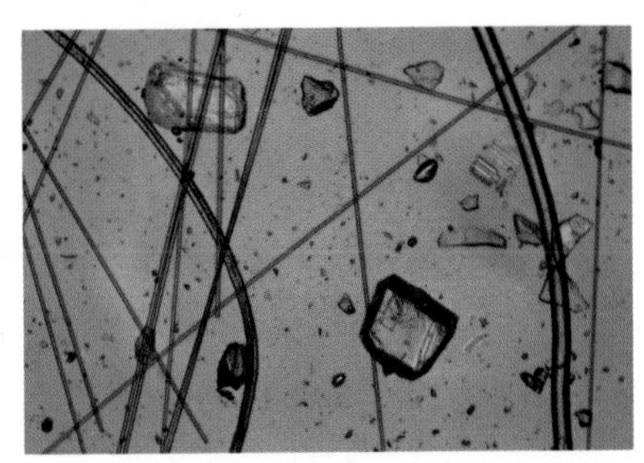

Figure 132. Crossed polars with one quarter wave plate.

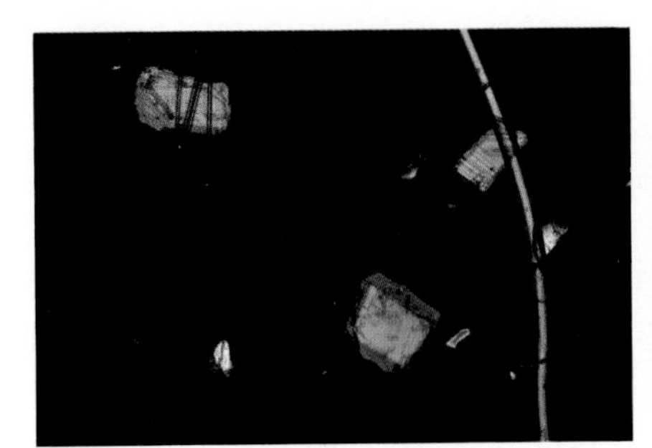

Figure 133. Crossed polars and crossed quarter-wave plates at 45°.

Figure 134. Slightly uncrossed polars and crossed quarter-wave plates at 45°.

Figure 135. 'Holey' polarizer, crossed analyzer, crossed-quarter wave plates at 45°.

Reprinted from
THE PARTICLE ATLAS
McCrone and Delly
Ann Arbor Science Publishers, Inc.

Table XXVI. Summary of optical and morphological properties of solids

	Isotropic		Anisotropic				
			Uniaxial		Biaxial		
	Glasses (supercooled liquids)	Cubic crystals	Tetragonal	Hexagonal	Orthorhombic	Monoclinic	Triclinic
		regular arrangement of lattice points					
	random arrangement of atoms; no faces, no angles	3 mutually perpendicular directions having identical spacings*	unique c axis normal to plane of 2 mutually perpendicular a axes having identical spacing*	unique c axis normal to plane of 3 a axes 120° apart and with identical spacing*	3 mutually perpendicular axes, a, b, c, with different spacing*and c < a < b	one axis, b, perpendicular to plane of 2 mutually oblique axes with all spacings* different and c < a	3 mutually oblique axes with different spacing*and c < a < b
	single refractive index for all directions		two refractive indices: ϵ, parallel to c; ω, for all directions perpendicular to c		three refractive indices, $\alpha < \beta < \gamma$: α, β, γ parallel to a, b, c, but not necessarily respectively	α, β or γ parallel to b; other two in plane of a and c	α, β and γ mutually perpendicular but oblique to a, b and c
extinction:			parallel or symmetrical		parallel or symmetrical (except on hkl faces)	parallel or symmetrical when observed in a - c plane; otherwise oblique	oblique
symmetry: minimum:	none	$4A_3$	A_4*	trigonal: A_3 hexagonal; A_6	$2A_2$**	A_2 or P	none
maximum:	none	$3A_4, 6A_2, 4A_3$ 9P, C	A_4, $4A_2$ 5P, C	trigonal; A_3, $3A_2$, 3P, C hexagonal; A_6, $6A_2$, 7P, C	$3A_2$, 3P, C	A_2, P, C	C

* spacing of lattice points
** may be "composite" axes

** may be "composite" A_2, sphenoids

F. PHOTOMICROGRAPHY OF MIXTURES OF ISOTROPIC AND ANISOTROPIC PARTICLES

Some of the complications of accurately representing both isotropic and anisotropic particles in the same photomicrograph are discussed in "Problems in Particle Photomicrography" (VI. F.). The basic difficulty is that the best illumination conditions for isotropic particles are the worst for anisotropic, and vice versa. In ordinary brightfield, glass fibers and undelustered nylon look the same — both appear — while glass fibers are not visible against the black field.

1. Slightly uncrossed polars

One answer to the problem is a compromise between these two kinds of illumination: the use of polars which have been uncrossed slightly by rotating one polar about 15° from the crossed position. This brightens the black field to a gray, against which colorless isotropic particles appear the same gray as the field and anisotropic ones show nearly the same interference colors as they do with crossed polars. These colors are diluted with the white light which is no longer completely stopped by the analyzer. This method was used in the first edition of The Particle Atlas. Several other ways of showing isotropic and birefringent particles under one set of illumination conditions are discussed below and are illustrated in Figures 130 through 135 (color plate).

Figures 127, 128 and 129 (color plate) illustrate the problem. A specially arranged particle preparation mounted in Aroclor® is illuminated with ordinary, unpolarized white light in Figure 127. The photomicrography provides no clue as to which substances are anisotropic. In Figure 128 one polar was used to illuminate the prep with plane polarized light. With one polar, anisotropic particles can show differences in contrast which are due to particle orientation. Two calcite crystals are oriented, one with its ω refractive index direction nearly parallel to the vibration direction of the polar and the other with an ϵ' index parallel to the polar vibration direction. The ω index direction bisects the acute angle of the calcite rhomb (see Figure 103). Since ω for calcite is 1.658, very near the index of the mountant (1.66), the first crystal appears at minimum contrast. The second crystal appears at maximum contrast because the index it is oriented to show approaches the ϵ value of 1.486. It is apparent that one could not hope to have every particle in a preparation favorably oriented to distinguish the anisotropic ones in a single rotation position of the microscope stage.

Figure 129, showing the preparation between crossed polars, illustrates the best view of the anisotropic particles at the expense of the isotropic ones, which disappear completely. In addition, any anisotropic particles that happen to be so oriented that they are at extinction or show an isotropic view (parallel to an optic axis) will not be seen at all. One way to avoid the disappearance of such particles is to use partly uncrossed polars, as described earlier in this section (Figure 130). The disadvantage is that there is no sure way to distinguish isotropic particles from birefringent ones at extinction.

2. Parallel polars

If the polars are set parallel instead of crossed, each anisotropic particle shows interference colors which are the complements of the colors it exhibits between crossed polars (Figure 131). The field is white instead of black, and isotropic particles are visible. There is, however, no way to distinguish between these and anisotropic particles showing isotropic views. Probably the chief disadvantage of using parallel polars is that small birefringences are hard to see against the bright background and hence are less sensitively detected than with slightly uncrossed polars.

3. Crossed polars with compensator

The use of crossed polars with a compensator is yet another way of avoiding the disappearance of isotropic particles but, like the two methods just discussed, it fails to differentiate anisotropic particles at extinction. In Figure 132 a single quarter-wave plate was used as the compensator.

4. Crossed quarter-wave plates

In Figure 133, the particles are shown between two quarter-wave plates which are placed between crossed polars. The quarter-wave plates are set at right angles to each other so that their slow directions are crossed, and at 45° to the polar vibration directions. The result is that all anisotropic particles show interference colors, even those oriented in extinction positions. The light emerging from the quarter-wave plate below the preparation is circularly polarized. When it passes through an anisotropic particle, it vibrates with equal amplitude in the two mutually perpendicular refractive index directions in the crystal, even when the crystal is so oriented that it would be at extinction if illuminated with plane polarized light. One component is retarded behind the other to produce an interference color as usual, but since the two components have equal amplitude they are equally bright for all rotation positions; thus no anisotropic particle lies at extinction. The retardation added to the system by the first quarter-wave plate is compensated by the second plate, so that the particles show exactly the same interference colors as they do when lying in the 45° position of rotation between crossed polars alone.

5. Crossed quarter-wave plates plus unpolarized light

Although the use of crossed quarter-wave plates differentiates anisotropic particles lying at extinction from isotropic particles, neither isotropic particles nor views down an optic axis of a uniaxial or biaxial substance would be ween under these conditions of illumination. To make them visible, it is necessary to put through the system a small amount of light which is not stopped by the analyzer. As mentioned earlier, this can be done by uncrossing the polars slightly. The combined devices of slightly uncrossing the analyzer and using crossed quarter-wave plates at 45° to the polarizer vibration direction were employed in most of the color photomicrographs in Volume II of Particle Atlas Two. Figures 61 and 134 illustrate the results (color plates).

A better way to supply light to illuminate isotropic particles is to make a small hole in the polarizer. This admits unpolarized light which can pass the analyzer to provide a gray field while allowing the polars to be kept strictly crossed to maximize intensity of the interference colors of the anisotropic particles. Crossed quarter-wave plates are used to eliminate extinction positions. Figure 135 illustrates the effect. A complication of this method is that a different sized hole is required for each microscope objective of different aperture. Probably the simplest solution is to make the hole in the polar as large as is needed for the lowest power objective to be used and, for the higher power objectives, reduce the diameter of the hole by superimposing metal rings of appropriate size on the polarizer.

See page 86 for another (final) solution to this problem as used in Volume V of The Particle Atlas.

X. DISPERSION STAINING

A. INTRODUCTION

Dispersion staining[54] is a particle identification technique based on the difference between refractive index dispersion of a particle and of the liquid medium in which the particle is immersed. Two different procedures are based on the use of stops in the objective back focal plane; both procedures give colored particle boundaries. An annular stop shows a color consisting of wavelengths near that at which particle and medium match in refractive index; a central stop shows colors complementary to those shown by the annular stop, that is, light of wavelengths refracted by the particle in that medium.

These effects make it possible to systematically identify transparent substances by their dispersion colors in known refractive index media. A system, based on Cargille[55] refractive liquids, already includes many common isotropic and anisotropic compounds[56]. Dispersion staining data have been plotted for these compounds, making it possible to "read" the dispersion staining colors shown by an unknown and to identify it from the graphs. The procedure has been applied to refractive index determination[57], identification of quartz in lung tissue[58], determination of toxic dusts[59], particle counting[60], identification of glass fragments[61], identification of asbestos[62-64], mineral characterization[65], identification of settled dust[66], fiber identification[67,68] and measurement of dispersion of the Kofler glass powder standards and Aroclors® [69].

54. Brown, K. M. et al., Microscope 13 311-22; 14 39-54 (1963).

55. R. P. Cargille Laboratories, Inc., 55 Commerce Road, Cedar Grove, NJ 07009.

56. McCrone, W. C., and J. G. Delly, The Particle Atlas, 2nd ed., Ann Arbor Science Publishers, 1973 and 1978.

57. Cherkasov, Y. A., Gos. Nauch-Tekhn. Izdat., Moscow 184-207 (1957). Translated by Ivan Mittin, Intern. Geol. Rev. 2 218-35 (1960).

58. Crossmon, G. C., Amer. Ind. Hygiene Quart. 18 341 (1957).

59. Crossmon, G. C., Occupational Health Review 16 3-7 (1964).

60. Thaer, A., Staub 38 30 (1954).

61. Grabar, D. G., and A. H. Principe, J. Forensic Sci. 8 54 (1963).

62. Julian, Y., and W. C. McCrone, Microscope 18 1-10 (1970).

63. McCrone, W. C., Environ. Health Perspectives 9 57-61 (1974).

64. McCrone, W. C., Microscope 26 (1978).

65. Crossmon, G. C., Anal. Chem. 20 976 (1948).

66. Grabar, D. G., J. Air Pollution Control Assoc. 12 560 (1962).

67. Forlini, A. L., and W. C. McCrone, Microscope 19 243-54 (1971).

68. McCrone, W. C., Microscope 25 25-30 (1977).

69. Goodman, R. A., Microscope 18 41-50 (1970).

B. THEORY

To produce dispersion staining colors, the particles and immersion liquids must have different dispersion curves which intersect sharply (Figure 136) in the visible light region. At λ_0, the wavelength at which both particle and liquid have the same refractive index, the particle-liquid preparation is optically homogeneous. At wavelengths on either side of λ_0, optical discontinuities exist between particle and liquid. At shorter wavelengths, e.g., λ_1, the index of the liquid, n_ℓ, will usually exceed the index of the particle, n_g; at longer wavelengths, the situation is reversed, $n_g > n_\ell$. As a result, light of wavelength λ_0, incident upon the particle, will pass through undeviated. Light of any other wavelength, e.g., λ_1 or λ_2, incident at any angle not normal to a surface, will be deviated according to the refractive index difference between the particle and the liquid at λ_1 or λ_2.

Figure 136 shows the optical basis for the effects observed. Suppose that λ_0 is 589 nm (yellow), λ_1 is 486 nm (blue) and λ_2 is 656 nm (red). Since the dispersion of a liquid is usually greater than that of a particle, it is possible to choose a liquid such that:

Liquid		Solid	
n_F (486 nm)	>	n_F	(blue)
n_D (589 nm)	=	n_D	(yellow)
n_C (656 nm)	<	n_C	(red)

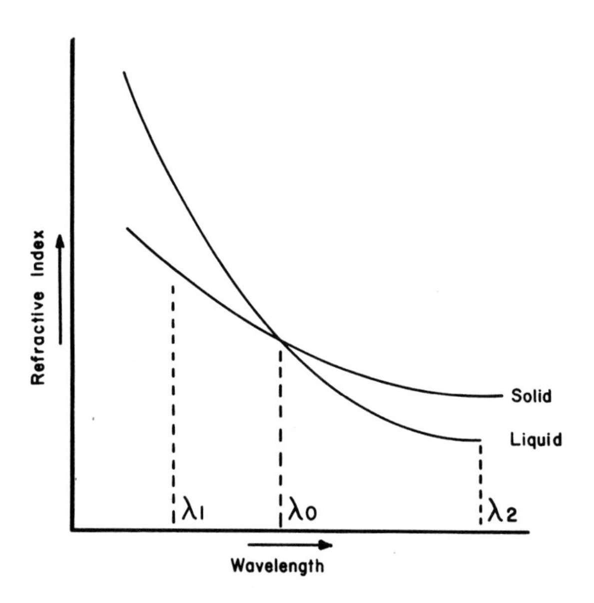

Figure 136. Dispersion curves for solid and liquid showing a match in refractive index at λ_0

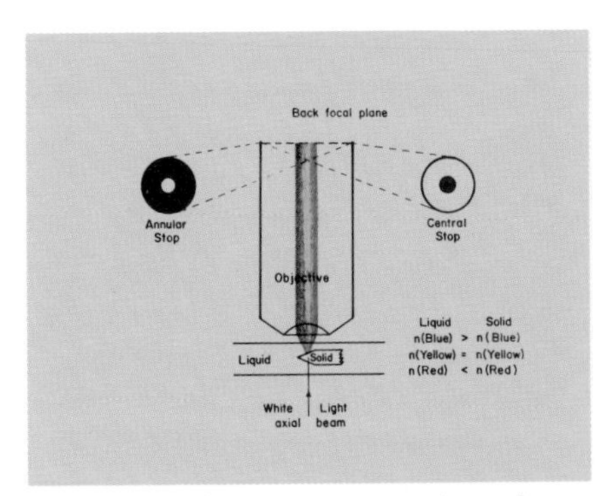

Figure 137. Schematic representation of annular and central stop dispersion staining.

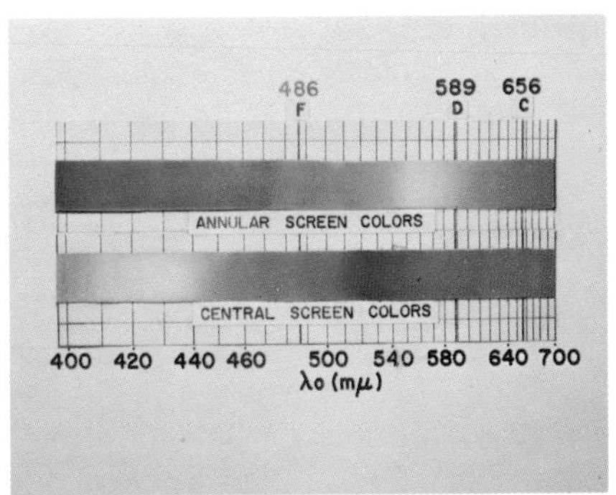

Figure 140. Annular and central stop colors.

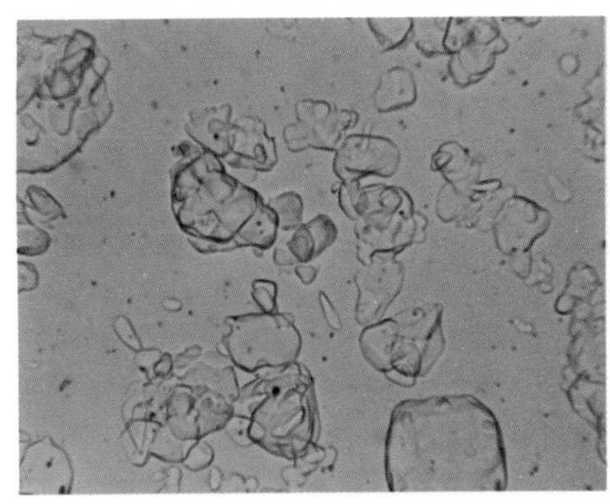

Figure 141. Sodium bromate in Cargille liquid 1.604; annular stop, 80X.

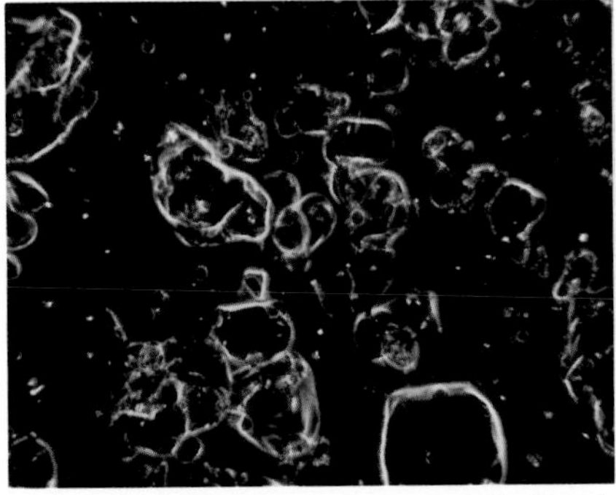

Figure 142. Sodium bromate in Cargille liquid 1.604; central stop, 80X.

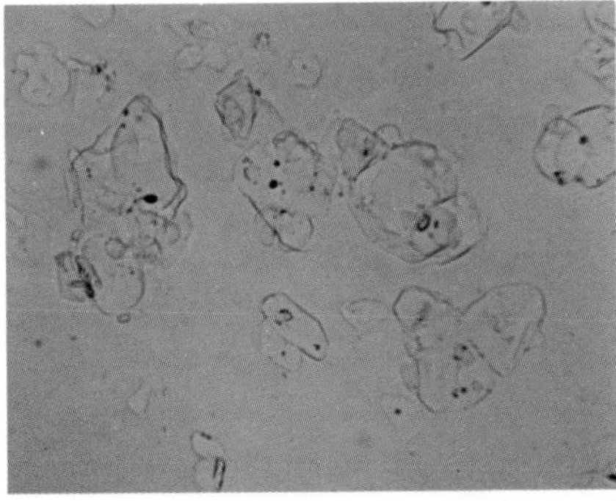

Figure 143. Sodium bromate in Cargille liquid 1.610; annular stop, 80X.

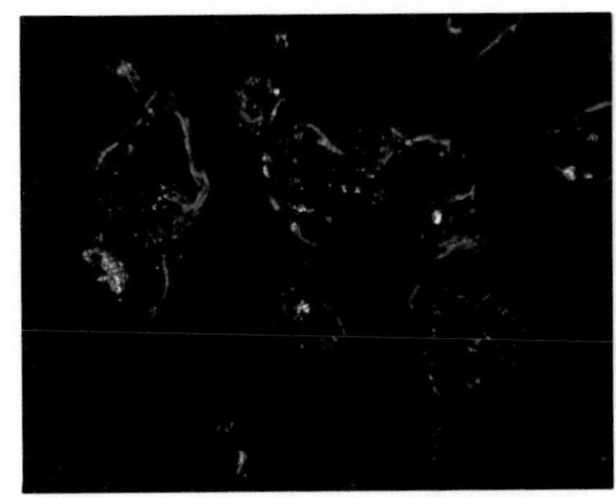

Figure 144. Sodium bromate in Cargille liquid 1.610; annular stop, 80X.

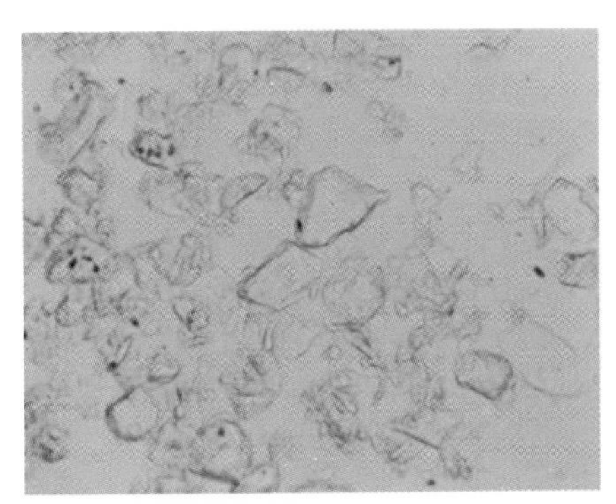

Figure 145. Sodium bromate in Cargille liquid 1.626; annular stop, 80X.

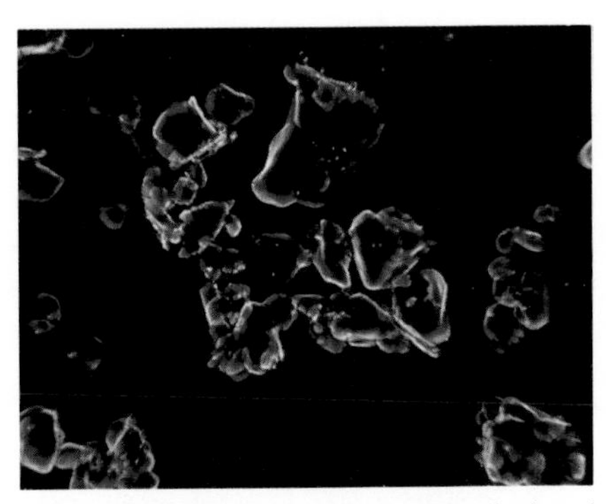

Figure 146. Sodium bromate in Cargille liquid 1.626; annular stop, 80X.

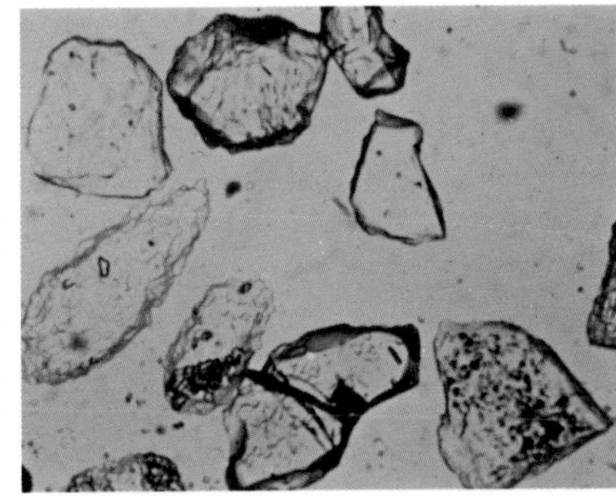

Figure 147. A mixture of almandite (blue), grossularite (green) and periclase (yellow) in Cargille liquid 1.740; annular stop, 80X.

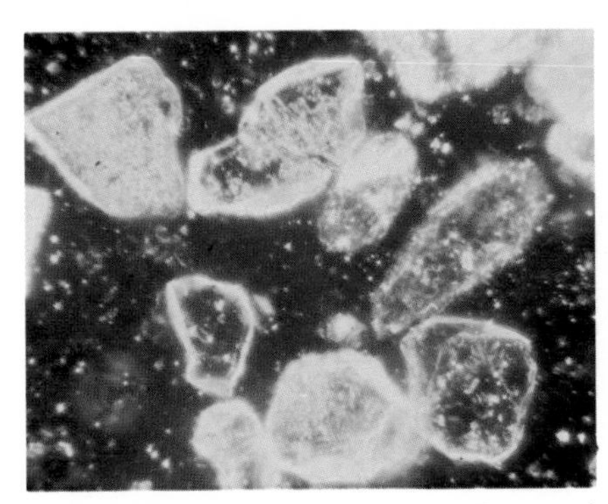

Figure 148. Same field as Figure 147, central stop, 80X.

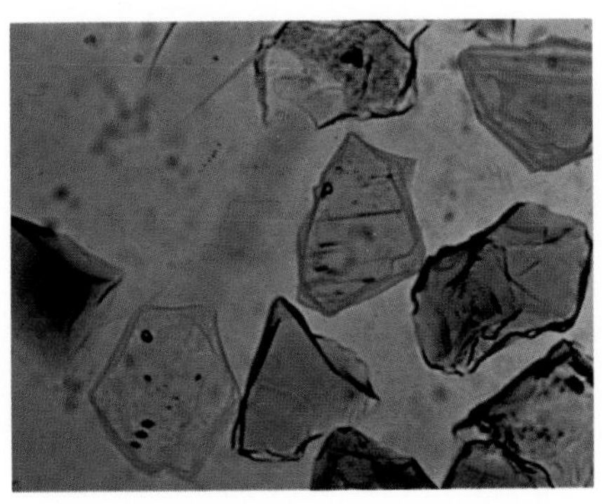

Figure 149. Tourmaline crystals mounted in Cargille liquid 1.630; annular stop, 80X.

Figure 150. Same field as Figure 149; central stop, 80X.

Reprinted from
THE PARTICLE ATLAS
McCrone and Delly
Ann Arbor Science Publishers, Inc.

Figure 137 (Color Plate) shows, in such a case, how the axial beam of white light is dispersed with short wavelengths refracted toward the liquid, long wavelengths toward the particle and wavelengths near 589 nm (n_D) very slightly or not at all. The net result, for λ_0 = 589 nm, is a blue Becke line moving outward and a red Becke line moving inward at the edges of the particle on focusing upward. A stop in the objective can be of a size, shape and position to stop or to transmit only the axial beam. A central opaque spot stops the axial beam allowing white light, minus λ_0 and those wavelengths close to it, to pass through. The annular stop, on the other hand, transmits only those wavelengths close to λ_0; hence the two stops produce complementary colors.

A particle-liquid system showing a match in refractive index for yellow light will show, with the annular stop, particles with yellow edges on a white field. The central stop with the same particle-liquid system, will show blue particles on a black field. It is very important to nearly close the substage iris for both annular and central stop dispersion staining. Furthermore, the two stops must be centered with respect to the image of the substage iris in the objective back focal plane.

Any angular deviation of light rays from their original direction will be indicated at the objective back focal plane by a corresponding linear displacement. An annular stop diameter of about 3 mm is required to cut off all the rays with large linear displacement or, conversely, a central stop diameter of about 4 mm is needed to cut off the rays with small linear displacement.

If the particle and liquid indices are too far apart to match at any visible wavelength, no dispersion colors will result. In this situation, the particles will show black edges on a white field with the annular stop. With the central stop, the same particles will show white edges on a black field.

In practice, even when λ_0 is much lower than 400 nm or much higher than 700 nm, the black outline with the annular stop will be tinted blue (low λ_0) or brown (high λ_0); with the central stop the white outline will, under the same circumstances, be tinted yellow (low λ_0) or blue (high λ_0).

The refractive index of an anisotropic particle will vary with crystal orientation and different dispersion colors will be obtained depending on the vibration direction of light. The two extreme orientations during rotation of the preparation with one polar may show two different colors if the indices in both positions match one index of the liquid in the visible light range. If one index of the solid is outside the range of the liquid, then for that orientation the particle will be black on a white field (annular stop) or white on a black field (central stop).

Table XXVII lists the colors observed with dispersion staining and both stops. The colors shown are a continuous series throughout the entire range of λ_0, therefore λ_0 can be estimated by comparing the colors produced, when the annular and central stops are used, with those listed in the table. With a little practice, one can estimate λ_0 to within ± 10-20 nm.

Table XXVII. Dispersion staining colors

Matching λ_0 nm	Annular stop colors	Central stop colors	
		in focus	Becke line
<420	blue-black	light yellow	faint gold + violet
430	blue-violet	yellow	" " "
455	blue	golden yellow	" " "
485	blue-green	golden-magenta	yellow + violet
520	green	red-magenta	violet + orange
560	yellow-green	magenta	blue violet + red orange
595	yellow	blue-magenta	blue + red
625	orange	blue	blue
660	orange-red	blue-green	green
>680	brown-red	pale blue	pale green

C. DISPERSION STAINING LIQUIDS AND OPTICAL ACCESSORIES

The application of the principles outlined so far to the identification of transparent particles is best considered under two main headings:

Selection of suitable liquids which cover the range of refractive indices under consideration and which, at the same time, possess the proper dispersion characteristics.

Choice of microscope optics to render dispersion colors visible.

1. Dispersion staining liquids

Numerous references describe particular applications of many dispersion staining liquids. Crossmon[58], in using darkfield dispersion staining to detect quartz, has employed, among other liquid media: cinnamaldehyde with butyl carbitol or ethyl phosphate; ethyl cinnamate alone and mixed with Aroclor 1242; styrene and α-bromonaphthalene with methylene iodide. Schmidt[70] includes a table listing many dispersion liquids. These cover a wide range of refractive indices which have been found suitable for characterizing various minerals.

Fortunately, commercially available liquids may be employed for the systematic application of an identification scheme on a routine basis. Sets of carefully standardized liquids varying regularly in intervals of 0.01 unit or less over a refractive index range of 1.300 to 2.11 are available from R. P. Cargille[55]; Grabar[61] has reported on their use. Each Cargille liquid label shows n_D^{25}, temperature coefficient of index and dispersion in terms of ν, a number related to dispersion by:

---

70. Schmidt, K. G., Z. Staub. 18 247 (1958); translation by A. L. Forlini, Proc. International Microscopy Symposium, Chicago (1960).

$$\nu = \frac{n_D - 1}{n_F - n_C} \tag{46}$$

where n_D is the refractive index when λ is 589 nm (yellow)

n_F is the refractive index when λ is 486 nm (blue)

n_C is the refractive index when λ is 656 nm (red).

With few exceptions the dispersion of these liquids has been found to be consistently as high as, if not higher than, the dispersion of other unrelated liquids which could be used for the same purpose. The advantages of having a complete set of liquids outweigh the disadvantages of slightly lower dispersion at some refractive indices than might be available with other liquids. Cargille, however, also markets a set of high dispersion liquids covering the range 1.50 - 1.645 to supplement the standard sets. Table XXVIII tabulates the dispersion values for both the standard and high dispersion sets. Note that a low value of ν corresponds to high dispersion.

2. Microscope optics for dispersion staining

To see dispersion staining colors, one must arrange the microscope to view the sample in either of the following ways:

1) by undeviated light for a brightfield image, termed annular stop, or

2) by deviated light for a darkfield image, termed central stop.

Of the two methods, the central stop provides certain advantages, notably in better resolution and in sensitivity to small particles. Note, in Figure 137, how the high angular aperture rays are utilized by the objective with the central stop but are screened out by the annular stop. This difference accounts for the higher resolution for the central stop.

It is important to realize that neither the central nor annular stop is designed to resolve particles. Both are designed to "resolve" colors. Matching the refractive indices of particles and liquid and using a condenser of NA of 0.05 is not a way to achieve resolution.

Experience has shown that there is no advantage in using higher magnification even with very small particles. To see the proper dispersion staining colors for even submicrometer particles requires a bright light source, clean slide and coverslip and a well adjusted central stop (very dark field of view).

The necessary microscope characteristics can be achieved by any of several methods described in the literature, but it must be emphasized that, whichever method is selected, axial illumination is of paramount importance. Crossmon[58] employs conventional darkfield techniques, and the use of phase contrast is described by Schmidt[70]. It is true that both darkfield and phase contrast systems give dispersion staining colors because both use a restricted condenser aperture. The important difference is that darkfield and phase systems use a high NA, hollow annular cone of illumination. The system advocated here is a very narrow axial cone of illumination. As noted, phase and darkfield give dispersion staining colors but they do not correspond to a narrow range of wavelengths around λ_0 and hence are more of a mixture and therefore fainter. Narrow axial illumination gives bright colors to which a definite value of λ_0 can be assigned. The axial system is particularly advantageous for anisotropic particles (95% of those encountered in practice). A hollow cone of illumination, as in phase or darkfield, will give a different λ_0 for each direction of light through the particle. This mixture of colors can approximate white or very faint dispersion staining colors.

Table XXVIII. Relative dispersion values for a few Cargille refractive index liquids in the range 1.45-1.70

Refractive index $n_D^{25°C}$	Dispersion value (ν)	
	Standard liquids	High dispersion liquids
1.45	56.0	—
1.46	55.3	—
1.47	50.5	—
1.48	46.6	—
1.49	43.4	—
1.50	40.7	26.2
1.51	38.4	24.5
1.52	36.4	23.0
1.53	34.7	25.4
1.54	33.2	24.8
1.55	31.9	24.2
1.56	30.7	23.7
1.57	29.6	23.3
1.58	28.7	22.8
1.59	27.8	22.4
1.60	27.0	22.1
1.61	26.3	21.6
1.62	25.6	21.3
1.63	24.7	21.0
1.64	22.9	20.7
1.65	21.5	—
1.66	20.4	—
1.67	20.1	—
1.68	19.8	—
1.69	19.5	—
1.70	19.2	—

Cherkasov[57] describes several methods which he has termed focal screening. His methods are useful because of the ease with which conventional microscopes can be adapted for them, and because they use a narrow axial cone of illumination.

For the annular stop an opaque disc with an accurately drilled central hole may be used. For a 10X objective, a hole two to three millimeters in diameter has proved suitable. In many objectives, the back focal plane is near the back surface of the rear lens element; hence the stop may be placed directly in contact with it. Difficulties may arise with other objectives in which the back focal plane is not readily accessible. Cherkasov reports that objectives equipped with adjustable iris diaphrams at the correct location for use with the universal stage give excellent results for annular stop dispersion staining. The easiest way to adapt an objective for a central stop is to place a dot of India ink, about four millimeters in diameter for a 10X objective, on a coverslip fitting the objective. An objective reserved exclusively for this work may have the India ink directly upon the back surface of the rear lens element. With a given stop, some trial and error may be necessary with either method before optimum screening, commensurate with resolution and visibility, is obtained.

McCrone Associates[71] has developed the dispersion staining objective shown in Figure 138 based on Cherkasov focal screening. This dispersion staining device has stops arranged on a metal disc in a turret. This plate can be rotated to orient either stop in position at the objective back focal plane. In a third position no stop appears and the device may be used as an ordinary objective with full aperture. Since this accessory has a standard RMS thread, it fits any microscope.

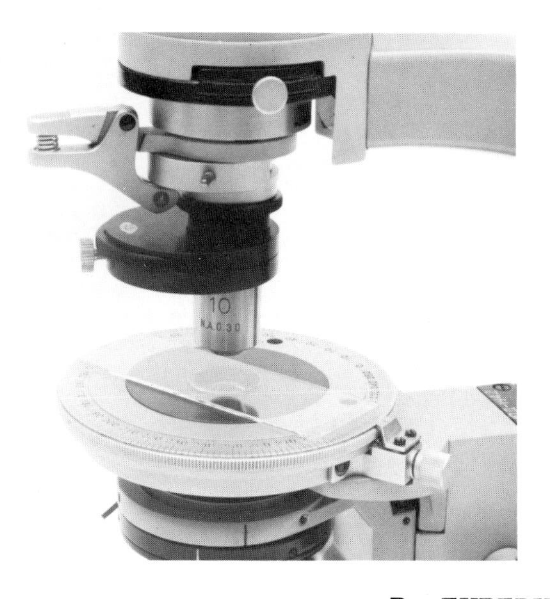

Figure 138. Dispersion staining objective

D. EXPERIMENTAL PROCEDURES

With the appropriate stop in the objective, the microscope is focused on a preparation and the substage condenser and diaphram are centered with respect to the objective. It is a considerable advantage to have a microscope with a centerable substage condenser. Even with a well-centered stop, any coverslip prep will act like a very small angle prism, hence the image of the substage iris will be displaced relative to the dispersion staining stop. One may, in the absence of a centerable substage condenser, be able to decenter the objective but this is an inelegant solution since it decenters the stage.

71. McCrone Associates, Inc., 2820 South Michigan Avenue, Chicago, IL 60616

If the stops are not centered, the dispersion staining colors will usually be different on opposite sides of the same particle. Often, moving the particle of interest about the field of view will change the colors. Any position of the particle in the field of view in which all borders of the particle are colored alike is the proper position for estimating λ_0. This procedure is often faster, easier and just as effective as centering the substage iris. In use, the substage condenser iris is closed to the point of optimum dispersion color and resolution. The image of the substage iris in the objective back focal plane should be just slightly smaller than the dispersion staining stop and concentric with it.

A readily appreciated application of dispersion staining is its ability to distinguish between components of a mixture. If morphological characteristics do not distinguish components differing only slightly in refractive index, they will appear similar when viewed microscopically by ordinary light. Dispersion staining causes each component to assume a different color which permits its identification and also, if required, facilitates particle size analysis or quantitative counting.

1. Plotting of analytical dispersion staining curves

To facilitate identification of unknowns, dispersion data for compounds of interest and their colors in suitable liquids must be tabulated in an analytically useful form. Figure 139 shows the dispersion relationship between a particle and a number of liquids differing in regular increments of refractive index (n_D) and having almost identical dispersion values. As the liquids increase in n_D, their dispersion curves intersect that of the particle at increasing wavelengths. The wavelength, λ_0, at which the dispersion curves intersect, determines the color seen by dispersion staining, and the particle, viewed with the annular stop, appears blue in liquid A, green in B, yellow in C and red in D according to Table XXVII. The central stop shows the complementary colors (yellow, reddish magenta, blue magenta and blue-white, respectively).

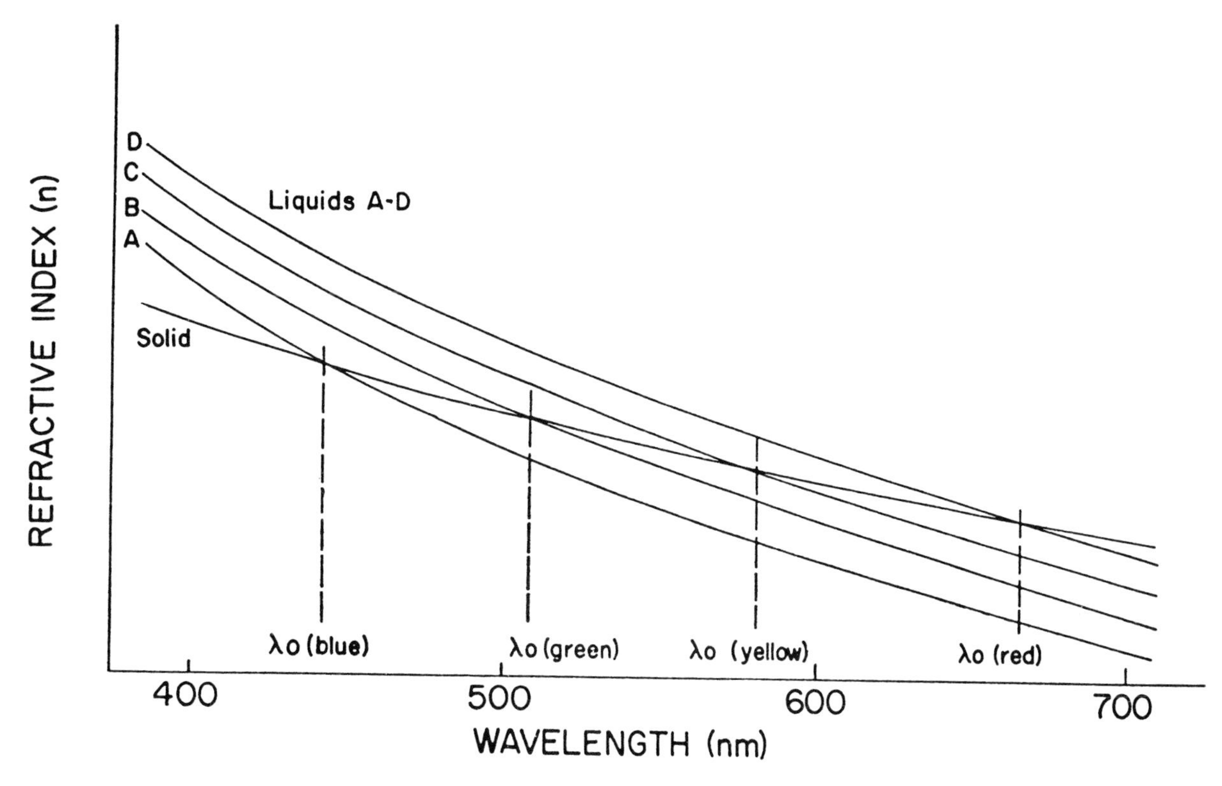

Figure 139. Dispersion curves for one solid and four different liquids

Figure 140 (Color Plate) shows these colors as a function of λ_0, the matching wavelength. The annular stop colors are identical with the visible portion of the electromagnetic spectrum. The central stop colors are, at each wavelength, the complementary color of the annular stop color just above.

Color Plate Figures 141-146 show the colors of sodium bromate particles mounted in three different liquids and viewed with both the annular and central stops. The resulting colors correspond approximately to the use of liquids A, B and D in Figure 139.

If the dispersion staining color of a particle in a given liquid is observed, λ_0 for that combination can be obtained from Table XXVII. When observations are made in a series of liquids, λ_0 for each liquid can be obtained and plotted against the n_D of the liquid used. A single curve, e.g., Figure 151 with linear plots of index versus wavelength describes the dispersion relationship for CsF. However, plotting the wavelength as $1/\lambda^2$ often yields a nearly straight line which is more convenient. A plot of this type (Figure 152) shows the same experimental data for CsF as shown in Figure 151.

There is a further very important difference between the true dispersion curves (Figure 153) and those used for analytical purposes (Figures 154-157). The usual dispersion curve is a plot of the actual refractive indices of the particles and liquids involved against wavelength. The analytical graphs in Figures 152, 154-157 show, as ordinate, the refractive index with sodium light of the Cargille liquids; the dispersion of these liquids is not given. To emphasize this point, the dispersion curve for each Cargille liquid is represented arbitrarily by a horizontal line at the n_D^{25} position for each liquid.

Figures 153-154 should make this important point more understandable. Quartz is anisotropic with two refractive indices, hence the dispersion staining colors vary with orientation of each crystal when polarized light (not crossed polars) is used. The two extreme colors for quartz in a given liquid are ϵ, parallel to the trigonal c axis and ω for all directions perpendicular to c. There is, then, a dispersion curve for each of these principal refractive index directions. Figure 154 is interpreted as follows: The ϵ and ω vibration directions of quartz show λ_0 = 486 nm (F) in Cargille liquids 1.547 and 1.538, respectively. Likewise, λ_0 = 589 (D) in 1.553 and 1.544, respectively; and λ_0 = 656 (C) in 1.556 and 1.547, respectively.

This does not mean that ϵ for quartz is 1.547 at 486 nm. It does mean that quartz and the Cargille liquid having its n_D^{25} = 1.547 have the same refractive index at 486 nm. That refractive index is found only by looking up the dispersion data for the Cargille liquids (Volume IV, Part I, Table V, Particle Atlas Two). Here, we find that the Cargille liquid whose n_D^{25} is 1.547 has a refractive index of 1.561 at 486 nm. The situation is similar for all λ_0's other than 589 nm (D). At 589 nm, of course, Cargille liquids have the index marked on each bottle.

Figure 154 was obtained from Figure 153 by looking up the actual refractive indices of the Cargille liquids at 486 and 656 nm and replotting the curves arbitrarily, making the Cargille liquid curves a set of horizontal lines. The quartz curves are then drawn in their proper positions relative to these arbitrary Cargille liquid lines. The dotted line in Figure 153 is the actual dispersion curve of Cargille liquid n_D^{25} = 1.553 and the analytical equivalent in Figure 154 of the straight horizontal line corresponding to Cargille liquid n_D^{25} = 1.553.

Obviously, Figure 154 is the simpler of the two and the easiest to use analytically. This is why all of the dispersion staining graphs are plotted in this way. We do not usually care what the actual index of quartz is at wavelengths other than 589 nm. Instead, we want to know which Cargille liquid, whose n_D^{25} is marked on the bottle, will give a specific λ_0 with quartz. You can see, however,

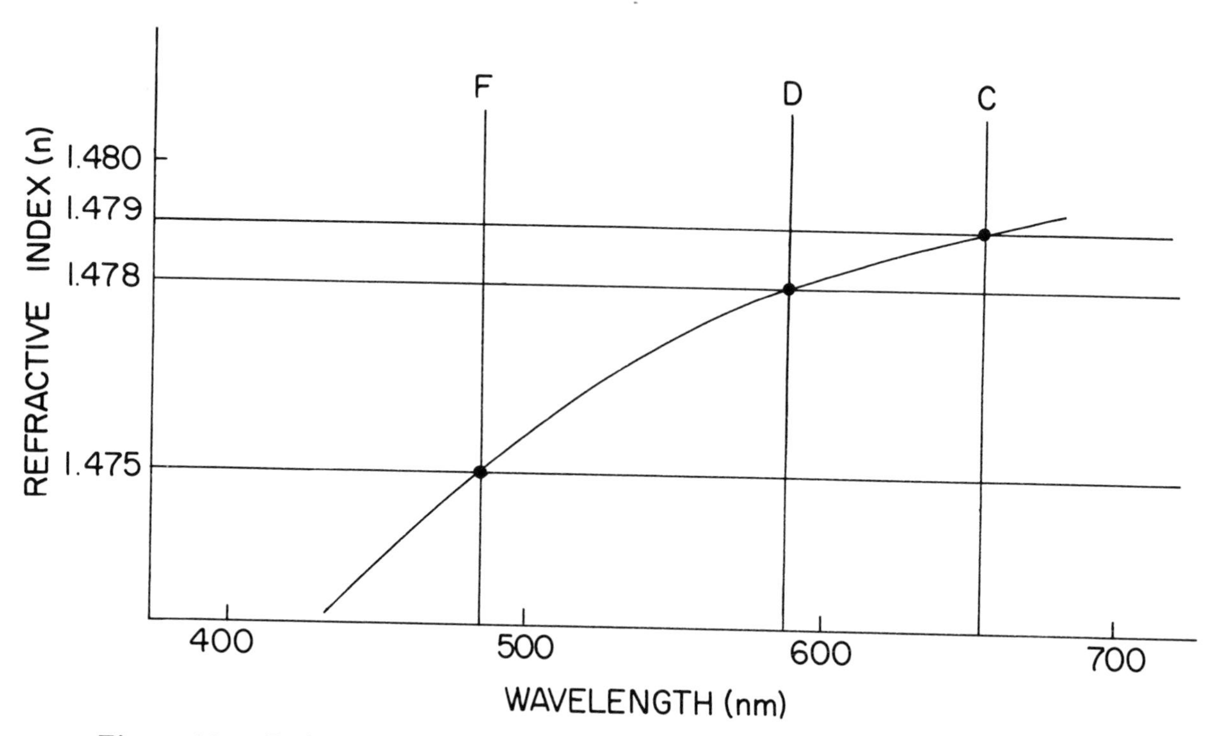

Figure 147. Refractive indices of the Cargille refractive index media matching CsF at various wavelengths (λ is plotted linearly)

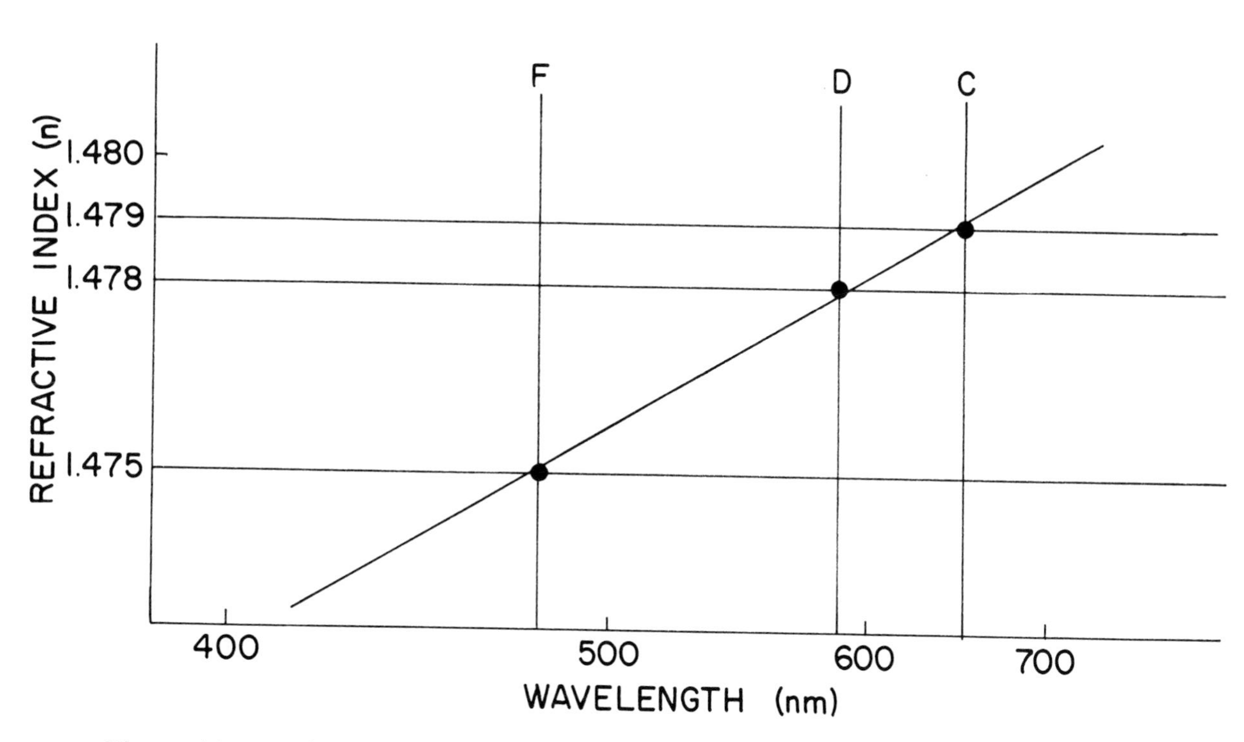

Figure 148. Refractive indices of the Cargille refractive index media matching CsF at various wavelengths (λ is plotted as $1/\lambda^2$)

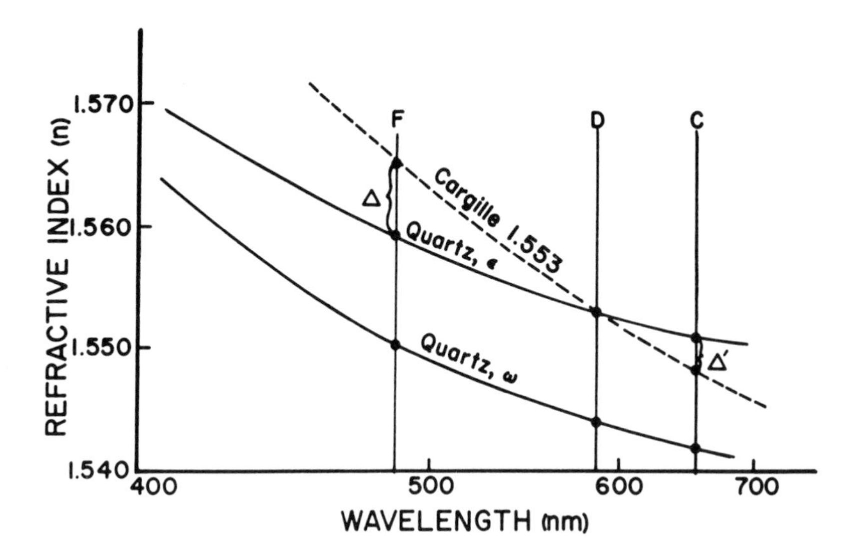

Figure 153. Dispersion curves for α-quartz (ϵ and ω) and for Cargille refractive index medium 1.553

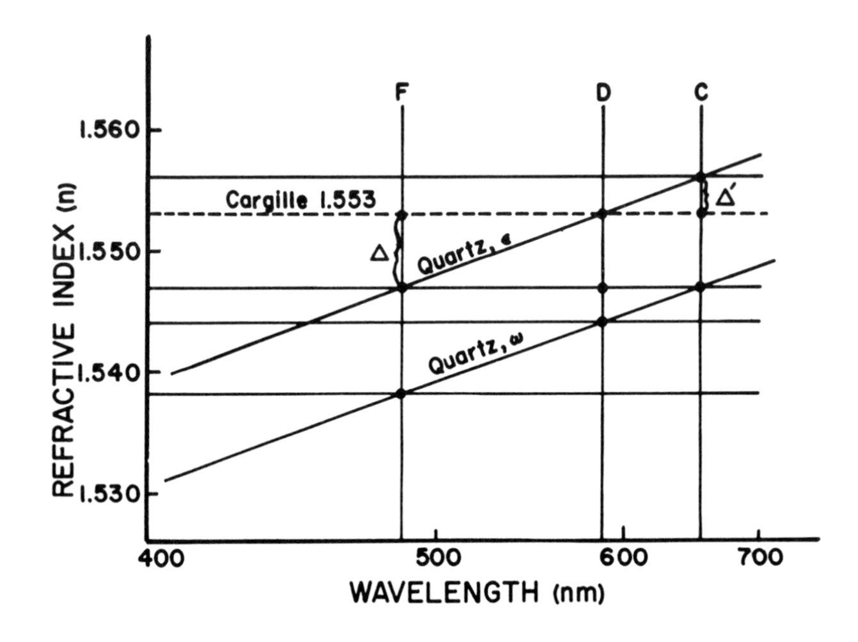

Figure 154. Dispersion staining curves corresponding to Figure 149

how to calculate dispersion staining data from actual dispersion data for any given substance. For example, to plot Figure 154 from actual literature values of dispersion for quartz, one takes ω or ϵ first, then finds that Cargille liquid whose index at, say, 486 nm is the same as the published value for quartz at 486 nm. One then plots not the actual index, n_F^{25}, but n_D^{25} of that same liquid against λ_0. Repeating this for the F, D and C lines gives a curve as in Figure 154.

No one should be disturbed by our talking about Cargille liquids 1.547, 1.553 and other odd numbers when Cargille makes no such liquids. One drop each, of the two adjacent liquids, 1.546 and 1.548 or 1.552 and 1.554 for the above examples, will simulate the intermediate liquids if they are ever needed. Normally, dispersion curves are measured by determining λ_0 in a variety of actual Cargille liquids. The plotted curve from these data often gives odd index values at the F, D and C wavelengths.

2. Examples of dispersion staining data

Figure 155 shows analytical dispersion staining curves for a group of common isotropic substances. If we look at the curve for sodium bromate, $NaBrO_3$, we see that this substance should show dispersion colors in Cargille liquids having refractive indices from about $n_D^{25} = 1.600$ to $n_D^{25} = 1.630$. With the annular stop, the colors shown should be blue for indices around 1.605, yellow around 1.617 and orange to red from about 1.620 to 1.630. These results are confirmed by the colors observed in Figures 141-146 (Color Plate).

Some particles have the same dispersion curve as a given Cargille liquid; in these cases the resulting dispersion staining curve will be a horizontal line at n_D for the matching liquid (Figure 155, KBr and RbBr). Such compounds show no dispersion staining colors in the standard set of Cargille liquids. They do, however, show dispersion staining colors in the high dispersion Cargille liquids.

Similar analytical dispersion staining graphs are shown in Figures 156-157 for uniaxial and biaxial substances.

In practice, an unknown particle is identified by mounting in a Cargille refractive index liquid and noting the dispersion staining color. If the selected liquid lies outside the dispersion range of the particle, no colors will be seen. Reference to the Becke line will then indicate whether a second preparation should be made with a liquid of higher or lower n_D. Even if the index of the liquid is too high or too low by less than about 0.03 to give λ_0 an indication of color is observed. A blue-black with the annular stop means λ_0 is < 400 nm and the liquid is too low, a brownish black means $\lambda_0 > 700$ and the liquid is too high. Note, however, that this rule would be reversed for $NaCrO_4 \cdot 4H_2O$ in Figure 157. Many chromates, sulfates, vanadates and the like have greater dispersion than the matching Cargille liquids. These compounds show a negative rather than a positive slope in their analytical dispersion staining curves. About 8% of the substances for which we have dispersion staining data have stronger dispersion than the matching standard Cargille liquids. When a color is observed, λ_0 is obtained by reference to Table XXVII or Figure 140 (Color Plate), or with a monochromator. A wedge interference filter graduated in a wavelength scale, such as that manufactured by Zeiss, is useful for this purpose.

With a monochromator, the particle disappears over a narrow range of wavelength using either the annular or central stop. The middle of this range is λ_0. With practice it becomes unnecessary to use the monochromator. Anyone using dispersion staining a great deal should expect to be able to draw a curve through his λ_0 data points with few, if any, points off by more than 5 nm. At first, however, estimated λ_0's will be off perhaps 20-30 nm especially at high and low values of λ_0 which are usually estimated low and high, respectively.

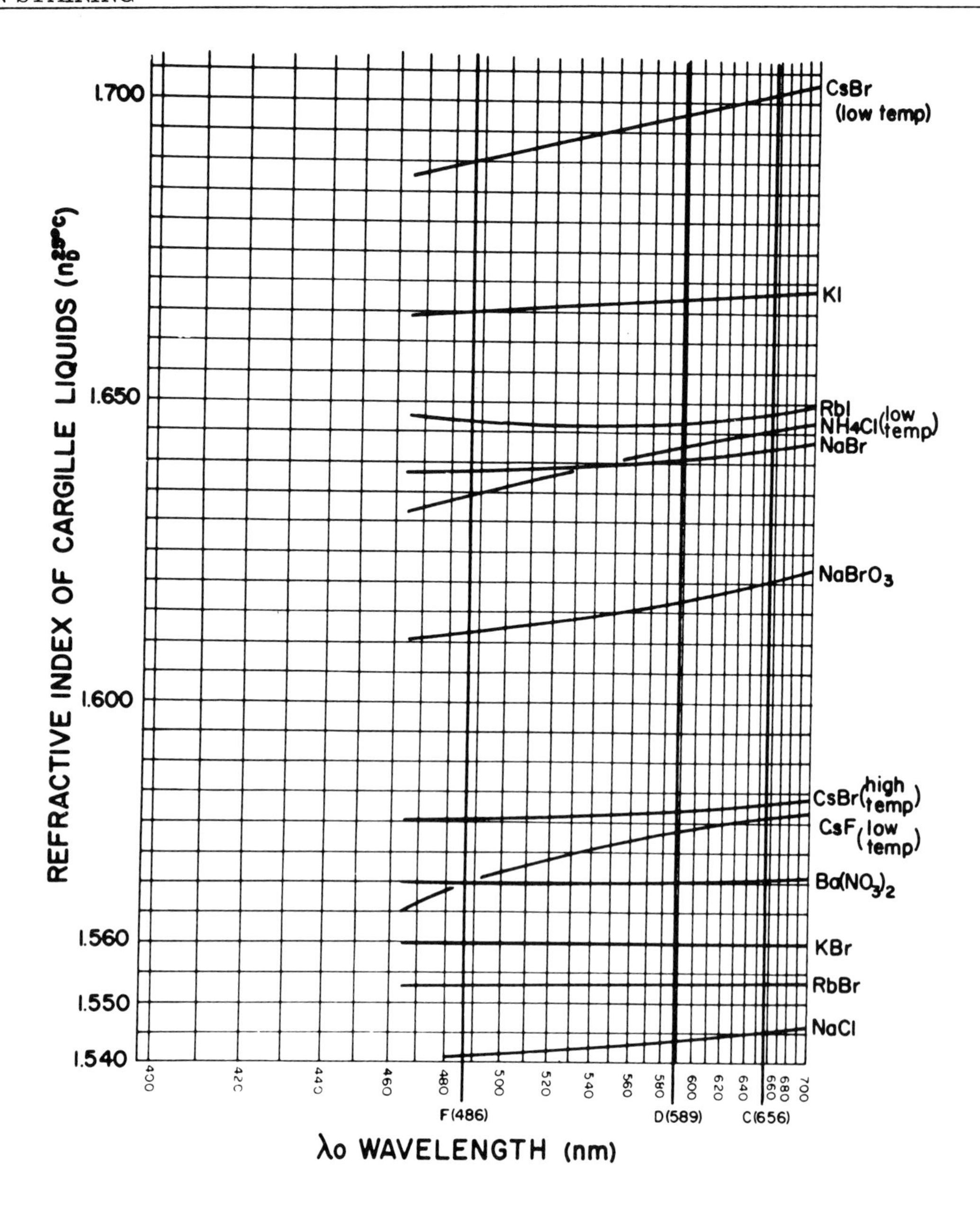

Figure 155. Dispersion staining curves for isotropic substances

3. Identification of isotropic unknowns

Returning now to identification of an unknown particle, we look for a point on the dispersion staining graphs for a known compound which represents n_D^{25} of the matching liquid and λ_0 as coordinates. A diagonal line on the graph passing through this point will indicate the substance thus identified. It is essential that the identification be confirmed by mounting in 1-2 other Cargille liquids to see that the proper λ_0 at each of those indices is obtained.

If observations are made at temperatures other than 25°C, suitable corrections, as indicated by the temperature coefficient supplied with each liquid, must be applied to n_D. A convenient way to change the refractive index of a given liquid is by varying the temperature. If, for example, trial shows that a given particle is outside the index range of the liquid medium and shows no color, and if a liquid has a higher index than the particle, the preparation can be heated to lower the liquid index. If the difference in index is not too great and if a good hot stage (e.g., Mettler or Kofler) is used, a match may be obtained at a definite higher temperature. The refractive index for the liquid at this higher temperature can then be calculated from the temperature coefficient of index given on each bottle of Cargille index medium. Strictly speaking, however, the calculated refractive index is a value for sodium light, and since we don't know precisely what dispersion the liquid has at temperatures other than 25° C, we should be cautious in using graphs like Figures 155-157 except for intersections at $\lambda_0 = 589$. In practice, we know that the dispersion curves are essentially parallel at different temperatures. This is another way of saying dn/dT may not vary significantly with wavelength. The variation of dn/dT with T at a constant λ is also very slight.

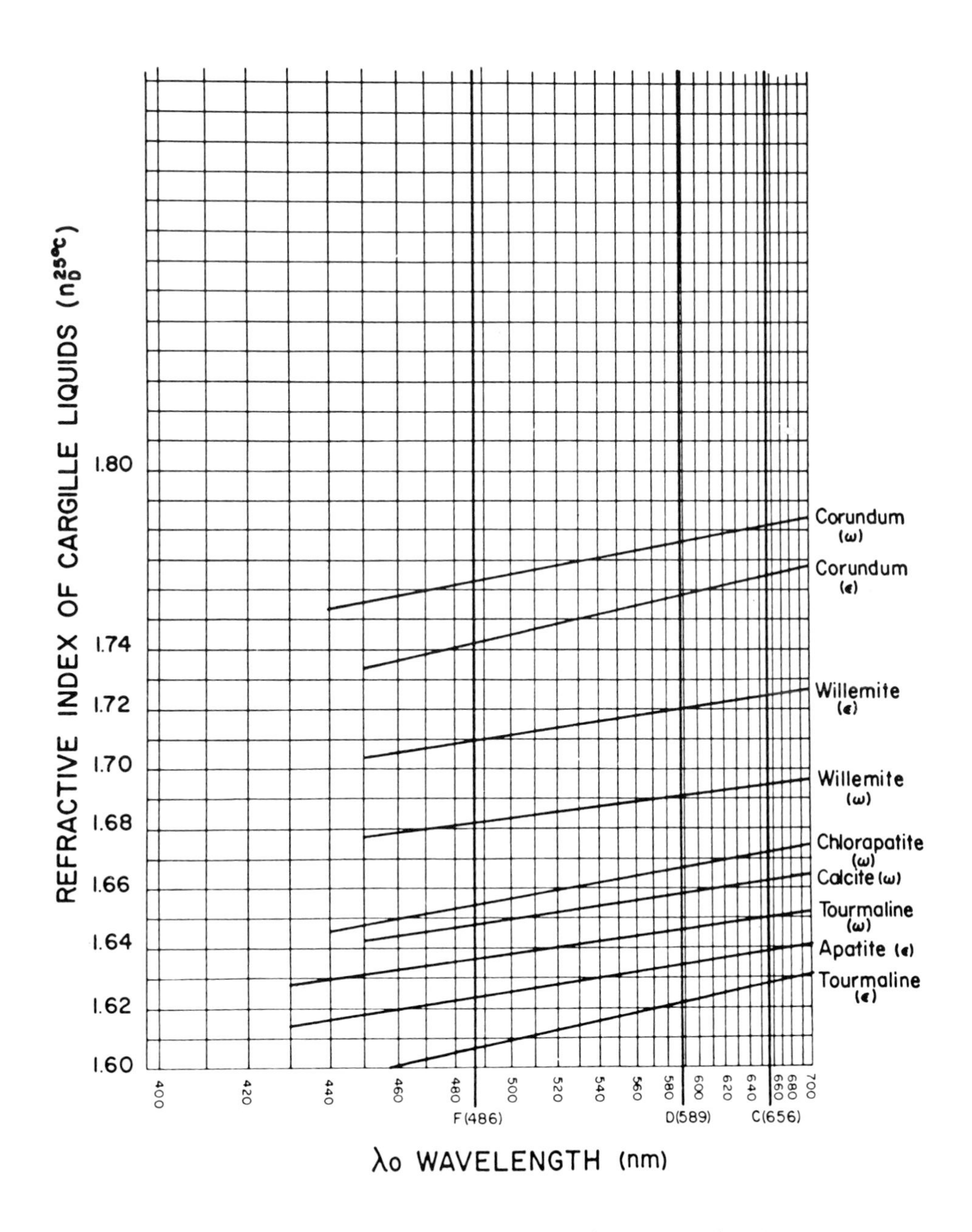

Figure 156. Dispersion staining curves for uniaxial substances

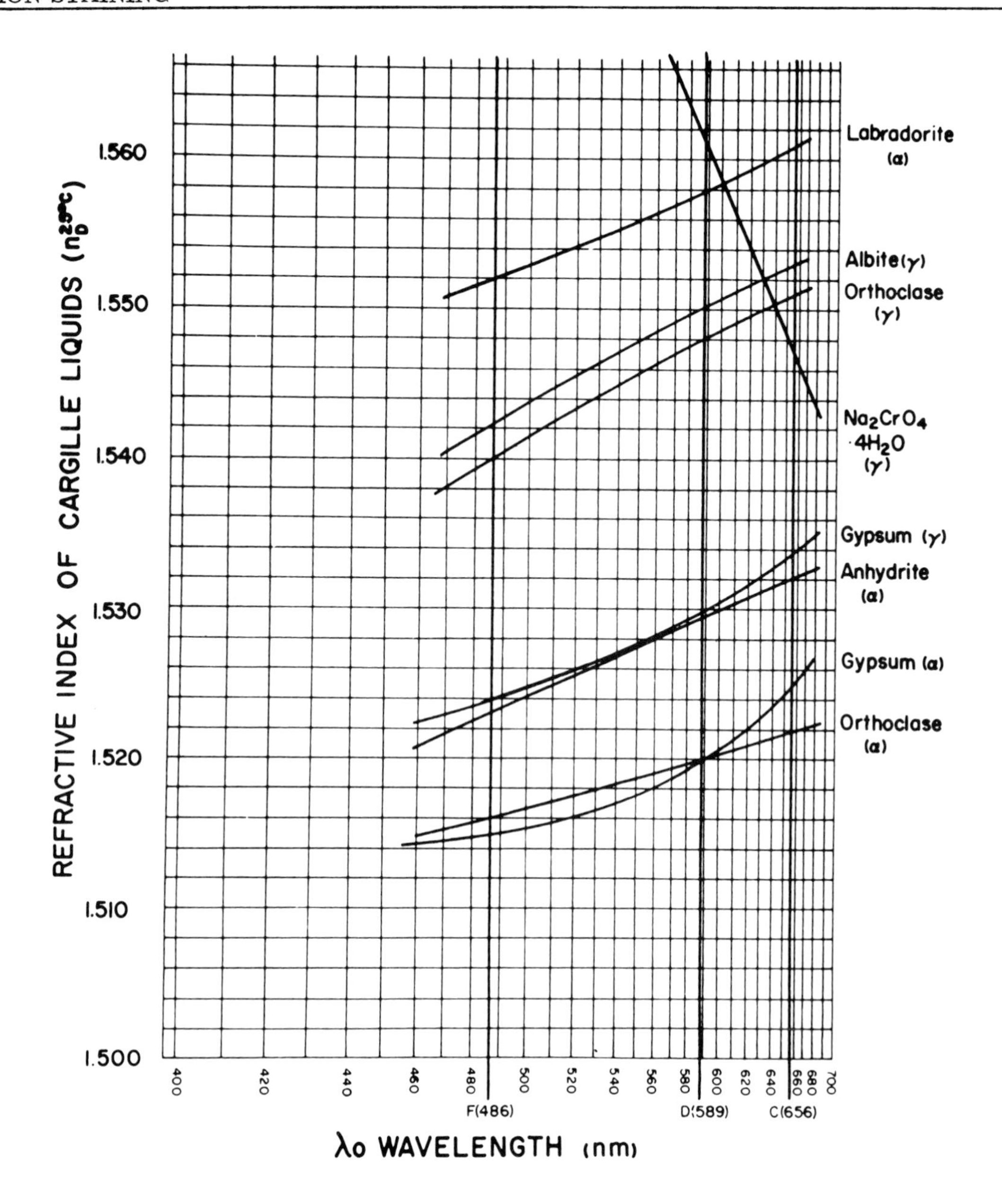

Figure 157. Dispersion staining curves for biaxial substances

Conversely, if the liquid has a lower index and if a cooling stage is available, the preparation may be cooled to increase the index of the liquid. Assuming a range of 75° C above and below room temperature, one can vary the index of most Cargille liquids at least 0.06 index units at a given wavelength. As the liquid-particle combination is heated, the index of the liquid decreases while the index of the particle changes so little, 0.00001/° C, we can ignore it. If we heat the preparation until the particle shows a measurable λ_0, then the calculated index at room temperature can be used with the graphs and tables to identify the substance.

Since an isotropic particle shows only one refractive index and can therefore exhibit only one color in a particular liquid at a given temperature, its single dispersion color serves as a means of rapid identification. Color Plate Figures 147-148 show particles of periclase and two closely related varieties of garnet: almandite and grossularite. The same field of view shows all three particles with annular stop (Figure 147, Color Plate) and central stop (Figure 148, Color Plate). Note, incidentally, that many very small particles of each component can be identified as such with the central stop but not with the annular stop.

4. Identification of anisotropic unknowns

Although some particles are isotropic, most are anisotropic. The latter either exhibits two characteristic refractive indices and are termed <u>uniaxial</u>, or they have three characteristic indices and are termed <u>biaxial</u> (see section IX).

For particles with more than one refractive index, the dispersion staining color depends on the crystal orientation with respect to the vibration direction of the polarized light used for illumination. Hexagonal crystals of tourmaline are shown in Color Plate Figures 149-150. In both figures the crystals are randomly oriented in Cargille liquid 1.630 and observed with one polar. Thus, a variety of refractive indices, ω, ϵ and ϵ', may be shown. The matching wavelength, λ_0, ranges from 445-690 nm in Cargille liquid 1.630 (Figure 156). This corresponds to colors ranging from blue (ω) to red (ϵ). For uniaxial crystals (hexagonal or tetragonal) the principal, or characteristic, refractive indices are the minimum and maximum values observed among all possible orientations of the crystal. These indices are termed ϵ and ω.

A uniaxial crystal may assume three general orientations:

1) The crystal may show ω in all directions.

2) It may show ω in one direction and ϵ at right angles to ω.

3) It may show ω in one direction and some intermediate index, ϵ', between ω and ϵ, at right angles to ω.

In ordinary unpolarized light, a given crystal shows a mixture of all the colors it would show during rotation of the stage with a single polar.

In a preparation of randomly arranged crystal fragments, all of these orientations will be seen, and in polarized light (one polar), various dispersion colors will result. The highest λ_0 will represent a crystal oriented to show ϵ; the lowest λ_0 a crystal oriented to show ω if the crystal is optically negative ($\epsilon < \omega$). If the crystal is positive ($\epsilon > \omega$), the highest λ_0 will correspond to ω. ω is easily identified because it is the one color all uniaxial crystals show in some position during rotation of the stage with polarized light (<u>one</u> polar).

Note in Figure 154 that quartz mounted in Cargille liquid 1.547 would show all λ_0's between 486 and 656 nm; in 1.544 all λ_0's between about 450 and 589 nm and in 1.540 all between about 410 and 520 nm. Thus, if a preparation is viewed by polarized light and the stage is rotated, most of the crystals will change their dispersion staining colors. The crystals which change color will all show a common λ_0, corresponding to ω, at some point during rotation of the stage. This indicates, first, that the crystals are indeed uniaxial; second, that the common color represents the ω index. If a given uniaxial crystal does not change color on rotation of the stage, the color shown is the ω color and the view is down the <u>c</u> axis. Note, however, that these rules are reversed if the solid has greater dispersion than the liquid. Color Plate Figure 150 shows tourmaline mounted in Cargille liquid $n_D^{25} = 1.630$ and viewed with the central stop. Tourmaline is uniaxial negative ($\omega > \epsilon$). The color common to each of the crystals (as the polarizer is rotated) and the lowest λ_0 shown is blue, which represents the ω index. The highest λ_0 is orange-red, which here represents the ϵ index. Color Plate Figure 149 shows the same field of view with the annular stop. The ϵ orientation is now blue and the ω orientation is orange.

Referring again to Figure 156, the analytical graph for a few uniaxial compounds, one finds two curves for corundum, representing the ϵ and ω indices. Thus, if an unknown is observed in a liquid of $n_D^{25} = 1.760$ and each crystal shows a common color corresponding to $\lambda_0 = 470$ nm, one can conclude tentatively that:

1) 470 nm corresponds to the ω index.

2) The unknown is uniaxial.

3) From the graph, the unknown may be corundum.

The unknown can be shown definitely to be corundum if λ_0 is noted for ϵ; this is done by looking for the color farthest from the ω color. This turns out to be about 610 nm in $n_D^{25} = 1.760$ and the graph (Figure 156) shows this to be ϵ for corundum.

It is, of course, possible that the two indices for a given compound are so far apart that a single liquid gives a dispersion staining color only with one of the two directions. In this case, the other direction will show black borders on a white field with the annular stop and white borders on a black field with the central stop. If the matching index is not too far outside the range of the mountant, the black (or white) outlines will be tinted red or blue depending on the relative indices of particle and mountant. For example, with the annular stop brownish-black borders mean the particle index is less that that of the mountant; if bluish-black the particle has the higher index.

If a crystalline material is not isotropic or uniaxial, it must be optically biaxial. Biaxial crystals have three principal refractive indices: α, β and γ. By convention, the indices are chosen so that $\alpha < \beta < \gamma$.

Again, in a preparation of randomly oriented biaxial crystals, one crystal may assume four general orientations:

1) any two of the three principal indices at right angles to each other;

2) one principal index and an intermediate index, α' or γ', in other directions;

3) intermediate indices in all directions; and

4) a single index, β, in all directions.

In general, the β index cannot be identified as such by simple observations unless a given crystal shows a dispersion staining color that does not change on rotation of the plane of polarized light. That color is λ_0 for β. If a large number of random orientations are obtained in a preparation, the lowest λ_0 will usually be close to the γ index, and the highest close to the α index. This will be reversed if the solid phase has stronger dispersion than the liquid.

As an example, assume that an unknown is examined with the annular stop in Cargille liquid $n_D^{25} = 1.524$, and as the polarizer is rotated the following observations are made:

1) A variety of colors is exhibited.

2) The crystals do not show a common color on rotation of the stage.

3) The highest $\lambda_0 \cong 640$ nm.

4) The lowest $\lambda_0 \cong 485$ nm.

One would draw these conclusions:

1) The unknown is biaxial.

2) 640 nm corresponds to the α index.

3) 485 nm corresponds to the γ index.

4) Referring to Figure 157, one would say that the material is gypsum.

Using the α index alone, one could not distinguish unequivocally between orthoclase and gypsum. A second preparation is mounted in Cargille liquid $n_D^{25} = 1.530$ in which gypsum (γ) is yellow by annular screening and orthoclase is outside the color range. This readily confirms the presence of gypsum. Wherever possible, a series of observations in adjacent liquids is always advisable for more dependable identification of an unknown.

Obviously much more information useful for identification is obtained with polarized light. It is both interesting and useful that removing the polar often makes an anisotropic substance behave as though isotropic. Note Figure 158, Color Plate, in which calcite shows quite different dispersion effects, depending on the vibration direction. Now look at Figure 159, Color Plate, which shows the same two calcite crystals but without any polar. Both are mounted in Cargille liquid $n_D^{25} = 1.654$. Both crystals in Figure 159 show what seems to be a single color in the absence of a polar. This effect seems to be characteristic of all uniaxial crystals, all biaxial crystals of low birefringence and all biaxial crystals with small optic axial angles (α and β or β and γ nearly equal, i.e., pseudo-uniaxial.) It is often useful to take advantage of this fact in examining mixtures. Looking for calcite in a mixture is much easier with unpolarized light since all of the crystals show a single color. A polar can always be inserted to confirm the identity of suspected particles. Note, however, that the liquid showing this effect must have an index matching the ω index of uniaxial crystals, be between α and β for optically positive biaxial crystals and between β and γ for optically negative biaxial crystals.

Since identification of the dispersion staining colors depends on careful color discrimination, the colors listed in Table XXVII can be regarded only as a guide. Each person might well make a comparable table before using the technique by observing a known material in a series of liquids. The Kofler glass powder standards sold by Arthur H. Thomas and William J. Hacker are useful for this calibration exercise. Their dispersion data has been determined by Lucy McCrone and published in a paper by Goodman[69]. Examination of such a series permits the analyst to prepare his own color series corresponding to Table XXVII or to draw a "spectrum" corresponding to Figure 140.

Dispersion staining can also be helpful in obtaining optical data which can be used with other crystal data tables. Winchell and Winchell[72] list many isotropic, uniaxial and biaxial compounds according to n_D^{25}, and in many cases include dispersion data, i.e., the refractive indices at 486 nm and 656 nm, respectively. When a complete plot of data is obtained on a compound, as in Figures 155-157, n_D for the compound can be read to ± 0.001 and, with more care, to ± 0.0002. From these plots the values of n_D^{25} for the Cargille liquids which match the crystal at the F and C wavelengths can also be obtained.

5. Checking a mixture for a suspected substituent

If the dispersion data for the suspected substance are known, the powdered sample is mounted in a single liquid chosen as follows:

1) If the suspected substance is isotropic, mount the sample in the Cargille liquid corresponding to $\lambda_0 = 540$ nm. Scan the field, using the central stop for magenta particles.

2) If the substance is uniaxial anisotropic, mount the sample in the Cargille liquid corresponding to $\lambda_0 = 540$ nm for the ω index. Proceed as in 1) but without polars.

3) If the substance is biaxial, mount the sample in the Cargille liquid corresponding to $\lambda_0 = 540$ nm for a line near but $> \alpha$ for a positive crystal, or near but $< \gamma$ for a negative crystal. Proceed as in 1) but without polars.

72. Winchell, A. N., and H. Winchell, The Microscopical Characters of Artificial Inorganic Solid Substances, Academic Press, New York, 1964.

In all three cases, rotate the stage with any magenta crystal found and check its behavior with one polar in place to confirm the identification.

6. Determination of refractive indices

Assume, for example, that we wish to measure the indices for quartz. This is best done with Cargille liquid $n_D^{25} = 1.560$ and a good hot stage, preferably the Mettler FP5 (or FP2).

Many ground quartz crystals are immersed in Cargille liquid $n_D^{25} = 1.560$ in the hot stage and observed with one polar. If there are 100 or so individual crystals in the field, rotation of the polarizer through 180° should disclose both ω and ϵ orientations recognizable as extreme colors. At 25° C, quartz crystals in Cargille liquid $n_D^{25} = 1.560$ will show all brownish-black crystal borders.

On heating, the index of the liquid decreases 0.0004 for every °C. The temperature should be set for and held at 35°C while the single polar is rotated and the crystals are observed for extremes in λ_0. TableXXIX shows the extreme in λ_0, corresponding to ϵ and ω, tabulated against successively higher temperatures.

If the experimental data in Table XXIX are plotted, the smooth curve gives values of n_D^{25} for the Cargille liquids that would match quartz ϵ and ω at λ = 486, 589 and 656 nm. The dispersion data for the Cargille liquids are used to convert n_D^{25} to n_F^{25} and n_C^{25}. That is to say, for example, that quartz ϵ has the same refractive index at 656 nm as the Cargille liquid whose $n_D^{25} = 1.556$. The refractive index of this Cargille liquid (and therefore quartz ϵ) is given by the dispersion tables for the Cargille liquids, i.e., 1.5510. This value for quartz ϵ at 656 nm compares very well with the literature value 1.5509. The other values of refractive index for quartz in Table XXX were obtained in this same way.

E. APPLICATIONS

To facilitate the use of dispersion staining, the known data are plotted in graphs and tabulated in two different ways. One set of tables lists the compounds by increasing n_D^{25} and the other lists the same compounds alphabetically; both tables include n_D^{25}, n_F^{25} and n_C^{25}.

The alphabetical tables are useful in locating a specific compound in the determinative graphs, and thus in predicting its dispersion staining color in any given liquid. The graphs and tables appear in Volume IV, Particle Atlas Two and, as an added help, complete dispersion and dn/dT data for all of the Cargille liquids, including the high dispersion set, have been included.

1. Counting analysis

Frequently, it is necessary to perform a counting analysis on a mixture of substances which appear similar in white light. A count made under these conditions is obviously difficult and may well be erroneous. Dispersion staining can cause the components to assume distinctive colors which simplify identification of the individual particles. Even when the components can be distinguished by other means, dispersion staining makes this analysis much easier.

Color Plate Figures 147-148 show a good example of differentiation of different substances. A mixture of almandite, grossularite and periclase mounted in a liquid of $n_D^{25} = 1.740$ is viewed with annular (Color Plate Figure 147) and central stops (Color Plate Figure 148). The three components show the following colors: almandite-blue, grossularite-green and periclase-yellow; the corresponding complementary colors are visible with the central stop. These colors would undoubtedly increase accuracy and speed in a counting analysis of the mixture. Ready identification of the particles would certainly render the work less fatiguing to a microscopist.

Table XXIX. Matching wavelengths, λ_0, for quartz at different temperatures

Temperature °C	λ_0 (nm) ω	λ_0 (nm) ε	n_D^T
35	>700	655	1.556*
40	>700	605	1.554
45	>700	570	1.552
50	>700	535	1.550
55	~700	510	1.548
58	650	495	1.547
60	630	490	1.546
65	586	468	1.544
70	554	440	1.542
75	530	~400	1.540
80	500	<400	1.538

* n_D^T is the refractive index, n_D^{25}, of the Cargille liquid at the temperature given, obtained from: $n_D^T = n_D^{25} - 0.00040\,(T-25)$.

Table XXX. Accuracy of refractive indices for quartz measured by dispersion staining

Wavelength (nm)	Refractive indices literature ε	literature ω	dispersion staining ε	dispersion staining ω	Error ε	Error ω
486	1.5590	1.5497	1.5587	1.5499	-0.0003	+0.0002
589	1.5533	1.5442	1.5532	1.5443	-0.0001	+0.0001
656	1.5509	1.5419	1.5507	1.5420	-0.0003	+0.0001

When quartz and gypsum are mixed, they can be easily differentiated with unpolarized light by mounting them in Cargille liquid n_D^{25} = 1.532. Both have relatively low birefringence as well as indices which are close to each other; hence, again unpolarized light should be used (Figure 160, Color Plate). Quartz crystals with the annular stop in Cargille liquid 1.532 are blue while gypsum is orange. With polarized light, quartz crystals will show a range of colors in the blue-green and gypsum a range of colors in the yellow-orange. In Cargille liquid n_D^{25} = 1.548, quartz shows a yellow color with the annular stop (Color Plate Figure 161); the blue component is laboradorite.

When the refractive indices of the two components are farther apart, it may be impossible to find a liquid in which both (or all) components show dispersion colors. In this situation, a liquid is chosen impârting a color to one of the components, generally that component which is isotropic or which has the lowest birefringence. This results in a single color for that component with unpolarized light. Other components will be dark-bordered with annular stop, white-bordered with the central stop.

2. Identification of mixtures

First, take the case in which one wishes to check for certain definite suspected substituents. To look for corundum, one first checks the appropriate graph in Volume IV, Particle Atlas Two or Figure 156 for the refractive indices of the liquids giving dispersion colors with both ϵ and ω for corundum. These range (Figure 156) from about 1.71 to about 1.79. When the compound is uniaxial and especially when the two curves (for ϵ and ω) are close together as they are for corundum, it is valid and easier to use unpolarized light. Under these conditions each corundum crystal shows a color which is a combination of all the colors it would show on rotation of the stage with polarized light. All corundum crystals will be orange-brown in 1.78 and blue in 1.75. There will be a greater tendency toward the ω color.

If the composition of a multicomponent mixture is completely unknown, it may be necessary to proceed by systematically mounting portions of the sample in liquids of successively higher index and looking for dispersion colors. The Cargille liquids to be used are:

1.400	1.526
1.412	1.556
1.426	1.590
1.442	1.630
1.460	1.680
1.480	1.74
1.502	1.82

If a hot-cold stage is available covering the range from -50° to +100° C, the following Cargille liquids will suffice:

1.400
1.470
1.556
1.660
1.80

Mounting a given mixture successively in these five liquids will cover λ_0 for all particles having indices from about 1.36 to about 1.88. Using the known temperature coefficient for any liquid medium, one can calculate the refractive index of the liquid at any temperature. Since the temperature coefficient of index is essentially nil for solids, the index of the particle at 25° C can be taken as equal to the index of the liquid for any temperature from -50 to +100° C. If, for example, a given solid shows λ_0 = 589 nm corresponding to its ϵ direction with the annular stop in Cargille liquid 1.606 at -40 °C, its refractive index, n_D^{25}, is:

$$= 1.606 + [25-(-40)]\, dn/dT$$
$$= 1.606 + 65\ (0.00041)$$
$$= 1.606 + 0.0267$$
$$= 1.633$$

Upon checking the graphs (Volume IV, Particle Atlas Two), we find that apatite fits this description.

3. Forensic applications

There are many applications of dispersion staining in forensic laboratories, particularly since proof of identity of two samples rather than final, absolute identification is often the problem. Two samples of glass, for example, one from the scene of a hit-and-run accident and one from the broken headlamp of a suspected vehicle, may be quickly compared and found to be different or identical with a high degree of confidence. At the same time, the results can be recorded as very convincing color photomicrographs (Color Plate opposite). The technique may be applied to most transparent substances: minerals (Color Plate Figures 163-165)[62, 63], plastics (Color Plate Figure 166)[73], fibers (Color Plate Figures 167-168)[67], narcotics and even plant and animal tissue[74]. The double exposures (Color Plate Figures 163-165 and 167-168) are a simple way of comparing λ_0 colors for the principal refractive index directions of fibrous substances. A group of parallel fibers is mounted in an appropriate Cargille liquid oriented with one index $n_\perp$ or $n_{||}$ parallel to the polar vibration direction. After one photomicrograph, the stage with fibers is rotated 90° and a second exposure is made on the same film frame. The procedure works, of course, only with the central stop, i.e., the darkfield position.

4. Submicrometer particles

Thaer[60], who has published extensively in the field of dust particle size determination by light microscopy, quotes a lower limit of one micrometer for the determination of quartz and silicate minerals in fine dusts. We have found that improved dispersion staining tools, particularly the central stop with its darkfield illumination and superior resolution, lower the limit, especially if a bright light source is used[69].

The central stop does, in fact, furnish a form of darkfield illumination. Just as particles much smaller than one micrometer are visible by scattered light with the usual darkfield illuminator, such particles also "scatter" colored light and are visible under the proper conditions with the central stop. Color Plate Figure 169 shows a few ppm of quartz in a mineral mixture. The magenta particles of quartz are readily recognized.

Another mechanism, diffraction, also gives a dispersion staining effect. Fortunately, both mechanisms reinforce each other since neither light scattering (diffraction) nor refraction will occur if the refractive indices are matched. Since scattering of blue light is more effective than scattering of light at other wavelengths, the dispersion colors are slightly stronger when scattered blue light is observed, viz., when the wavelength at which $n_{particle}$ matches n_{liquid} is in the yellow or red and the central stop is used. Another reason for using blue or blue magenta as the central stop dispersion staining color is the sensitivity of the eye to this color range compared to yellow.

---

73. McCrone, W. C., ASTM Special Tech. Publication No. 348, 125-30 (1963).

74. McCrone, W. C., J. Roy. Microscop. Soc. 83, 217-19 (1964).

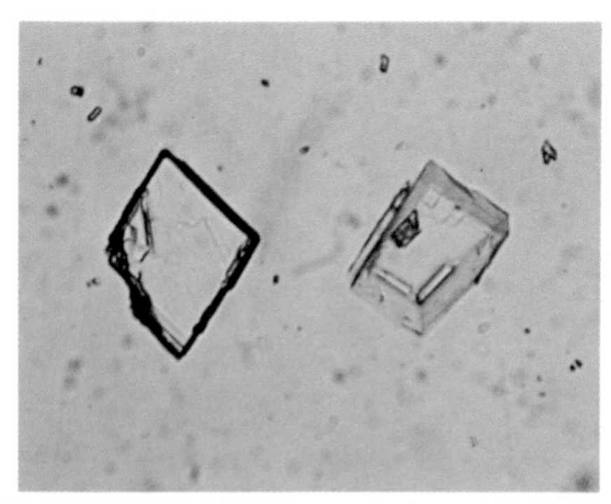

Figure 158. Two calcite crystals mounted in Cargille liquid 1.648; annular stop, polarized light, 80X.

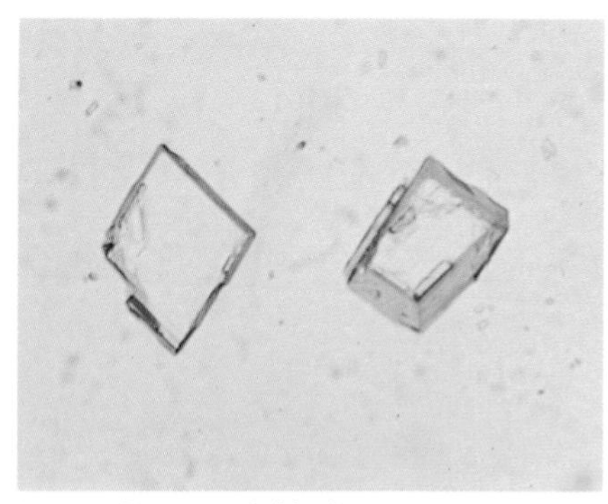

Figure 159. Same field of view as Figure 158; annular stop, unpolarized light, 80X.

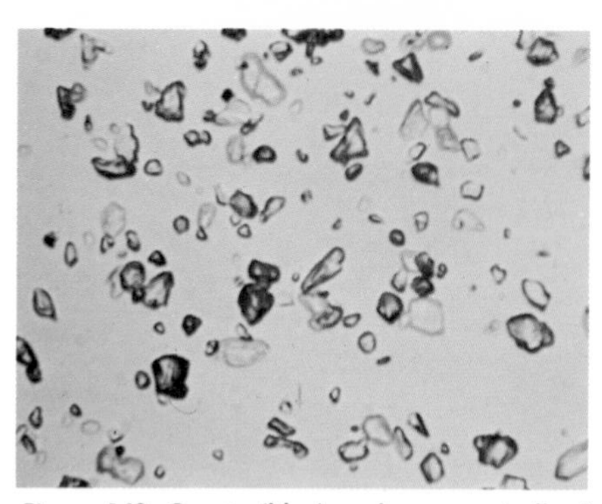

Figure 160. Quartz (blue) and gypsum (yellow) mounted in Cargille liquid 1.532; annular stop, unpolarized light, 80X.

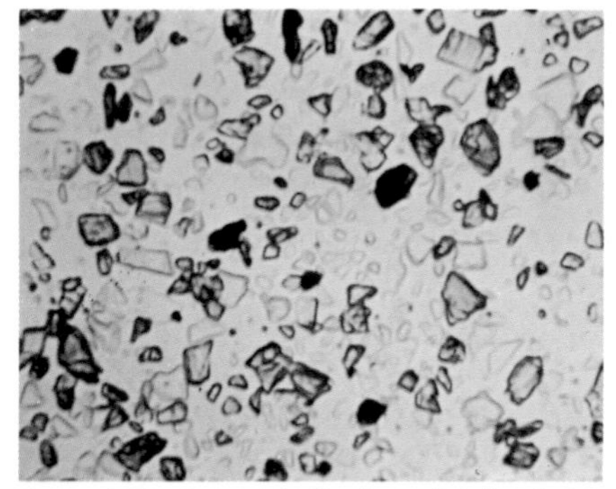

Figure 161. Quartz (yellow) and labradorite (blue) mounted in Cargille liquid 1.548; annular stop, unpolarized light, 80X.

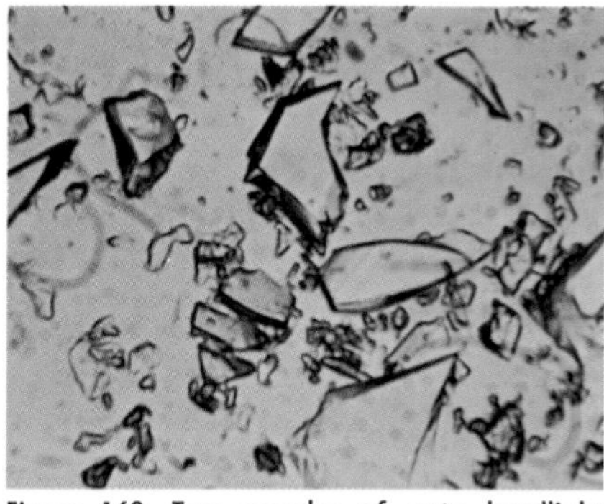

Figure 162. Two samples of auto headlight glass mounted in Cargille liquid 1.520; annular stop, 80X.

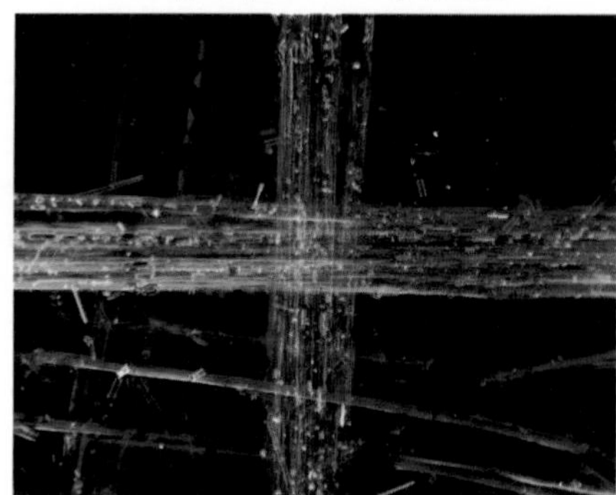

Figure 163. Double exposure of chrysotile asbestos fibers mounted in Cargille liquid 1.570; central stop, 80X.

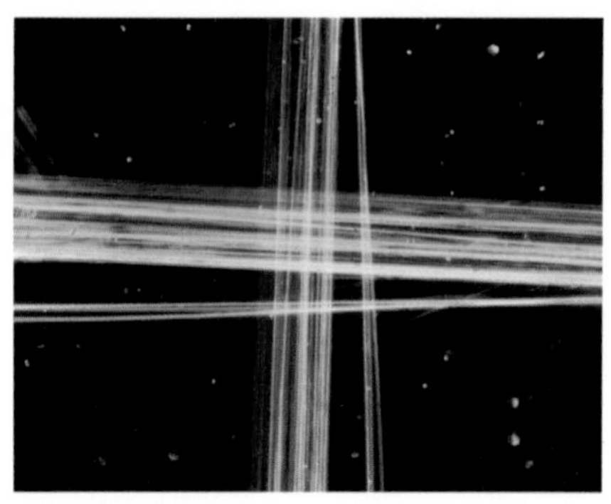

Figure 164. Double exposure of amosite asbestos fiber mounted in Cargille liquid 1.670; central stop, 80X.

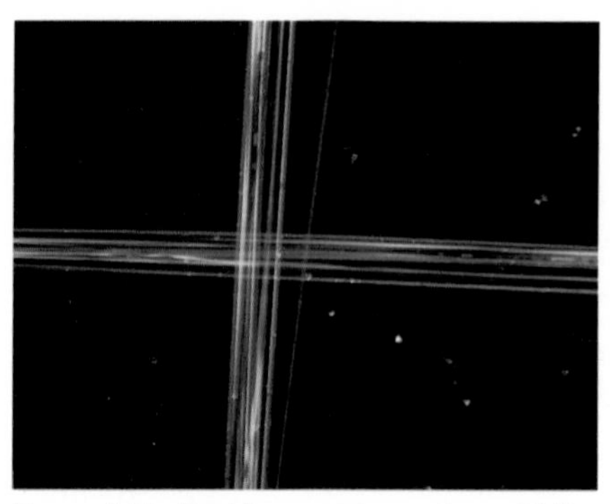

Figure 165. Double exposure of crocidolite fibers, mounted in Cargille liquid 1.690; central stop, 80X.

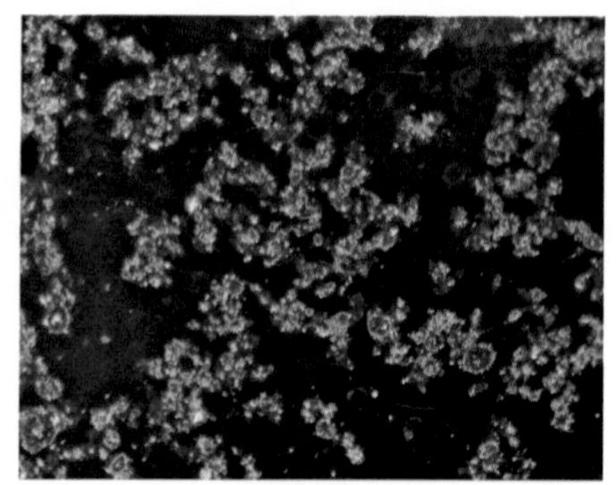

Figure 166. A phenolformaldehyde resin in aqueous potassium mercuric iodide; central stop, 80X.

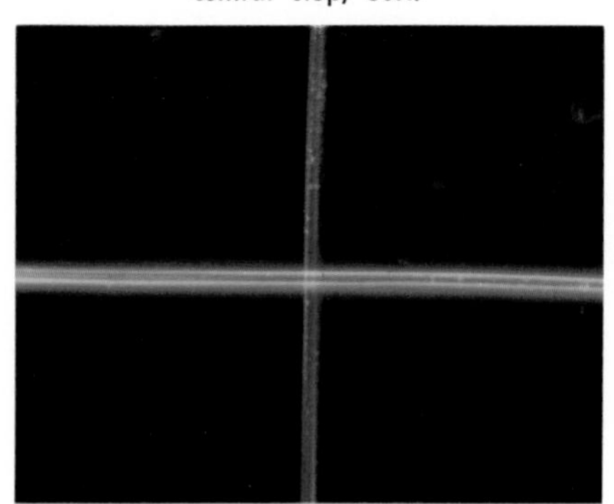

Figure 167. Double exposure of viscose rayon fibers mounted in Cargille liquid 1.530; central stop, 80X.

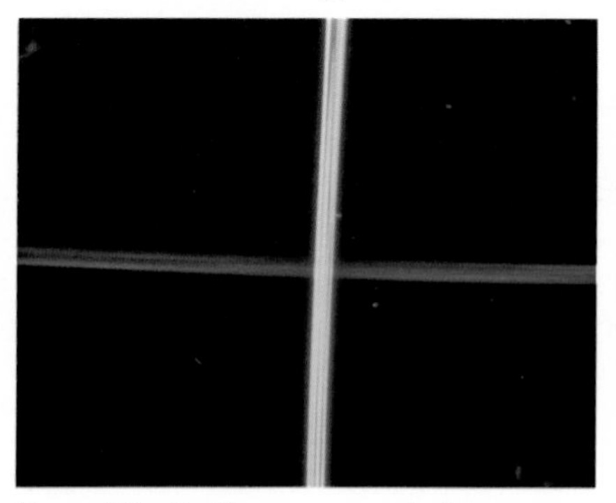

Figure 168. Double exposure of viscose rayon fibers mounted in Cargille liquid 1.550; central stop, 80X.

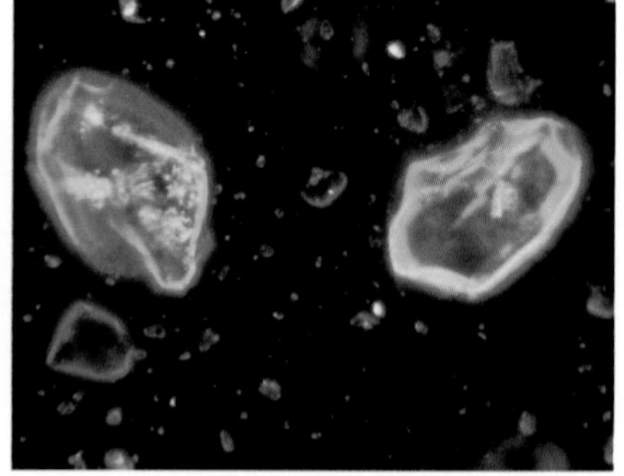

Figure 169. Mine atmosphere mixture with trace amounts of finely divided quartz in Cargille liquid 1.536; central stop, 80X.

Reprinted from
THE PARTICLE ATLAS
McCrone and Delly
Ann Arbor Science Publishers, Inc.

ADDITIONAL REFERENCES ON DISPERSION STAINING
(arranged by date)

Christiansen, C. (Experiments on determination of the refractive index of white powders by the liquid immersion method), Ann. der Physik und Chemie (Wiedemann's Annalen) 23, 298-306 (1884).

Lord Rayleigh, Paper of Christiansen effect, Phil. Mag. 20, 358 (1885).

Ambronn, H., "Farbenerscheinungen an der Grenzen farbloser Objecte in Mikroskop.," Bericht Gesell. Wiss. Leipzig, Math-Phys. Kl. XLVIII, 134-140 (1896).

Wright, F. E., "Oblique illumination in petrographic microscope work," Am. J. Sci. 35, 63-82 (1913).

Smithson, E., "The application of phase contrast microscopy to mineralogy and petrology," Mineral Mag. 28, 384 (1942).

Winchell, A. N., and W. B. Meek, "Birefringence-dispersion ratio as diagnostic," Am. Mineral. 32, 336-43 (1947).

Crossmon, G. C., "Optical staining of tissue," J. Opt. Soc. Am. 38, 417 (1948).

Dodge, N. B., "The darkfield color immersion methods," Am. Mineral. 33, 541-9 (1948).

Emmons, R. C., and R. M. Gates, "Use of the Becke line colors in refractive index determination," Am. Mineral. 33, 612-18 (1948).

Douglas, P. C., and G. C. Crossmon, "Examination of textile fibers by the dispersion staining method," Textile Res. J., 644-6 (1949).

Crossmon, G. C., "The dispersion staining method for the selective coloration of tissue," Stain Technol. 24, 61-5 (1949).

Crossmon, G. C., "Dispersion staining with phase contrast microscope accessories: the microscopic identification of quartz," Science 110, 237-8 (1949).

Crossmon, G. C., "Determination of free silica by dispersion staining microscopical methods," Am. Ind. Hyg. Quart. 12, 118-21 (1951).

Thaer, A., "Contribution towards light microscopical mineral determination of fine dust, especially in coal mining," Staub 38, 555-70 (1954).

Crossmon, G. C., and W. C. Vandemark, "Microscopic observations correlating toxicity of beryllium oxide with crystal structure," A.M.A. Arch. of Ind. Hyg. and Occupational Med. 9, 481-7 (1954).

Thaer, A., "Microscopic evaluation of dust specimens in German mineral coal mining," Glückauf 91, 29-37 (1955).

Schmidt, K. G., "New measurements for the phase contrast method," Staub 43, 47-9 (1956).

Schmidt, K. G., "New working sheets for the determination of refractive index and dispersion determination of minerals of powdery consistency," Z Staub 18, 247 (1958).

Crossmon, G. C., "Chemical microscopy in the optical industry," ASTM Special Tech. Pub. No. 257, 29-38 (1959).

Wilcox, R. E., Proc. of 1960 Microscopy Sym., 160-5, McCrone Research Institute (1960).

Crossmon, G. C., "Instrumentation for the microscopy and identification of dust," Encyclopedia of Instr. for Ind. Hyg., 281-3 (1961).

Radczewski, O. E., "The differentiation of minerals by optical staining in the limiting dark field," Report of the Deutsche keramische Gesellschaft 38, 381-98 (1961).

Thaer, A., "Methods and instruments for microscopic fine dust analysis," Leitz-Mitt. 2, 17-23 (1961).

Crossmon, G. C., "Dispersion staining microscopy as applied to feed preparations," Offic. Proc. of the 10th Ann. Meeting, Am. Assoc. of Feed Microscop. (1962).

McCrone, W. C., and J. S. Martin, "Identifying colorless transparent particles by microscopy," Res./Develop., 26-31 (1964).

Crossmon, G. C., "New developments in phase and dispersion staining microscopy for the examination of dust samples," Am. Ind. Hyg. Assoc. J. 25, 25-7 (1964).

Wilcox, R. E., "Immersion liquids of relatively strong dispersion in the low refractive index range (1.46-1.52)," Am. Mineral. 49, 683-8 (1964).

Grabar, D. G., "Thermal-optical effects in acrylic fibres," Microscope 4, 209-14 (1964).

Kirchgessner, W. G., and A. R. Gaisser, "Application of dispersion staining to fiber and plastic examination and identification," Textile Res. J. 35, 78-80 (1965).

Kantz, M. R., "A new application of dispersion staining," Microscope 14, 422-4 (1965).

Crossmon, G. C., "Macroscopic dispersion staining," Microscope 14, 498-503 (1965).

Jones, F. T., "Dispersion and refractive index determinations by means of an annular wedge interference filter," Microscope 14, 440-7 (1965).

Wright, H. G., "The petrographic use of the spindle stage and focal plane screening in the determination of the optical variation of some zoned minerals," Mineral Mag. 35, 656-60 (1965).

Crossmon, G. C., "Some macro and microscopic applications of dispersion staining as applied to industrial hygiene and air pollution," Microchem. J. 10, 273-85 (1966).

McCrone, W. C., and R. G. Draftz, "Think small," Heating Piping Air Conditioning (January 1966).

Kirchgessner, W. G., "Comparative dispersion staining using a modified comparison microscope," Microscope 15, 511-15 (1967).

Crossmon, G. C., "Techniques," Bausch & Lomb Focus 39, No. 2, 19-22 (Fall 1968).

Beyer, H., and G. Schöppe, "Application of the colour immersion method in phase contrast and dark field in the analysis of mineral dust," Jena Rev., No. 4, 228-33 (1969).

Dodd, J. G., "Observations with a schlieren microscope," Microscope 17, 1-14 (1969).

Wiemann, M. R., Jr., F. C. Besic, and O. Keller, "Application of dispersion staining to the study of refractive index of enamel," J. Dental Res. 51 (4) (July-August 1972).

McCrone, W. C., "Systematic identification of particles using dispersion staining," Powder Technol. 2, 366-7 (1968/9).

Crossmon, G. C., "Techniques," Bausch & Lomb Focus 40, No. 1, 11-14 (Spring 1969).

Forlini, A. L., "Expanded and revised tables for the determination of unknowns," Microscope 17, 29-54 (1969).

Julian, Y., and W. C. McCrone, "Identification of asbestos fibers by microscopical dispersion staining," Microscope 18, 1-10 (1970).

Goodman, R. A., "Expanded uses and applications of dispersion staining," Microscope 18, 41-50 (1970).

Forlini, A. L., and W. C. McCrone, "Dispersion staining of fibers," Microscope 19, 243-54 (1971).

McCrone, W. C., and I. M. Stewart, "Asbestos," Am. Laboratory, pp. 13-18 (April 1974).

McCrone, W. C., "A new dispersion staining objective," Microscope 23, No. 4 (1975).

McCrone, W. C., "Determination of n_F, n_D and n_C by dispersion staining," Microscope 23, No. 4 (1975).

Dodd, J. G., and W. C. McCrone, "A schlieren eyepiece," Microscope 23, 89-92 (1975).

O. Goldberg, "An obvious illuminator for dispersion staining," Microscope 24, 291-4 (1976).

McCrone, W. C., "Dispersion staining," Microscope 26 (1978).

XI. HOT STAGE METHODS

Hot stage methods were developed in the United States[75] somewhat later than the Mikro Methoden of the Koflers[76] although apparently independently. Since these two schools of thought developed along different lines with different results, it is interesting to make a systematic comparison. Both groups were familiar almost from the start with the publications of Otto Lehmann who, in his volume titled Die Kristallanalyse, dated 1891[77], described many of these techniques, including the mixed fusion (Kontaktmethode) for qualitative determination of the phase diagram (as a basis for establishing identity or lack of identity of two compounds).

Although Lehmann's efforts in this field were almost completely ignored for nearly fifty years, they have now been extended to many applications in organic and physical chemistry. A partial list of these follows:

A. APPLICATIONS

1. Determination of purity
2. Analysis of mixtures
3. Characterization and identification of fusible compounds and mixtures
4. Determination of composition diagrams
   a. Two-component systems
   b. Three-component systems
5. Investigations of polymorphism
6. Measurement of physical properties
   a. Molecular weight
   b. Rates of crystal growth
   c. Crystal morphology
   d. Crystal optics
7. Study of boundary migration
8. Study of kinetics of crystal growth
9. Correlation of physical behavior with crystal properties (e.g., thermal stability of decomposable compounds with lowering of melting point, change in refractive index of the melt etc.)

A fundamental difference between the two groups has resulted from the dependence of the Kofler group on the highly precise Kofler hot stage. Most of the early work in this field in the United States had as its objective the development of simple rapid techniques applicable with modest technical background and rudimentary equipment. Most observations were made, therefore, on preparations melted over a microflame as they cooled on the bare microscope stage. The Kofler group, in general, make their observations during heating, the United States group during cooling of the preparation.

---

75. McCrone, W. C., Fusion Methods in Chemical Microscopy, Interscience, New York, 1957.

76. Kofler, L., and A. Kofler , Thermo-Mikro-Methoden zur Kennzeichnung Organischer Stoffe und Stoffgemische, Wagner, Innsbruck, 1954.

77. Lehmann, O., Die Kristallanalyze, Engelmann, Leipzig, 1891.

The Mettler hot stage, introduced in 1967, has increased the use of hot stage microscopy, partly because of its precision but mainly because of its ease of use[78-81].

B. SCOPE

Before further comparing the two approaches, we should consider the following list of properties determinable by hot stage methods to appreciate the wide scope of these methods.

1. During heating
   - a. Sublimation
     - (1) Degree of sublimation
     - (2) Nature of sublimate
       - (a) Liquid globules
       - (b) Liquid plus crystals
       - (c) Badly-formed crystals
       - (d) Well-formed crystals of stable form
         - [1] Profile angles
         - [2] Forms, habit, system
           - [a] Isotropic or anisotropic
           - [b] Uniaxial or biaxial
         - [3] Extinction
         - [4] Sign of elongation
         - [5] Principal refractive index or indices
         - [6] Conoscopic observations
           - [a] Optic axial plane
           - [b] Optic axial angle
           - [c] Dispersion
           - [d] Sign of double refraction
         - [7] Pleochroism
       - (e) Well-formed crystals of unstable form
         - [1] - [7] Same as (d) [1] - (d) [7]
   - b. Polymorphic transformation
     - (1) Temperature at first discernible transformation
     - (2) Monotropic transformation
     - (3) Enantiotropic transformation
       - Transition temperature
   - c. Loss of water (or solvent) of crystallization
     - Temperature at first discernible loss
   - d. Decomposition
     - (1) Temperature at first discernible decomposition
     - (2) Color of decomposition products

---

78. McCrone, W. C., "Applications of hot stage microscopy," Lab. Management (January 1968).

79. McCrone, W. C., "Applications of thermal microscopy," Mettler Technical Information Bulletin 3003 (1968).

80. Woodard, G. D., "Calibration of the Mettler FP2 hot stage," Microscope 18, 105-8 (1970),

81. Julian, Y., and W. C. McCrone, "Accurate use of hot stages," Microscope 19, 225-34 (1971).

2. During melting
   a. Melting point
   b. Refractive index of the melt
   c. Temperature coefficient of refractive index of the melt
   d. Dispersion of refractive index of the melt

3. During cooling
   a. Supercooling of melt
      (1) Supercools slightly
      (2) Readily supercools
          Mobility of supercooled melt at room temperature
   b. Crystal growth rate as a function of temperature
   c. Form of crystal front
      (1) Badly-formed crystals
          Same as 1 [2] - 1 [7]
      (2) Well-formed crystals
          Same as 1 [1] - 1 [7]
   d. Polymorphic transformation
          Same as 1 (a) - 1 b (3)
   e. Gas bubbles
   f. Shrinkage cracks
   g. Mechanical twinning
   h. Anomalous polarization colors
   i. Characterization of unstable polymorphs
          Same as 1 [1] - 1 [7], 3a - 3c, 3e - 3h
4. Meltback (leaving some crystalline material unmelted as seed)
          Same as 1, 2 and 3, omitting 1a, 2b, 2c and 3a
5. After cooling to room temperature
   a. Supercooled liquid
      (1) Refractive index
      (2) Temperature coefficient of refractive index
      (3) Dispersion of refractive index
      (4) Mobility of the supercooled melt
   b. Stable crystalline phase
          Same as 1 [1] - 1 [7]
   c. Unstable crystalline phases
          Same as 1 [1] - 1 [7]
6. Mixed fusion with reference standards
   a. Composition diagram
      (1) Solid solution
          Types I, II, III (Roozeboom)
      (2) Complete immiscibility
      (3) Simple eutectic
         (a) Melting point of eutectic
         (b) Eutectic composition
      (4) Addition compound
         (a) Melting points of addition compound (or compounds) and eutectics
         (b) Composition of addition compound (or compounds) and eutectics
   b. Crystallographic properties of crystals growing into zone of mixing
          Same as 1 [1] - 1 [7]
7. Recrystallization from a high-boiling solvent on a microscope slide
          Same as 1 [1] - 1 [7]

On cursory examination, this outline seems to include an abundance of quantitative physical properties which would serve to uniquely characterize any possible compound. Unfortunately, however, too many compounds show only a few or none of these properties. Some, like hexachloroethane, sublime completely before melting; others, like sucrose, decompose completely before or during melting or crystallize to give badly-formed, very small crystals without crystallographic character. Another difficulty is that while one group of compounds may grow well from the melt to give a number of quantitative characteristics, a second group may give an equal number of numerical characteristics but of a different type (e.g., refractive index of the melt, temperature coefficient and dispersion of the refractive index as compared with profile angles, extinction angles and optic axial angles). Obviously the properties to be chosen for classification purposes must be measurable on all compounds to be included.

The requirements for the ideal physical property to characterize all compounds studied might be listed as follows. Each property should:

1. be measurable with high accuracy and precision on all compounds under study;
2. be easily and quickly determinable without extraordinary background and training;
3. be numerically expressible;
4. vary widely from compound to compound;
5. be tabulated readily so that an unknown can be identified quickly.

C. CHARACTERIZATION AND IDENTIFICATION METHODS

Consideration of the properties determined by hot stage methods as listed above shows that few, if any, properties satisfy all of these criteria. For example, anomalous polarization colors, dispersion of refractive index and pleochroism defy ready tabulation; profile angles do not vary greatly from compound to compound; the presence of gas bubbles, shrinkage cracks etc. cannot be expressed numerically; conoscopic observations cannot be made with rudimentary background and equipment; and none of them can be determined on all fusible compounds. Even with this limitation individual compounds can be quickly recognized (Figure 170).

The Koflers [76] have been especially ingenious in the manner in which they have surmounted these difficulties. A few listings from their identification tables illustrate their approach to this problem (Table XXXI).

In these tables, which include more than 1,000 compounds, the primary tabulating characteristic is the melting point or the best possible substitute for the melting point for those compounds which decompose or sublime before melting. Then, since most compounds are stable somewhat below the melting point, the Koflers have introduced the eutectic melting point with two standard compounds. The eutectic melting point is as easy to measure as the melting point of the pure compound and nearly as valuable diagnostically.

This gives three numerical constants characteristic of each compound, yet the Koflers, in order to make identification even more certain, have developed a very clever technique for measuring the refractive index of the melt. This involves the use of a set of glass powder standards covering the range from 1.34 to 1.67 in increments of about 0.01. The refractive index measurement is made by determining the temperature at which the glass powder standard has the same index as the melt. This temperature is, of course, quite unique although the compound must be very pure before the determination is made. The Koflers have, of course, taken care of this requirement by developing the absorption method of purifying compounds by soaking up the eutectic melt in filter paper as the temperature rises. The unmelted residue just before complete melting is the pure compound.

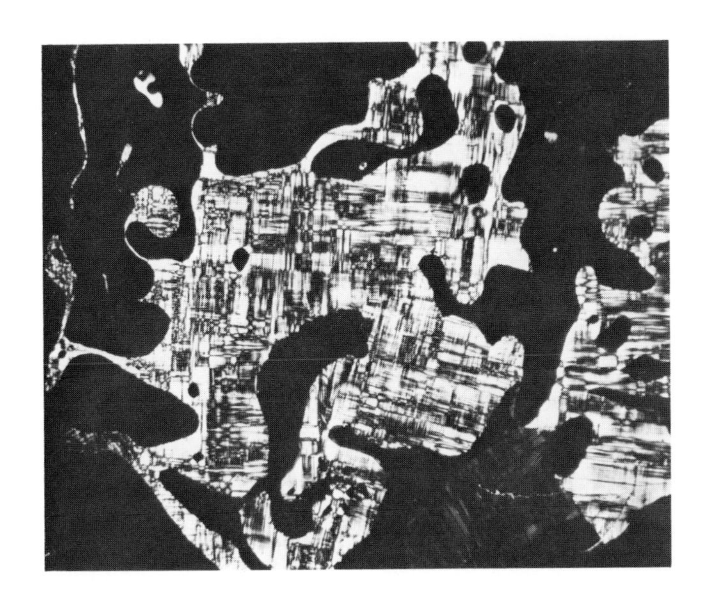

Figure 170. Fusion preparation of distearyl ethylenediamine

Table XXXI. Excerpt from Kofler's tables for the identification of organic compounds

Melting point, °C	Compound	Eutectic temperature with Phenacetin	Benzanilide	Refractive index of melt	Temperature, °C	Special characteristics
151-153	α-Benzo-inoxime	113	133	1.5609	147-149	sublimes at 120°C to give kernels, rods and droplets; solidifies to a glass; on rewarming gives a spherulitic mosaic
153	Diphenyl-thiourea	116	133	1.6128 1.6010	155 167	sublimes at 130°C to give needles and plates
150-158	β-Nitroso-α-naphthol	111	129	decomposes		brownish green; sublimes at 130°C to give needles, rods and, later, droplets; melts sluggishly with decomposition
151-158	Boldin	116	130	decomposes		brown; gives a brown melt which solidifies to a glass; with 50% phenacetin, n=1.5700 at 113-115°C; 1.5609 at 124-126°C

Occasionally when the compound decomposes it is possible to at least bracket the refractive index between two of the glass powder standards (e.g., α-benzoinoxime) or, in some cases, to determine the refractive index in the usual manner, not on the pure compound but on an accurately weighed mixture of that compound and a standard (e.g., boldin with phenacetin).

More recently the Kofler group has added UV absorption data to the tables as a substitute for refractometry of decomposing substances[82,83].

Characterization and identification of fusible compounds and their mixtures have developed quite differently in the United States than in Europe. Most of the value of this means of identification has come in the study of relatively small groups of compounds: substituted aminoquinolines[84], sterols[85], high explosives[75], hexachlorobenzenes[86] etc. In these cases the analyst has worked with all of the compounds almost daily so that he becomes familiar with each by sight. Identification is usually made on the basis of some outstanding morphological or optical characteristic such as anomalous polarization colors, unique shrinkage cracks or gas bubbles, odd crystal habit, transformation mechanism etc. Obviously such characteristics cannot be recorded in tabular fashion and only with difficulty in descriptive terms. The analyst must remember each characteristic, perhaps with the help of photomicrographs or a tabulated suggestion as to the proper key property of each. Despite these limitations there seems to be no difficulty in applying this method to the analysis of groups of compounds up to at least fifty in number. In one research program on high explosives[75] the analyst had no difficulty remembering the key properties of a group of nearly seventy high explosives and mixtures thereof. As a result, any one of that group could be recognized unequivocally in one or two minutes. This included sample preparation time. Actual observation time through the microscope averaged probably under ten seconds for conclusive identification and often a glance alone sufficed.

This system has the serious limitation that only an analyst trained in the recognition of each compound can make the analysis although a second man may learn the distinctive characteristics very quickly. Also, the analyst must maintain his ability by constant examination of compounds in the system; otherwise he may forget the key properties. In spite of these limitations, however, the method has obvious application to research and analysis. The problem of general analysis of a number of compounds by fusion properties assembled into tables analogous to the Kofler tables has also been considered. Rather than melting points and refractive index of the melt, however, crystallographic properties would be utilized as identifying characteristics.

The use of crystallographic properties has the advantage that present tabulations[87,88,89,90] of data can be used directly but it has the serious disadvantage, absent from the Kofler system,

---

82. Kuhnert-Brandstätter, M., R. Hoffmann and M. Senn, Microchem. J. 7, 357 (1963).

83. Kuhnert-Brandstätter, M., Thermomicroscopy in the Analysis of Pharmaceuticals, Pergamon, Oxford, 1971.

84. Goetz-Luthy, N., J. Chem. Educ. 26, 159 (1949).

85. Gilpin, V., Anal. Chem. 23, 365 (1951).

86. Arceneaux, C. J., Anal. Chem. 23, 906 (1951).

87. Winchell, A. N., and H. Winchell, The Microscopical Characters of Artificial Inorganic Solid Substances, 3rd ed., Academic Press, New York, 1964.

88. Crystallographic data, Anal. Chem., monthly, March 1948-March 1961; The Microscope and Crystal Front since 1963.

89. Winchell, A. N., Optical Properties of Organic Compounds, 2nd ed., Academic Press, New York, 1954.

90. Kordes, Ernst, Optische Daten, Verlag Chemie, Gmb H., Weinheim/Bergstr., 1960.

that the analyst must have a more extensive background in crystallography. It is possible, however, to determine both optical and morphological properties by hot stage methods with the advantage that the unknown need not be purified before analysis. The general method involves successive steps, as outlined above, in the classification of properties determinable by hot stage methods. The method is suitable for compounds which decompose if they are recrystallized on the microscope slide from thymol, nitrobenzene, benzyl alcohol, Aroclor® or a similar high-boiling liquid having high surface tension. This technique is, by the way, an excellent means of obtaining otherwise unobtainable, unstable high-temperature polymorphs; it is for this reason that, in the identification scheme, the original crystals of unknown usually are not melted completely so that they remain as seeds during cooling.

Laskowski[91] characterizes polynuclear aromatic compounds by tabulating the melting points of all solid phases formed during mixed fusion with 2, 4, 7-trinitrofluorenone. Since each system includes at least one eutectic, an addition compound, a second eutectic or a transition temperature (peritectic reaction) and the compound itself there are at least four easily determined melting points to use for identification purposes. This scheme approaches the Kofler procedure based on eutectic melting points with two standard compounds and the refractive index of the melt for each compound; however, the 2, 4, 7-trinitrofluorenone method is restricted to substances forming addition compounds with that reagent.

D. POLYMORPHISM

Finally, the use of hot stage methods for the study of polymorphism (Figure 171) and of two- and three-component phase diagrams (Figure 172) has been developed to a high degree[75,76,92,93]. Most organic compounds possess several polymorphic forms and the phase diagrams for organic systems are fully as complex as those obtained for metals. The discovery of quasieutectic syncrystallization, in which the supercooled melt crystallizes — apparently as eutectic — over a wide range of binary composition, was made as a result of microscopical study of crystallization from the melt. Observing the stabilization of lattices, unstable for a pure compound, by forming suitable solid solutions is a curious and important operation difficult, if not impossible, by classical macromethods.

E. DETERMINATION OF PURITY

Purity is usually determined by semiquantitative observation of crystallization as the melt cools. The amount of eutectic melt remaining after crystallization indicates the amount of impurity. The absence of eutectic melt is possible only with pure compounds, when the impurity shows solid-solid solubility or in the trivial case when the two melts are completely immiscible. Observation of the melt itself before crystallization will detect the latter as well as the presence of higher-melting, undissolved solid components. The use of the melting point depression in known two-component systems, the change of optical properties in isomorphous mixtures, the use of refractive index differentiation of impurity from the major component, the change of crystal habit and changes in the rate of crystal growth have all been used to indicate purity. Some of these techniques can be made quantitative and become useful, then, as methods for the analysis of mixtures.

---

91. Laskowski, D. E., D. G. Grabar, and W. C. McCrone, Anal. Chem. 25, 1400 (1953); 30, 542 (1958).

92. Grabar, D. G., J. P. Hession, and F. C. Rauch, Microscope 18, 241 (1970).

93. McCrone, W. C., Microscope 18, 257 (1970).

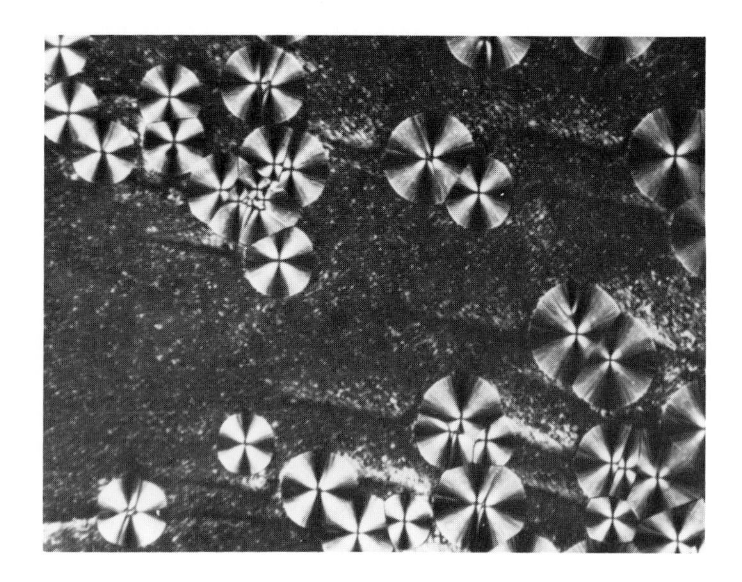

Figure 171. Polymorphic transformation in cholesteryl acetate

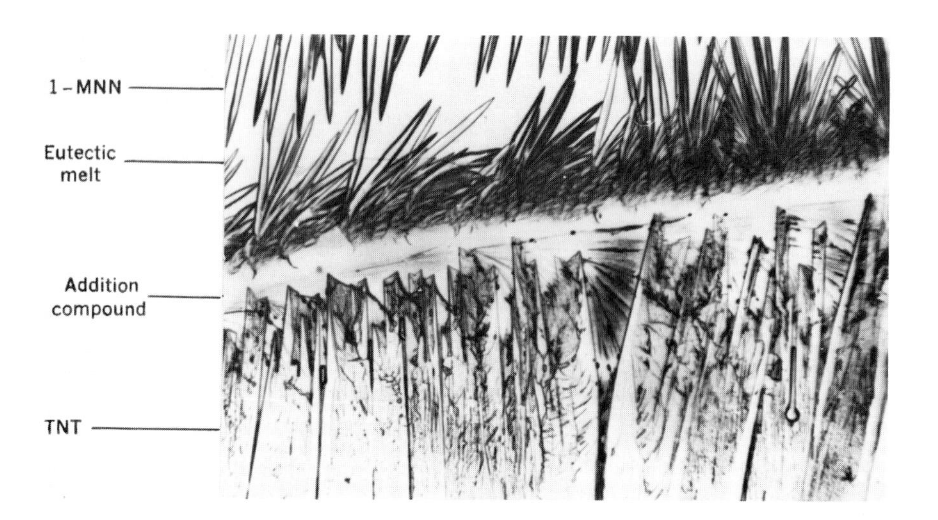

Figure 172. Addition compound formed in fusion preparation of TNT and mononitronaphthalene

F. ANALYSIS OF MIXTURES

Mixture analysis has been carried out by measuring the optic axial angle or other optical property in isomorphous systems[94], by measuring the rate of crystal growth of the principal component of the mixture[95] and by determining the melting point depression. The first and third of these methods are suitable only for binary mixtures. The method based on rate of crystal growth is suitable

94. Bryant, W.M.D., J. Amer. Chem. Soc. 60, 1394 (1938).

95. McCrone, W. C., A. Smedal, and V. Gilpin, Anal. Chem. 18, 578 (1946).

for polycomponent mixtures if the impurities, no matter how many or even whether they are known, are always present in the same ratio. A given impurity may increase or decrease the rate of growth of the major component; hence the method has to be carefully standardized in terms of the nature of the impurities.

G. DETERMINATION OF COMPOSITION DIAGRAMS

The qualitative technique for determining composition diagrams is often used as a means of proving lack of identity or identity of a given pair of compounds[75]. In this test two compounds are fused and allowed to crystallize in contact across a narrow zone of mixing so that the entire composition diagram is shown in the composition gradient. Lack of any discontinuity during growth (including growth rate) across this gradient indicates identity of the two components. A discontinuity in the rate of growth indicates a difference in purity (one discontinuity) or nonidentity (two discontinuities).

H. MECHANISM OF CRYSTAL GROWTH

The use of rate of crystal growth in determining the amount of p, p'-DDT in technical DDT has created an interest in learning more about the mechanism of crystal growth by fusion methods. The results have been published in part[75, 84]. Additional work has shown that the rate of crystal growth of a given compound in a binary mixture is directly related to the viscosity of the mixture. It is hoped also that additional work on the kinetics of crystal growth may result in the calculation of heats of fusion from crystal growth rate data.

Boundary migration[96] is a phenomenon occurring in a few organic compounds which show anisotropy of elasticity; it permits them, in a sense, to be under unidirectional stress. As a result, crystals of such compounds show unidirectional growth into and through each other. This is strictly an orientation effect such that one particular face grows into a different face of another identical crystal in contact with it. This is a most interesting effect which can be observed and studied only by hot stage methods.

I. MISCELLANEOUS APPLICATIONS OF HOT STAGE METHODS

The use of hot stage methods to measure other physical properties of fusible materials has also been studied. For example, the decomposition of organic compounds on heating can be followed quantitatively and the kinetics of decomposition determined by fusion techniques. Decomposition will result in a lowering of the equilibrium melting point and a change in refractive index of the melt. Either property can be used, depending on the temperature at which the study is to be made and the melting point of the compound under study. The density of cast high explosives is related to the manner in which they crystallize since gases dissolved in the melt and shrinkage cracks lower the density. Microscopical hot stage methods are, therefore, a logical means of studying this problem.

Finally, hot stage methods have been pushed to high temperatures[97, 98, 99] and to the study of inorganic systems[97, 99, 100]. This is done by direct application of hot stage methods at high temperatures by means of special hot stages and furnaces, and with auxiliary lens systems to prevent damage to the microscope optics. It is also done indirectly by using low-melting organic reagents, such as 8-hydroxyquinoline, to give characteristic precipitates with the various inorganic ions[94].

---

96. McCrone, W. C., Disc. Far. Soc. 5, 158 (1949).
97. Bauman, H. N., Jr., Bull. Amer. Ceram. Soc. 277, 267 (1948).
98. Cech, R. E., Rev. of Sci. Instr. 21, 747 (1950).
99. Grabar, D. G., and W. C. McCrone, J. Chem. Educ. 27, 649 (1950).
100. West, P. W., and L. Granatelli, Anal. Chem. 24, 870 (1952).

XII. MICROCHEMICAL TESTS

A. INTRODUCTION

In the day of such sensitive microanalytical instruments as the electron microprobe, the scanning electron microscope with energy dispersive detector and the ion microprobe, one might assume that microchemical tests under the light microscope would have been superseded. On the contrary, the routine analysis of subnanogram samples[101] has sparked interest in any and all means of ultramicroanalysis. Microscopical microchemical tests have several assets:

1. They detect beryllium and lithium, most anions and many organic compounds; for these the electron probe and SEM are insensitive or incapable.

2. They are inexpensive, sensitive, reliable and quick.

3. In a few fields, like drug detection, they are widely used and the literature has grown fairly continuously.

In Chamot's day it made sense to use an analytical scheme based entirely on microchemical tests, and he devised a fairly complete scheme for inorganic ions[102]. Today instrumental methods are faster, more quantitative and sometimes more sensitive[103] — they should be used when an unknown must be completely analyzed. Microchemical tests are best used to check for specific entities.

The most generally useful microchemical tests are based on morphology, not of the original particles but of precipitates formed by adding a reagent to an aqueous solution of the sample.

B. INORGANIC QUALITATIVE ANALYSIS

1. Chamot's System

Basically, each chemical element or ion can be precipitated from a drop of solution by one or more of a group of specific reagents, and each gives a characteristic crystalline precipitate. Most of the tests require only a minute or two and not only is identification reliable but, with practice, a fairly accurate estimate of the percentage of that element can be made.

The test is usually performed by mixing a small drop of the reagent with a small test drop on the same microscope slide. When the mixture has stood for a few seconds or minutes, characteristic crystals of the resulting compound can be observed. The specific reagent and conditions for each precipitation can be found in volume II of The Handbook of Chemical Microscopy[102]. Some of the most useful of these tests are summarized in Table XXXII.

In general, the reagent is dissolved in a small drop of water and made to run quietly into contact with a test drop containing the unknown. Occasionally it is best to add the reagent crystals directly to the test drop or to draw a reagent drop of solution across a dried residue from the test drop. The details of each test should be studied in Chamot and Mason's book before any actual testing.

---

101. Bayard, M. A., Microscope 15, 26 (1967).

102. Chamot, E. M., and C. W. Mason, Handbook of Chemical Microscopy, Vol. II, Wiley, NY, 1940.

103. McCrone, W. C., Particle Analyst 49, Ann Arbor Science Publishers (1968).

Later important contributors to inorganic microscopical qualitative analysis include Schaeffer[104] and Keune[105]. Feigl's spot tests[106] are also useful in special cases although these tests are not visually microscopical.

Table XXXII. Microscopical tests for inorganic ions

Ion	Reagents
Sodium	Uranyl acetate, zinc uranyl acetate
Potassium	Chloroplatinic acid, perchloric acid
Ammonium	Chloroplatinic acid, iodic acid
Beryllium	Chloroplatinic acid, potassium oxalate
Calcium	Ammonium carbonate, sulfuric acid
Strontium	Ammonium carbonate, iodic acid
Barium	Ammonium carbonate, potassium ferrocyanide
Magnesium	Ammonium hydroxide and phosphate
Zinc	Oxalic acid, potassium mercuric thiocyanate
Copper, lead	Zinc, potassium mercuric thiocyanate
Mercury	Potassium bichromate, lead
Aluminum	Ammonium bifluoride, cesium sulfate
Tin	Zinc, cesium chloride
Arsenic	Ammonium molybdenate, cesium sulfate and potassium iodide
Chromium	Silver nitrate, lead acetate
Uranium	Thallous nitrate and ammonium carbonate
Fluoride	Sodium fluosilicate
Iron	Potassium ferrocyanide, potassium thiocyanate
Carbonate	Silver nitrate, calcium acetate
Nitrate	Silver nitrate, nitron* sulfate
Cyanide	Silver nitrate, ammonium sulfide and ferris chloride
Chloride	Silver nitrate, thallous nitrate

* "Nitron" is the common name for diphenylenedianilohydrotriazole.

2. Scanning for Individual Ions

Special procedures have been developed for scanning a sample microchemically for particular ions. These tests are based on the use of gelatin-coated microscope slides or a similar substrate. This technique, first proposed by Seely[107], requires the gelatin or other substrate to be impregnated with a reagent specific for the test ion. Particles of the unknown are made to impinge on the gelatin surface where, if they are soluble at the pH of the gelatin, they dissolve and may react chemically with the reagent in the gelatin.

---

104. Schaeffer, H. F., Microscopy for Chemists, pp. 114-57, Van Nostrand, New York, 1953; reprinted by Dover Publications, New York, 1966.

105. Keune, H., Bilderatlas zur Qualitativen Anorganischem Mikroanalyse, VEB Deutscher Verlag für Grundstoff industrie, Leipzig, 1967.

106. Fiegl, F., Spot Tests in Inorganic Analysis, 5th ed., American Elsevier, New York, 1958.

107. Crozier, W. D., and B. K. Seely, Proceedings of First National Air Pollution Symposium, p. 45, Stanford, California, Stanford Press, 1950.

Assume, for example, that the gelatin is impregnated with silver nitrate and the particles impinged on the gelatin surface consist partly, at least, of sodium chloride. On standing for a short time, the sodium chloride particles or, for that matter, the particles of any soluble chloride dissolve and begin to react with the silver nitrate. The result is a halo of crystalline silver chloride at the site of each chloride-containing particle. These halos have a diameter, when fully developed, of eight to ten times the size of the original particle; hence they are easily observed microscopically.

This test has been developed for a variety of cations and anions and could be applied to almost any ion. Different substrates, such as a Millipore filter, have been used by Lodge and others. Lodge and Fanzoi[108] have given the following directions for applying this test to the detection of soluble calcium compounds:

> Purified Eastman Kodak pigskin gelatin (2.5 g), in 8 ml of glycerol and 5 ml of 30 percent aqueous ammonium ferrocyanide, is heated in an oven at 85° C until the gelatin dissolves. Microscope slides are then coated as described by Pidgeon[109]. After several days, particles containing soluble calcium react with the ferrocyanide in the gelatin to form white microcrystalline halos consisting of large single crystals.

This procedure has the advantage that it more directly associates the test ion with the particles since only those particles containing the test ion in a soluble form will produce a halo. It also, when calibrated, indicates the size of the reacting particle and gives visible halos with particles at or even below the resolving power of the light microscope. It is, therefore, particularly useful for the characterization of suspended particles (1 μm) in contrast to the larger settled dust particles which can, in most cases, be better handled by morphology.

The most serious disadvantage of the gelatin procedure is the required solubility of the particle containing the test ion. If the test ion as combined in the parent particle is less soluble than the precipitate formed by that ion and the reagent, no halo will form. It is, of course, possible to increase the rate of reaction by storing the slides in a moist atmosphere. The solubility may also be enhanced by exposing the slide plus particles to moist hydrochloric acid or ammonium hydroxide vapors. All in all, the procedure is quite useful, particularly when the particles to be tested are below 1 μm or when they cannot be identified by morphology.

C. ORGANIC QUALITATIVE ANALYSIS

Much of the emphasis in the development of microcrystal tests for organic compounds has been placed on methods for drugs. Fulton's monumental effort[110] is a compendium of such tests. Less complete, but easier to use, is the English translation of Behrens-Kley[111] which is oriented towards organic compounds more generally.

---

108. Lodge, J. P., Jr., and H. M. Fanzoi, "Extension of the gelatin method for the detection of micron sized particles," Anal. Chem. 26, 1839 (1954).

109. Pidgeon, F. D., "Controlling factors in identification of microscopic chloride particles with sensitized gelatin films," Anal. Chem. 26, 1832 (1954).

110. Fulton, C. C., Modern Microcrystal Tests for Drugs, Wiley-Interscience, New York, 1969.

111. Behrens, T. H., and P. Kley, Organische Mikrochemische Analyse, L. Voss, Leipzig, 1922; English translation by R. Stevens, Microscope Publications Ltd., 1969.

A different approach to organic microchemistry was advanced by Laskowski, who found reagents which form molecular addition compounds with various classes of organic compounds[112-114]. The individual compound is then identified by physical data, such as melting points of addition compounds and eutectics.

As noted earlier, Fiegl's spot tests[115] may be useful in special cases.

D. MINIATURIZATION

Most of the microchemical methods considered thus far require milligrams or, at best, micrograms of sample. The impregnated gelatin slides can work for individual 1 μm particles, however. Other means of reducing the quantity of sample needed are discussed below.

The electron microscope may be used to observe the characteristic morphology of a reaction product[116,117]. Although identifying optical properties such as color birefringence and refractive index are lacking, electron diffraction can be applied to identify reaction products.

Instead of depending on increased microscope resolution to observe smaller amounts of product, one can reduce the size of the system in which the reaction occurs. Microchemical tests have been successfully scaled down to include picogram particles. Several approaches are possible but all involve a means of restricting either the drop size or the area of precipitation. By doing this electrolytically with very fine-tipped cathodes, Brenneis[118] succeeded in detecting 50 pg of copper or silver. Rachelle[119] localized his precipitate and controlled the size of the droplets by placing two very small droplets of reagent and test solution close together in a drop of oil. The two drops coalesce to give a visible precipitate.

By treating the microscope slide with silicone oil, reagent drops can be kept from spreading so that aqueous or organic liquids can be confined to very tiny droplets[120]. Precipitation from aqueous solutions on the smallest scale requires control of humidity, however. This has been achieved by using small cells which can be flushed with moist or dry air at will. Picogram quantities of lithium and beryllium have been detected, and the method is general[121].

---

112. Laskowski, D. E., D. G. Grabar, and W. C. McCrone, Anal. Chem. 25, 1400 (1953).

113. Laskowski, D. E., and W. C. McCrone, Anal. Chem. 26, 1497 (1954); 30, 542 (1958).

114. Laskowski, D. E., and O. W. Adams, Anal. Chem. 31, 148 (1959).

115. Fiegl, F., Spot Tests in Organic Analysis, 7th ed., Elsevier Publishing, New York, 1954.

116. Wiesenberger, E., "Experiments for the performance of tests under the electron microscope," Mikrochim. Acta 1957, 506.

117. Tufts, B. J., and J. P. Lodge, Jr., "Chemical identification of halide and sulfate in submicron particles," Anal. Chem. 30, 300 (1958) (electron microscopy).

118. Brenneis, H. J., Mikrochemie 9, 385 (1931).

119. Benedetti-Pichler, A. A., and J. R. Rachelle, Ind. and Eng. Chem., Anal. Ed. 12, 233-41 (1940).

120. Laskowski, D. E., Anal. Chem. 37, 174 (1965).

121. McCrone, W. C., Microscope 19, 235 (1971).

E. QUANTITATION

Benedetti-Pichler[119] felt that the sensitivity limit for microchemical tests with the light microscope is in the femtogram range but at this level the precipitated crystals would have to be very fine needles or very thin plates to be visible. For most quantitative analysis more material is necessary. Benedetti-Pichler provides a very complete treatment of ultramicrochemical analysis in his book, Identification of Materials Via Physical Properties, Chemical Tests and Microscopy, Academic Press, 1964.

Klimeš and Janák perform photometric analysis of picogram samples using silica gel particles as cells in which the color reactions take place[122]. Well-established color reactions, such as dimethylglyoxime for nickel, have been used at the picogram level.

122. Klimeš, L., and J. Janák, Microchem. J. 13, 534 (1968).

XIII. MORPHOLOGICAL ANALYSIS

A. INTRODUCTION

If you can recognize a robin or a banana by sight, you should be able to recognize single particles of coal, cement or gypsum when the microscope magnifies them to an equivalent size. This is the thesis of those who identify small particles by microscopy. No other microanalytical method identifies such small samples of such diverse substances so quickly.

Single particles of almost any substance — animal, vegetable, mineral, industrial by-product, raw material, corrosion product, flyash — can be identified in a few seconds to a few minutes by a microscopist trained in particle identification. The particle must be at least 2 to 3 micrometers in diameter and 10 picograms or 10^{-11} gram in weight for identification by light microscopy, or 200 to 300 Angstroms in diameter and 10 attograms or 10^{-17} gram in weight for identification by electron microscopy (Figure 68, Chapter VII).

Who needs methods this sensitive? Industrial hygienists, for one; particles ingested into the human lung are nearly all smaller than 5 μm in diameter. Harmful particles like asbestos, quartz, lead and beryllium compounds found in lung tissue are seldom larger than 5 μm and these must be recognized and identified. Many particles in air and water pollution samples are smaller than 5 to 10 μm and must be identified. The particles clogging fine fuel jets or synthetic fiber spinnerets are often this small. Many particles smaller than 10 μm can cause serious troubles if they get into magnetic recorder tape or photographic film emulsions. Many solutions, such as ampouled drugs or bottled beverages, are expected to be optically clear, and very tiny particles can scatter light or form visible residues. Finally, as electronic systems become more and more microminiaturized, the more often small particles will cause failures.

The need for identification of small particles is continually increasing. This is especially true since chemical identification of a particle usually also identifies the source of that particle. A variety of such problems has been solved by morphological analysis.

In one tragic case, a clogged jet fuel line on an aircraft caused a fatal accident. Examination of the clogged jet showed tiny short lengths of glass fibers from a protective filter placed in the fuel line supplying the jet aircraft. In another case, noisy magnetic tapes were found to be due to fine dust particles; the fine dust particles, in turn, were found to be magnetic tape and coating abraded from the edges of the tape by rough metal wind-up spools. One final example from the many that might be chosen was the presence of very tiny metal particles in photographic film emulsion, each of which caused precipitation of an easily visible spot of metallic silver.

B. EQUIPMENT REQUIRED

What steps must be taken to identify these tiny samples? Since nearly all are invisible to the naked eye, they must be made visible by using an appropriate microscope. To be able to identify any of the thousands of particles that may be present requires a good polarizing microscope and a trained microscopist. The equipment required may cost as little as $1000 although any microscopist skilled in this highly specialized field probably deserves more sophisticated equipment as a fringe benefit. A total cost of perhaps $8,000 is more realistic.

In addition, few laboratories are clean enough to permit work of this kind. The laboratory itself is dirty; the air in the room is dust-laden; fingers carry particles; slides and coverslips are never clean. In short, there are millions of particles in any laboratory just as large and just as visible microscopically as the particle to be identified. It is essential that the sample be opened, examined and identified under conditions such that contamination is avoided. This requires a clean bench if not a clean lab. Figure 173 shows a laboratory arrangement for two microscopists. Each of three benches is fully enclosed with a polymethacrylate hood. The entire back wall is the support for a full-size absolute filter. Laminar flow of cleaned air toward the microscopist prevents entrance of dust-laden room air. Each microscopist has a stereo binocular microscope and a polarizing microscope, the latter fitted with a vertical illuminator.

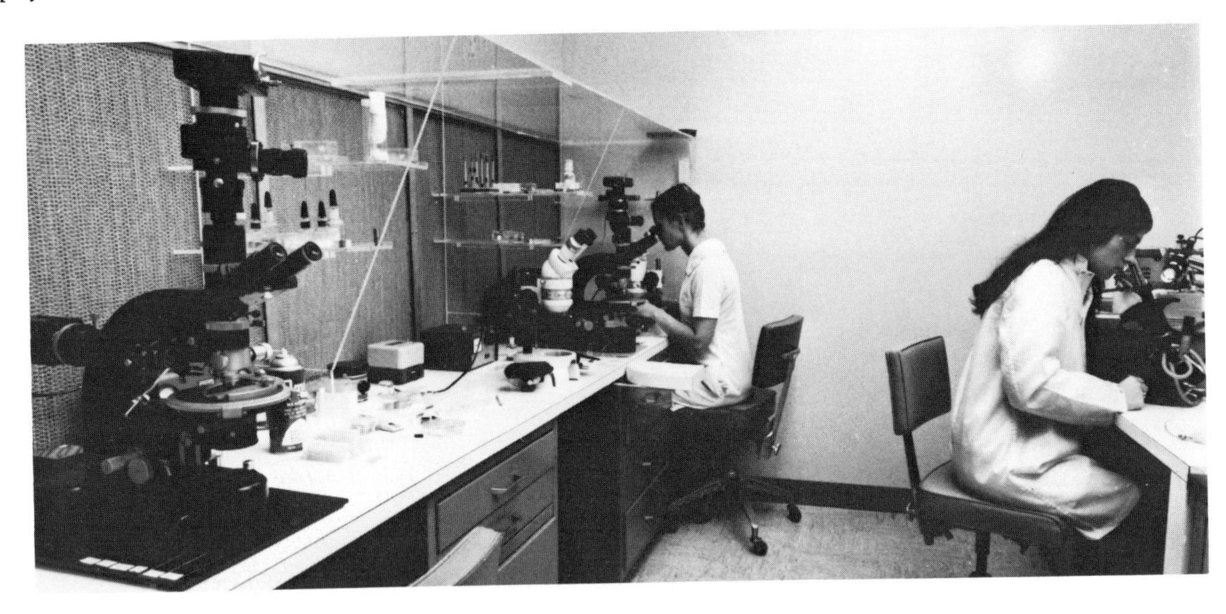

Figure 173. Clean benches for particle microscopy

C. IDENTIFICATION BY MORPHOLOGY

Now, what properties can be measured on small particles that will aid identification? First is shape: Is it a very flat sheet of mica, highly fibrous asbestos, a pitted fiber of wood pulp, conchoidally fractured quartz or ground glass, or a perfect sphere of flyash from a pulverized coal boiler? Next, size may be important for some substances like pollens, starches and fibers. Color is very important, and both transmitted and reflected light must be used. The color should also be associated with degree of transparency. Crossed polars should be used to differentiate between isotropic substances such as glass, pumice, fluorite or sphalerite and birefringent substances such as quartz, talc, nylon, fur fibers and starch. The degree of birefringence often permits recognition of limestone, foraminifera, Orlon®, quartz and so forth. The refractive indices are usually not determined precisely, although an estimate of the value relative to the mounting liquid is often useful. For example, calcite and siderite can be quickly and easily differentiated when mounted in the usual mounting medium, Aroclor®, which has a refractive index of 1.66. Particle identification with these techniques is called morphological analysis.

Difficulties begin when the microscopical examination fails to identify the particle. Then, the microscopist must have recourse to electron microscopy, x-ray diffraction, electron microprobe and neutron activation.

D. AIDS TO IDENTIFICATION

To be able to identify a given particle, the microscopist must either have seen one before, or, at least, a good photomicrograph of it. His aim is to associate such microscopical characteristics as shape, surface, transparency, color, birefringence and refractive index with the name of that substance. A photomicrographic atlas is an ideal solution to the problem because a microscopist can remember the morphological characteristics of only so many different substances. If he can remember a large number, he is still the only one who can make the analyses. An atlas supplements the microscopist's memory and helps other microscopists learn to identify particles. Numbers of photomicrographic atlases have been published on specific classes of materials. These include wood[123], papermaking fibers[124,125], fur[126], hair[127], textile fibers[128,129], pollens[130,131], cement clinkers and slags[132] and boiler effluent[133].

A more general work, The Particle Atlas[134], includes representative substances from these and other categories. Now in a greatly enlarged second edition it contains more than 1400 full-color photomicrographs and a detailed description of 1000 substances. Also included are careful descriptions of the techniques necessary for collecting, manipulating and characterizing particles as well as analytical tables for the identification of specific particles after a simple preliminary classification. With this atlas, an interested person willing to invest the necessary time could become expert in identifying the common particles found anywhere in the world.

---

123. Forest Products Research Laboratory, An Atlas of End-Grain Photomicrographs for the Identification of Hardwoods, Bulletin 26, Her Majesty's Stationery Office, London, 1953.

124. Armitage, F. D., An Atlas of the Commoner Paper Making Fibres, An Introduction to Paper Microscopy, Guildhall, London, 1950.

125. Carpenter, C. H., et al., Papermaking Fibers, Technical Publication 74, State University College of Forestry, Syracuse, New York, 1963.

126. Brevoort, H. L., Fur Fibers as Shown by the Microscope, Royal Microscopical Society, London, 1886.

127. Lochte, T., Atlas der menschlichen und tierischen Haare, Schöps, Leipzig, 1938.

128. Luniak, B., Identification of Textile Fibers, Pitman, London, 1953.

129. Mauersberger, H. R. (ed.), Matthews' Textile Fibers, 6th ed., Wiley, 1954.

130. Hyde, H. A., and K. F. Adams, An Atlas of Airborne Pollen Grains, St. Martin's, New York, and Macmillan, London, 1958.

131. Kremp, G. O. W., Morphologic Encyclopedia of Palynology, University of Arizona, Tucson, 1965.

132. Konovalov, P. F., B. V. Volkonskiǐ and A. P. Khashkovaskii, Atlas of Microstructures of Cement Clinkers, Refractory Materials and Slags, Gos. Izd. Lit. po Stroit, Arkhitekt. i Materialam, Leningrad, 1962.

133. Hamilton, E. M., and W. D. Jarvis, Identification of Atmospheric Dust by Use of the Microscope, Central Electricity Generating Board, London, 1963.

134. McCrone, W. C., and J. G. Delly, The Particle Atlas, 2nd ed., Volumes I-IV (1974), Volumes V-VI (1978), Ann Arbor Science Publishers.

GENERAL REFERENCES

Benedetti-Pichler, A. A., Identification of Materials via Physical Properties, Chemical Tests, and Microscopy, Academic, New York, 1964.

Bunn, C. W., Chemical Crystallography, 2nd ed., Oxford, 1961.

Burrells, W., Industrial Microscopy in Practice, Morgan and Morgan, Hastings-on-Hudson, New York, 1964 and Fountain, London, 1961.

Chamot, E. M., and C. W. Mason, Handbook of Chemical Microscopy, vol. I, 3rd ed., Wiley, New York, 1958.

Cosslett, V. E., Modern Microscopy, G. Bell and Sons, London, 1966.

El-Hinnawi, E. E., Methods in Chemical and Mineral Microscopy, Elsevier, Amsterdam, 1966.

Hallimond, A. F., The Polarizing Microscope, 3rd ed., Vickers, York, 1970.

Hartshorne, N. H., and A. Stuart, Crystals and the Polarizing Microscope, 4th ed., Edward Arnold, London, 1970.

Johannsen, Albert, Manual of Petrographic Methods, 2nd ed., 1918, Facsimile reprint by Hafner, New York, 1968.

Loveland, Roger P., Photomicrography; A Comprehensive Treatise, 2 vols., Wiley, New York, 1970.

Michel, K., Die wissenschaftliche und angewandte Photographie, Springer, Vienna, 1967.

Needham, George H., The Practical Use of the Microscope Including Photomicrography, Thomas, Springfield, Illinois, 1958.

Rogers, A. F., Introduction to the Study of Minerals, 3rd ed., McGraw-Hill, New York, 1937.

Shillaber, C. P., Photomicrography, Wiley, New York, 1944.

Wallis, T. E., Analytical Microscopy, 3rd ed., Little, Brown and Co., Boston, 1965.

White, G. W., Introduction to Microscopy, Butterworths, London, 1966.

XIV. LABORATORY EXERCISES

MICROSCOPY NOTEBOOKS

A good microscopy notebook is a valuable record of observations, drawings and other experimental data. Like any other well-kept laboratory notebook, it forms an official record of your work and is an original source from which analyses, research reports, papers for publication and patentable inventions are drawn. You should keep your microscopy notebooks in such a way that you, or another microscopist, will understand what you entered on each page, not only tomorrow but five years from tomorrow — otherwise your notebook will not be worth retaining, either as part of your own research records or as part of your company's.

Type of notebook. By far the most satisfactory type of notebook is permanently bound, measures about 5 x 8", and is bound along one of the 5" sides. Such a notebook can be used on the bench next to the microscope while occupying a minimum of usable space. Such notebooks can be made by sawing larger (8 x 10") notebooks in half, or can be ordered from scientific notebook dealers*; they are not normally available from stationers in the U.S., although they are in Britain.

Drawings. Accurate drawings of microscopic objects are valuable not as records only, but as a means of forcing the microscopist to observe an object accurately. A good drawing is seldom simply an exact reproduction of the image seen at one setting of the microscope focus — it is more often an accurate interpretation of the object, arrived at by scanning the field to choose typical, or atypical, parts of the preparation and by racking the fine focus up and down to observe the images at various depths within the object.

The most common mistakes of the beginner are to make drawings too small, too fuzzy in outline, too black in shaded areas, and to make beautiful drawings of such artifacts as air bubbles, dust particles or extraneous fibers. Many of the prettiest drawings by a beginner are of artifacts on the top of the coverslip.

These mistakes are particularly fatal in the study of crystal morphology. Crystals should be drawn large enough so that you need to use a straightedge to make them look pretty (a microscope slide makes a handy straightedge). The edges directly visible from the top should be drawn with solid lines; those underneath should be drawn with dashed lines. Shading is not particularly helpful in most crystal drawings; if it is used at all, the best kind is illustrated in the drawings of the flattened and distorted octahedra on page 351 of Chamot and Mason's Handbook of Chemical Microscopy, Vol. I, 3rd ed. If you are not skillful at drawing, you may find it helpful to use a drawing camera until you have developed greater proficiency in representing what you see through the microscope. Remember that the interfacial angles are very important; 90°, 120° and 60° ∠'s should be ± 1-2°, oblique ∠'s really so.

Photomicrographs. Photomicrographs are sometimes to be preferred to drawings and sometimes are not. Whenever prints of photomicrographs are available, they should be pasted into the notebook. Be sure to record the conditions of illumination, magnification etc.

Table of contents. Although an alphabetical index to a microscopy notebook would be a very nice thing to have, few microscopists take the trouble to make one. A table of contents is perhaps 90% as useful, and it requires little extra effort to make such a table in the front of your notebook as you go along. It is recommended that you get into the habit of doing so.

Representative microscopy notebook pages are shown on the following page.

---

* McCrone Associates, Inc. now markets a microscopist's notebook.

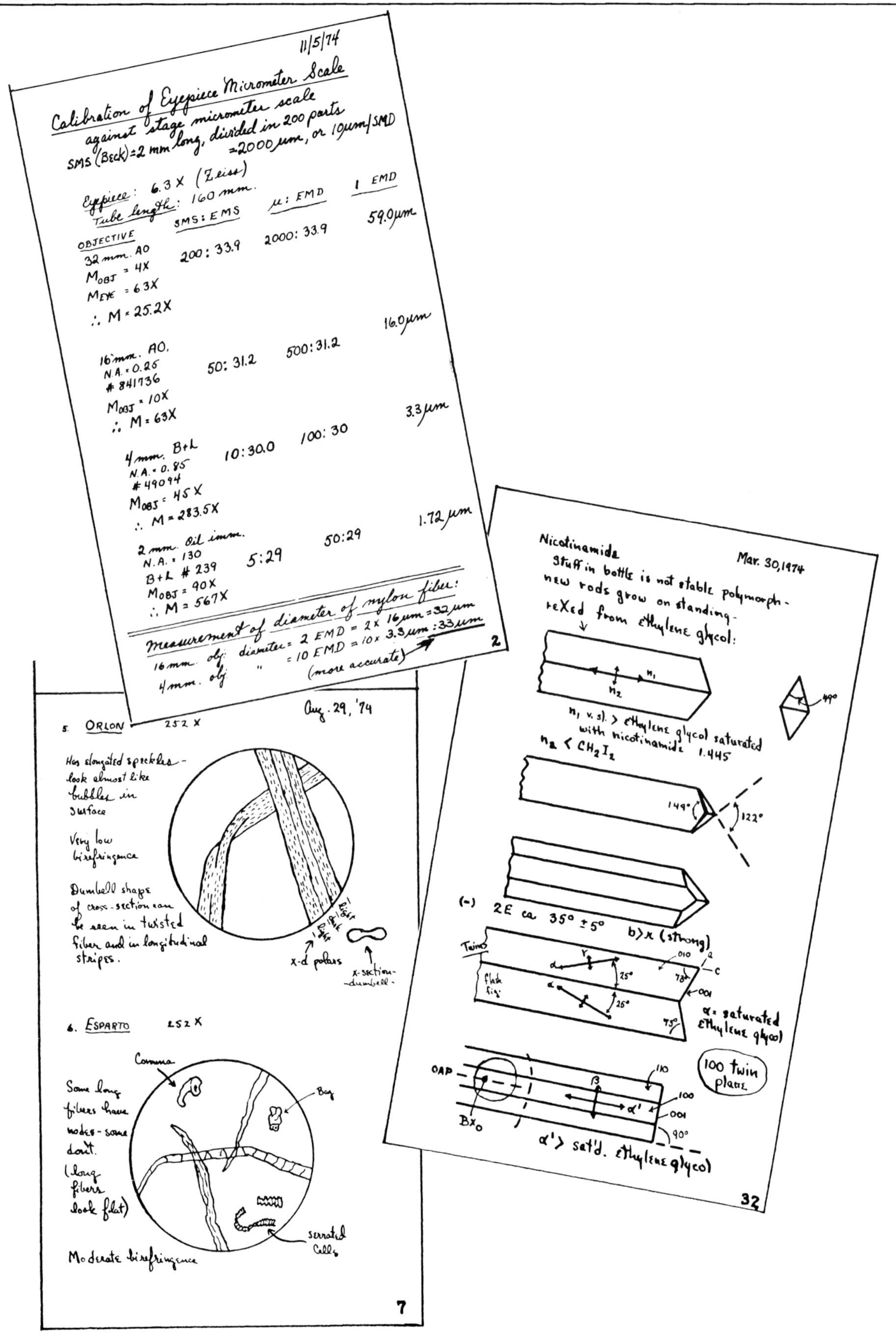

11/5/74
Calibration of Eyepiece Micrometer Scale
against stage micrometer scale
SMS (Beck) = 2 mm long, divided in 200 parts
= 2000 µm, or 10 µm/SMD
Eyepiece: 6.3X (Zeiss)
Tube length: 160 mm.
OBJECTIVE
SMS : EMS
µ : EMD
1 EMD
32 mm. AO
M OBJ = 4X
M EYE = 6.3X
∴ M = 25.2X
200 : 33.9
2000 : 33.9
59.0 µm
16 mm. AO.
N.A. = 0.25
# 841736
M OBJ = 10X
∴ M = 63X
50 : 31.2
500 : 31.2
16.0 µm
4 mm. B+L
N.A. = 0.85
# 49094
M OBJ = 45X
∴ M = 283.5X
10 : 30.0
100 : 30
3.3 µm
2 mm. Oil imm.
N.A. = 1.30
B+L # 239
M OBJ = 90X
∴ M = 567X
5 : 29
50 : 29
1.72 µm
Measurement of diameter of nylon fiber:
16 mm. obj. diameter = 2 EMD = 2 x 16 µm = 32 µm
4 mm. obj. " = 10 EMD = 10 x 3.3 µm = 33 µm
(more accurate)
2
5. ORLON 252 X
Aug. 29, '74
Has elongated speckles - look almost like bubbles in surface
Very low birefringence
Dumbell shape of cross-section can be seen in twisted fiber and in longitudinal stripes.
x-d polars
x-section - dumbell -
6. ESPARTO 252 X
Comma
Bug
Some long fibers have nodes - some don't.
(long fibers look flat)
serrated cells
Moderate birefringence
7
Nicotinamide
Mar. 30, 1974
Stuff in bottle is not stable polymorph - new rods grow on standing.
reXed from ethylene glycol:
n1
n2
n1 v. sl. > ethylene glycol saturated with nicotinamide 1.445
n2 < CH2I2
149°
122°
(-) 2E ca 35° ± 5°
b > α (strong)
Twins
Flash fig.
25°
75°
α = saturated ethylene glycol
100 twin plane
OAP
Bxo
β
α'
110
100
001
90°
α' > sat'd. ethylene glycol
32

FAMILIARIZATION WITH THE POLARIZING MICROSCOPE

Each student should become familiar with his microscope; learn what each part is for, what parts are to be moved, which require factory adjustment, what to clean and what not to clean. Follow the outline and ask questions when necessary.

A. Substage condenser

1. Remove the substage condenser. Can the top element be removed? Does it swing out? Clean the condenser elements after checking with the instructor.

2. Is there an iris diaphram?

3. Can the condenser be centered with respect to its mount on the microscope?

4. Can the condenser be raised and lowered?

5. Is there a stop to prevent the condenser from going too high? Can it be adjusted upward and downward?

6. Is there a keyed position for the condenser in its uppermost position?

7. If there a filter holder on the condenser?

   a. Is it removable?
   b. Can it be rotated?
   c. Is it a Polaroid filter or calcite?
   d. What is the vibration direction? Note how to set it for the desired E-W vibration direction.

B. Microscope stage

1. Is the stage movable? Can it be locked in place?

2. Is there a mechanical stage? Any stage clips?

3. Can the stage be centered? What is the accuracy in reading the angular position of the stage?

C. Objectives

1. What type of nosepiece do you have: rotating or "quick-change"?

2. Are the objectives centerable? If so, how is this accomplished?

3. What objectives are furnished?

4. Remove and examine each objective; check the front lens surface for dirt, scratches etc.; examine the back of the interior lens system for dust, cement separation etc. Use a hand magnifier to focus on each lens element. Report any positive findings to the instructor before proceeding. Replace objective.

5. Copy the markings on each objective. Explain what they mean.

D. Bodytube

1. Can the bodytube be raised and lowered?

2. Is there a fine and coarse adjustment? Is the fine adjustment graduated?

3. Can the coarse or fine adjustment tension be adjusted? Do you detect any backlash in the fine adjustment?

4. Is the bodytube adjustable in length?

5. Remove the ocular. Is there anything in the light path between objective and ocular?

   a. Is there an analyzer? Polaroid? Calcite? Dusty?
   b. Can the analyzer be rotated? Can it be removed from the light path?
   c. Is there a Bertrand lens? Can it be removed from the optical path?

6. Can the bodytube be removed from the stand?

7. Is there a compensator slot in the bodytube? A dust protector for the compensator slot?

E. Oculars

1. Remove and examine the oculars; record and translate the markings.

2. Hold the ocular up to the light. Is there a scale, crosslines or other object in focus?

3. Examine the ocular for scratches, blemishes and dust. Check with the instructor before attempting to clean it.

4. Can the top lens be focused? Is there an orientation pin to position the ocular in the bodytube?

5. Can the ocular be used as a hand magnifier?

F. Illuminating system

1. What type of transmitted light illuminator do you have?

a. Wattage ________ Voltage ________
b. Type of filament.
c. Is there a condensing lens? Focusable?
d. Is there a field iris?
e. A filter holder?

2. How is the light from the illuminator brought to the substage condenser of the microscope?

3. Is the substage mirror universally adjustable?

4. Describe each side of the mirror.

5. Can the bulb of the illuminator be centered in the lamp housing?

6. Do you have a separate illuminator for top lighting?

a. Wattage ________ Voltage ________
b. Type of filament.
c. Is there a condensing lens? Focusable?
d. Is there a field iris?

QUANTITATIVE MICROSCOPY

A. Size estimation

Using any preparation of fibers, estimate the average fiber diameter based on the assumption that the field of view is a picture being observed at a normal reading distance of 10 inches. Use each of the objectives and record the estimated average diameter in micrometers at the going rate of 1000 micrometers (μm) to the millimeter (mm) and 25.4 mm to the inch.

	Objective	Estimated average fiber diameter (μm)
1.		
2.		
3.		

Now, observing the stage micrometer scale which is usually one millimeter long and graduated into one hundred 10 μm divisions, estimate the total diameter of the field of view of each objective.

	Objective	Size of field of view (μm; mm)
1.		
2.		
3.		

B. Calibration of ocular micrometer

Each state micrometer has a scale with divisions 100 μm (0.1 mm) apart and at least 1 or 2 of these divisions are subdivided into 10 μm (0.01 mm) divisions. These are the standard against which the arbitrary divisions in the micrometer ocular are to be calibrated. Each objective must be calibrated separately by noting the correspondence between the stage scale and the ocular scale with each objective. Starting with the lowest power objective, focus on the stage scale and arrange the two scales parallel and in good focus. It should be possible to determine the number of ocular divisions equal exactly to some definite portion in micrometers of the stage scale.

The calibration consists, then, of calculating the number of micrometers per ocular scale division. To make the comparison as accurate as possible, a large part of each scale must be used. Let's assume that with the 10X objective 6 large divisions of the stage scale are equal to 38 divisions of the ocular scale. This means that 38 ocular scale divisions (osd) are equivalent to 600 micrometers, hence:

$$
\begin{aligned}
1 \text{ osd} &= 600/38 \\
&= 15.8 \ \mu\text{m}
\end{aligned}
$$

This means that when that micrometer ocular is used with that 10X objective each division of the ocular scale is equivalent to 15.8 μm and it can be used to make an accurate measurement of any object on the microscope stage. A particle, for example, observed with the 10X objective and measuring 8.5 divisions on the ocular scale, is 8.5 (15.8) or 135 μm in diameter.

This calibration must be repeated for each of the objectives on your microscope.

A convenient way to record the necessary data and to calculate μm/osd is by means of a table.

Objective	No. ssd = No. osd	μm = No. osd	μm = 1 osd
example	6 = 38	600 = 38	15.8 μm
1.			
2.			
3.			

C. Measurements of size using the ocular micrometer scale

You now have a calibration for each objective with the particular ocular and tubelength used. Now remeasure the fibers whose diameters you estimated above, using each of the objectives, and complete the following table:

Objective	Fiber diameter (μm)	
	Estimated	Measured
1.		
2.		
3.		

D. Particle size frequency determination

The calibrated objectives can now be used to estimate or measure the average particle size of a particulate sample. Using a prepared slide of a mixture of ThO_2 and SiO_2 beads and the 10X objective, estimate the particle size range (minimum to maximum) and average size in micrometers. Record these values:

ThO_2 + SiO_2 (Prep No. ___):

Estimated average particle size: ______μm (ThO_2); ______μm (SiO_2)
Estimated range of sizes: ______μm (ThO_2); ______μm (SiO_2)

Using the 16-mm objective and the calibrated micrometer ocular, proceed with an accurate particle size determination of the spheres whose size you have already estimated. The first step is choice of a partner such that at least one of you can read and write while the other should be able to measure particles rapidly and accurately as the preparation is moved slowly past the ocular scale. Both of you should be reasonably honest in measuring and recording scientific data. One person should measure several hundred particles of the ThO_2 (high contrast in the medium) while the partner records data, then reversing roles several hundred SiO_2 particles should be measured.

The particles are measured (Martin's diameter) and recorded as the whole number of osd which most closely fits the diameter of the particle at best focus. Measure all particles whose centers fall within the osds of the scale as the slide is moved in a straight line (parallel to the lines that represent the scale divisions). The slide is moved with a mechanical stage, by rotation of the microscope stage rendered off-axis (by means of the centering screws on either the objective or the stage) or by hand. As the successive diameters are measured, they are recorded by one partner as a tally opposite each size class (e.g., size class 8: 1111 1111 1111 1111 1111 1111 1111 1111 111 etc.). Your data should be recorded in the following table. The size class is indicated as ocular micrometer divisions rather than μm to simplify later calculations.

Sample ______________________; Observer ______________________

Objective ______________________; Recorder ______________________

Calibration factor: μm = 1 osd

Size Class, d (osd)	Number of particles in each size class, n	Total
1		
2		
3		
4		
5		
6		
7		
8		
9		
10		
11		
12		
	Grand Total	______

A size frequency determination should now be made on the second component of the mixture using the other microscope and reversing the observer and recorder. Only the total particles in each size class need be copied in the other notebook.

After totaling the numbers for each size class, the data can be handled statistically in a variety of ways of which the following is one of the simplest. Carry out the necessary operations by filling in the following table:

Sample ______________________________; Calibration factor 1 osd = μm

d	n	nd	nd^2	nd^3	nd^4
1					
2					
3					
4					
5					
6					
7					
8					
9					
10					
11					
12					
Totals					

Calculate the various particle size averages as follows:

Sample: ______________________________

Average particle size with respect to:

number = $\bar{d}_1 = \Sigma nd/\Sigma n =$ = osd = μm

surface = $\bar{d}_3 = \Sigma nd^3/\Sigma nd^2 =$ = osd = μm

weight = $\bar{d}_4 = \Sigma nd^4/\Sigma nd^3 =$ = osd = μm

The specific surface, S_m, can also be calculated by the following formula if the surfaces of the particles are smooth.

$S_m = 6/\bar{d}_3\rho$, where the density, ρ, of SiO_2 is 2.2 and of ThO_2 is 5.6, therefore:

$$S_m = 6/\bar{d}_3\rho = \quad m^2/gram$$

Record the corresponding data from your partner's data on the other component of the SiO_2 - ThO_2 mixture.

Sample: ______________________________

Average particle size with respect to:

number = $\bar{d}$, = μm

surface = $\bar{d}_3$ = μm

weight = $\bar{d}_4$ = μm

Specific surface, S_m, = $m^2/gram$

E. Counting analyses

Make a count of the total numbers of SiO_2 and ThO_2 spheres in your preparation. Count all particles whose centers fall within the field of view; count sufficient fields to give a total of at least 500 particles. You may wish to use the 5X objective.

Sample: SiO_2 - ThO_2 spheres

number of spheres of SiO_2 and ThO_2 in successive fields

	Field	1	2	3	4	5	6	7	Totals
ThO_2									
SiO_2									

Grand Total________

Now calculate the percentages of SiO_2 and ThO_2 in the mixture

% ThO_2 by number = $\Sigma\, ThO_2/(\Sigma\, ThO_2 + \Sigma\, SiO_2) \times 100$

Volume of average ThO_2 particle = $\pi(\bar{d}_4)^3/6$

=

= μm^3

Convert to cm^3 by x 10^{-12} = x 10^{-} cm^3

Weight of average ThO_2 particle = average volume x ρ

= x 10^{-} grams

Volume of average SiO_2 particle = $\pi(\bar{d}_4)^3/6$.

=

= μm^3

Convert to cm^3 by x 10^{-12} = x10^{-} cm^3

Weight of average SiO_2 particle = average volume x ρ

= 10^{-} grams

Percentage ThO_2 by weight in the mixture =

$$= \frac{\text{weight of average } ThO_2 \text{ particle}}{\text{weight of average } ThO_2 \text{ particle} + \text{weight of average } SiO_2 \text{ particle}} \times 100$$

CRYSTAL MORPHOLOGY

There are a variety of ways to recrystallize compounds on a microscope slide: sublimation, precipitation by a metathetical reaction, by evaporation of solvent from a saturated solution, by cooling a warm saturated solution.

A. Sublimation

This is an excellent way to obtain well-formed crystals of substances having sufficient vapor pressure. This includes many, if not most, organic compounds, e.g., iodoform, hexamethylene tetramine, caffeine, benzoic acid, phthalic anhydride, tetrachlorobenzene and quinone; and even a few inorganic compounds, e.g., mercuric iodide and arsenic trioxide.

The procedure is very simple; place a few milligrams of the sample near one end of a microscope slide and cover with a coverslip. Holding the opposite end of the slide between thumb and forefinger, slowly heat the sample over a micro flame, moving the slide slowly across the flame and back to cause uniform heating. As soon as any sublimate appears, stop heating and allow the slide to cool before observing the underside of the coverslip for crystalline sublimate.

The same procedure can be used to separate a volatile organic compound from a mixture of extraneous materials as, for example, caffeine crystals from a general settled dust sample collected near a coffee roaster or from coffee itself.

1. Sublime hexamethylene tetramine (cubic: rhombic dodecahedra) as described above. Draw representative crystal views.

2. Sublime iodoform (hexagonal basal pinacoids) using very low heat such as the heat from your illuminator for a minute or two.

B. Precipitation by a metathetical reaction

If the desired compound can be prepared metathetically ($AB + CD \rightarrow AD + CB$) and if it is more insoluble than any of the other salts a crystalline precipitate will result. If very insoluble, e.g., $CaCO_3$, $AgCl$, $BaSO_4$ and PbI_2, very dilute solutions of the reactants must be used. More soluble precipitates require more concentrated solutions of the reactants, e.g., NH_4ClO_4, K_2PtCl_6 and $CaSO_4$. Most microchemical tests involve this latter type because it is easier to obtain larger recognizable crystals of the more soluble "insoluble" compounds.

This procedure is also very simple. Prepare two small aqueous drops close to each other (but not touching) on the same slide. Keep in mind how much of each reactant you use so that correction can be made if a second try is needed — more solute if no precipitate results and less if the precipitate is too fine. Once the solute has dissolved in each drop join them gently together using the glass tip and let flow and diffusion do the mixing. This will give a variety of concentrations, some of which may be more conducive to large well-formed crystals.

Try this procedure using the following reactions:

a. NH_4ClO_4 (60)* + KI (38) $\rightarrow$ $KClO_4 + NH_4I$

b. $Pb(NO_3)_2$ (25) + 2KI (38 $\rightarrow$ $PbI_2\downarrow + 2KNO_3$

* number of vials in wooden reagent block.

C. Evaporation of a solvent from a saturated drop

1. — given an already saturated solution to start

Saturated solutions of a number of salts have been prepared and are available at each bench, e.g., NaCl, $Pb(NO_3)_2$, K/alum, $NH_4H_2PO_4$, $NaNO_3$, NH_4ClO_4, $BaCl_2 \cdot 2H_2O$, borax and $CuSO_4 \cdot 5H_2O$. Shake the chosen bottle and pipette a small droplet onto a clean slide. Observe microscopically for good crystals. If none are apparent use the glass tip to scrape the slide surface within the drop and to crush any large crystalline aggregates. Once good crystals of sufficient size are present a coverslip may be used to prevent rapid evaporation. This is not usually necessary since the crystals last long enough in a deep drop (not spread in a thin layer) for several representative crystals to be drawn — 5-10 minutes.

A dry drop should be resuscitated by adding a small drop of H_2O and using the glass tip to stir and crush the crystals. The drop will soon again be saturated.

2. — given dry powders and pure water

The preceding paragraph suggests the procedure to be used. One places a small drop of water at the center of the microscope half-slide and adds to the center of the drop a few crystals of the desired solute. A glass rod drawn out to a 0.5-1.0 mm tip is used as a spatula, a stirring rod and a crystal-crusher. Be sure to hold the glass rod straight up and perpendicular to the slide during stirring and crushing since otherwise the drop will spread out. After crushing the crystals and stirring the drop, observe the solution process under the microscope. If all of the crystals dissolve, add an additional increment with repetition of the crushing and stirring. Repeat, in fact, until the crystals do not all dissolve.

Then, keeping the slide at room temperature, observe the edges of the drop until a heavy crust of crystals develops. This crust should be pushed into the center of the drop, carefully using the glass rod so as not to enlarge the drop. On repeating this operation and crushing the crust inside the drop the time soon comes that the drop is saturated and well-formed crystals appear. If you have done a good job and the drop is deep, these crystals will slowly grow and develop for many minutes before they protrude through the upper surface of the drop and become useless for further study.

When the drop dries completely, a small drop of water can be added at the center of the dried crystals and by stirring and crushing a new batch of crystals soon appears. To be sure the drop is small enough it is transferred from the dropper to the glass tipped rod for transfer to the slide. Too large a drop will not stay on the glass tip. Remember only crystals completely covered with solution are worthy of study.

a. Recrystallize the following compounds from a drop of water on a microscope slide. Do, at least, one starting with dry crystals. Be sure all crystals are completely covered with solution and draw several representative views of each.

NaCl	$NH_4H_2PO_4$
$Pb(NO_3)_2$	$NaNO_3$
K alum	NH_4ClO_4
	borax
	$CuSO_4 \cdot 5H_2O$

D. Cooling a saturated warm drop between slide and coverslip

Organic compounds cannot be easily recrystallized from the usual low boiling organic solvents in drop sizes on a microscope slide. If, however, higher boiling solvents are used, good crystals can often be obtained. Best results are obtained by slight warming of a saturated drop containing a slight excess of solid solute. Solvents found to be useful for this purpose include:

thymol	Cargille liquids
aniline	benzyl alcohol
tricresyl phosphate	nitrobenzene

Recrystallize a small amount of TNT from a drop of benzyl alcohol saturated with TNT at room temperature, using a slide and coverslip preparation. Heat slide gently on your illuminator to partly dissolve the crystals, and observe the prep as it slowly cools to room temperature. If the rounded crystals do not fill in with plane faces and angular edges, the preparation was not heated sufficiently; on the other hand, if the crystals grow rapidly with needlelike projections (skeletal crystals), the preparation was heated too much. In either case, if the preparation is reheated to the proper temperature, perfect crystals of TNT will be formed on cooling.

CRYSTAL OPTICS

A. Measurement of refractive indices

1. Refractive index of sodium chloride

Mount crushed fragments of NaCl (n_D^{25} = 1.544) in three Cargille liquids chosen from the three ranges given:

1.510 - 1.520
1.542 - 1.546
1.570 - 1.580

Note the degree of contrast and the movement of the Becke line on focusing up and down. What is the effect of closing the substage iris? Why? What is the best position of the iris for best Becke lines?

2. Analysis of an unknown mixture (No. 22 in the reagent block)

This mixture contains one or more of the following (and nothing else):

Potassium alum	n_D^{25}	= 1.456
Sodium chloride	"	= 1.544
Potassium bromide	"	= 1.559
Potassium iodide	"	= 1.667

Choose one single mounting liquid to tell all components present. Estimate the quantitative percentage of each component.

3. Identify the isotropic unknown (No. 21 in reagent block) by measuring its refractive index.

Use an orange filter, half slides and coverslip fragments (see note below). Mount successive small increments of the unknown powdered sample in small amounts of Cargille refractive index liquid under small fragments of coverslip. Check the Becke line and remount until n is determined to ±0.001 — note room temperature and, if necessary, make an appropriate temperature correction for the index of the liquid (dn/dT is on each bottle). Identify the compound from the tables in Winchell's Handbook of Chemistry and Physics or this manual, p. 133.

Note: Small pie-shaped fragments of coverslip may be prepared by placing a clean coverslip on a book, tablet or notebook and striking the center with a blunt-pointed instrument.

B. Refractive indices of anisotropic crystals

1. Calcite

Examine the prepared slide of calcite (20) which is mounted in Aroclor® 5442. The refractive indices are:

Aroclor® 5442	n_D^{25}	= 1.662
Calcite (ω)	"	= 1.658
" (ϵ)	"	= 1.486

Using a single polar, note the effect of rotating the stage on the contrast shown by a number of individual crystals of calcite. Do they all behave similarly? Roll a well-formed crystal into a position looking down the A_3 asis and note the indices on rotation of the stage. Explain.

2. Na_2SO_3 (No. 44)

Determine the ω refractive index of Na_2SO_3

3. Uniaxial unknown

If the crystals have recognizable morphology, then the ϵ index can be located parallel to c, and ω perpendicular to c. Carefully recrystallize the uniaxial unknown, No. 51 in the reagent block, from a drop of water on a slide and allow the drop to go slowly to dryness. Using tiny fragments of coverslip and very tiny amounts of index liquid, determine ϵ for the unknown by mounting successive crystals around the periphery of the dried drop in Cargille index liquids. Repeat the process for ω. Identify the compound by consulting the literature or the table on page 134.

4. Biaxial crystals

To measure the principal refractive indices for biaxial crystals you must (a) have crystals with recognizable morphology, (b) have an interference figure, or (c) determine α by mounting crystals, randomly oriented, in successive lower index liquids until no crystal in any position shows a lower index. Repeat for γ going the other way.

Determine α for gypsum by lowest n method.

Determine γ for gypsum using morphology. (See p. 164, Winchell.)

C. Dispersion staining

1. Adjustment of the objective

Remove the 10X (or the 40X) objective from its centerable mount and replace it with the dispersion staining objective (DSO). The desired screening of the back focal plane of the DSO is controlled by the rotating plate. Three positions are possible: ○ = full aperture, ● = annular stop, ⊙ = central stop. Set the plate at full aperture and focus on the calcite prep from the slide set. Make certain that you have Köhler illumination (page 214) and that the stage is centered, then with the Bertrand lens in rotate the central stop into position. It does not much matter if the stop is not in sharp focus but you may be able to sharpen it temporarily by sliding the ocular up out of the bodytube. This could be done during adjustment of the aperture diaphram, then lowered for observation of dispersion staining colors.

Next close the substage iris and observe whether the outline is concentric with the central stop. You must be able to close the iris completely behind the central stop. Ideally, you should be able to center the substage condenser to achieve centration. A slightly tipped coverslip will affect centration and centration should be effected and observations made with a flat coverslip. One can tolerate a modest amount of decentration of iris and stop but the proper dispersion staining colors will be observed only if the particle is observed on the optical axis of the stop. If you look at the field of view (orthoscopically, i.e., no Bertrand) with the central stop in place as you close the substage iris a dark spot with fuzzy edges should appear in the center of the field. The spot grows larger as the iris is closed further. If the stop is not centered the dark spot will not be centered but a particle placed in the center of the spot (in the field of view) will show proper colors.

Hopefully, the center of stage rotation and the center of the dark area in the field of view will coincide. A microscope may have centerable condenser, stage and/or objective. A microscope for dispersion staining should ideally have a centerable stage. If you have unusual difficulty in centering the DSO stops with your substage iris you should remove the top lens of your condenser.

If you have arranged the illumination carefully so that the field diaphram is focused and centered in the field of view it can be closed to define only a central area and the contrast will be greatly improved due to decreased glare. The substage iris can also be opened slightly to give increased brightness of the colors. There is also always some glare from dirt particles on the slide, coverslip and condenser top lens; these should always be cleaned.

B. With the annular stop in place examine the calcite prep (which is mounted in Aroclor®) using one polar only. Note the effect of opening and closing the substage iris on the dispersion staining color for calcite (ω). Note the effect of rotation of the polar on the dispersion staining color.

What is λ_0 for calcite (ω)? Why don't you get a color for calcite (ϵ)?

Repeat, using the central stop. Note that to produce the best colors the substage iris may require a setting slightly different from the optimum for the annular stop.

Mount a few quartz crystals in a Cargille liquid having $n_D^{25} = 1.532$ and observe at room temperature. Explain.

C. Plot the dispersion curve for NH_4Cl on a dispersion staining chart (next page). Mount a few particles of NH_4Cl in each of six or eight Cargille liquids in the dispersion staining range, and plot λ_0 for each prep, after applying whatever temperature correction is necessary for the liquid.

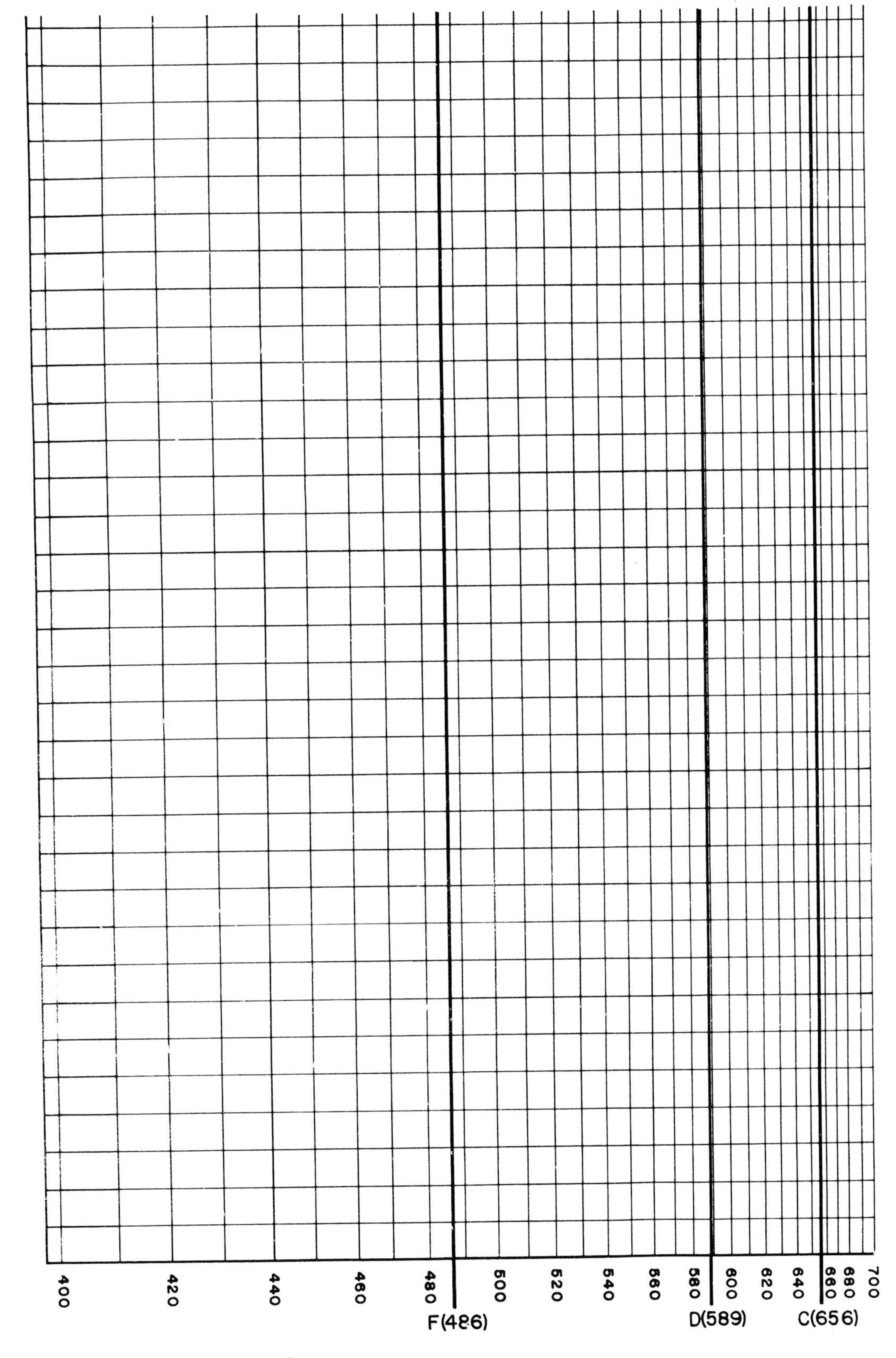
REFRACTIVE INDEX OF CARGILLE LIQUID ($n_D^{25°C}$)
400
420
440
460
480
500
520
540
560
580
600
620
640
660
680
700
F(486)
D(589)
C(656)
λo WAVELENGTH (nm)

D. Birefringence, use of compensators, interference figures

1. Polarization colors, effect of thickness

Prepare a wedge-crystal of thymol by melting a few crystals under a coverslip one side of which is supported on the edge of a second slip. Press down the unsupported edge of the first slip and seed the melt near that point. Maintain pressure until crystal growth has proceeded about 1/4 the coverslip diameter.

Examine this preparation between crossed polars, ⊕. Note the sequence of interference (polarization) colors as a function of thickness. Compare with the Michel-Lévy chart. Assuming the thymol layer at the supported edge is the same as the coverslip, 0.17 mm, what is the birefringence of the thymol?

Some compounds show anomalous interference colors (but not thymol) because the birefringence varies with wavelength. Repeat the above experiment with picric acid, with benzil.

2. Polarization colors, effect of birefringence

Compare: dynel (No. 65), cultivated silk (No. 61), viscose rayon (No. 14), wool (No. 15), human hair (No. 87) and nylon (No. 12) between crossed polars. To a first approximation, each has a circular crossection so measure the diameter, assume an equal thickness, estimate the retardation and read the birefringence from the Michel-Lévy chart.

3. Extinction

Rotate the stage while observing any of the above fibers between crossed polars. Note the variation in intensity of the interference colors. Each should be dark (extinction) when parallel to each crossline, <u>i.e.</u>, the vibration direction of either polar; each should be brightest 45° from any of the four extinction positions. Fibers and any other substances behaving in this manner are said to show parallel (or straight) extinction.

Study fiberglass (No. 16) in the same manner. Such substances are, of course, isotropic and show neither interference colors nor extinction.

Study cotton (No. 10) between crossed polars. Cotton, like a few other fibrous substances, <u>e.g.</u>, sisal and some wood fibers, show interference colors but no extinction. Such behavior is unusual and usually infers some form of polycrystallinity (overlapping of crystals, some at extinction but always some bright). Cotton is a helical arrangement (60° helix) of cellulose crystals with those on top bright when those on the bottom are at extinction and vice versa.

Study a good calcite (No. 20) rhomb between crossed polars. Its extinction is said to be symmetrical.

Study a well-formed single-crystal rhomb of gypsum (No. 21) between crossed polars. Here, extinction is said to be oblique.

4. Sign of elongation

Examine a prepared slide of viscose rayon (No. 14) for two fibers crossing at right angles. Note the interference colors at the center of this intersection (both fibers in the brightness position) — explain the black cross.

Now, look for two fibers crossing nearly parallel; note the interference colors of each fiber alone and where they overlap. Explain this observation.

Referring to a single viscose rayon fiber, find one showing an area of first order red (Red I); any red you see will be Red I. With this fiber rotated into an upper left-lower right position insert the Red I compensator (it is a single crystal plate with a retardation of 530 nm, Red I, and a slow component vibration direction crosswise of the plate). What, now, is the retardation of the same part of the fiber that was Red I before insertion of the compensator. Now, rotate the stage 90° watching the same area of the fiber. What happened and why?

A fiber having its slow component (high refractive index) vibration direction parallel to its length is said to have a positive sign of elongation. Most fibers are positive but a few are negative.

Determine the sign of elongation for Orlon (No. 13) and nylon (No. 12).

5. Interference figures

Fuse a few crystals of $NaNO_3$ (No. 43, reagent block) between slide and coverslip. It is high melting (>300°C) so place them near one end of the slide to decrease chance of breaking the slide by uneven heating. After cooling, note the interference colors of the various grains. Find one showing the lowest retardation throughout stage rotation and observe its interference figure — leave that grain in the center of the field of view, switch to the 40X objective, rack the objective down until the air gap is about 0.5 mm, open the aperture diaphram, insert the Bertrand lens, defocus the lamp filament to even the illumination of the back focal plane and focus the interference figure to the largest circle by racking the bodytube up or down very slightly. You should see a black cross with concentric circles of interference colors, a typical uniaxial interference figure.

Note and explain the results on inserting Red I and a 1/4 λ compensator.

Determine the sign of double refraction: +, $\epsilon > \omega$; -, $\omega > \epsilon$.

Observe the interference figure for mylar film (No. 30). Note the sign of double refraction on this typical biaxial figure: +, $\gamma - \beta > \beta - \alpha$; -, $\gamma - \beta < \beta - \alpha$. Explain why this pseudo biaxial crystal shows three refractive indices.

Measure the optic axial angle 2E and any dispersion apparent at the optic axes.

6. Pleochroism

Mount a few crystals of tetracyclone in Aroclor® 1260. Observe and explain the color changes upon rotation of the stage, using one polar. Which colors correspond to which refractive index directions?

MICROCHEMICAL TESTS

A. Inorganic cations

1. Test for lead using a more active metal: prepare a dilute solution of lead nitrate; add several small bits of magnesium or zinc metal and cover with a coverslip; observe and draw the resulting lead metal dendrites.

2. Repeat with a dilute solution of silver nitrate.

3. Test for lead in a dilute $Pb(NO_3)_2$ test drop by adding a dilute solution of potassium iodide: observe and draw the thin yellow hexagonal plates of lead iodide. Long white needles which may appear and "consume" the PbI_2 are the double salt of PbI_2 and KI.

4. Test for zinc in a dilute solution of any zinc salt by adding a small potassium mercuric thiocyanate crystal; observe and draw the "drusy crosses" that result.

5. Test for potassium using chloroplatinic acid: mix dilute drops of potassium iodide and chloroplatinic acid*; observe and draw the orange octahedra.

6. Test for sodium using zinc uranyl acetate and draw representative crystals.

B. Inorganic anions

Test for phosphate using silver nitrate: prepare very dilute drops of ammonium dihydrogen phosphate and silver nitrate and mix together. Observe digestion of the light yellow "amorphous" precipitate into dendrites, stars, crosses and, especially, three-pointed dendrites.

C. Organic compounds

Test for polynuclear aromatics using picric acid; melt a few milligrams of picric acid on a slide under a coverslip. After crystallization flip off the slip and sprinkle phenanthrene over the surface; note the red spots characteristic of the reaction between phenanthrene and picric acid on warming to the eutectic mp (about 100°C).

* Try to avoid contamination of the chloroplatinic acid (and other chemicals, for that matter) by using the glass tip only after cleaning it with moist Kleenex.

POLYMER FIBERS AND FILMS

A. Examine prepared slides of longitudinal mounts of various vegetable, animal and synthetic fibers.

Note: 1. morphological characteristics, e.g., uniformity of size, fiber tips, size of lumen if present, probably crossection shape, surface markings etc.;

2. optical properties, e.g., general magnitude of n's, birefringence, extinction, sign of elongation.

B. Examine prepared slides of rayon crossections.

Note: 1. morphological characteristics, including those of skin and core;

2. optical properties.

C. Examine undrawn nylon and then stretch pieces 1X, 2X and 3X their original lengths.

1. Observe each between crossed polars with white and with monochromatic light; examine necking region. Explain observations in terms of molecular rearrangement during drawing.

2. Measure n's of drawn and undrawn nylon. Explain results (work in groups of three or four and combine data).

D. Prepare fusion prep of Dacron or mylar (both are polyethylene terephthalate).

1. Crystallize slowly on hot bar just below m.p. and observe crystal habit and number of nuclei.

2. Crystallize rapidly on a hot bar at about 120° and compare with 1.

3. Melt prep and quench by placing on cold metal block and compare with 1 and 2. Reheat prep slowly at increasing temperature on hot bar and interpret observations.

4. Repeat 1, 2 and 3 with nylon and compare results with those for Dacron.

What are indications in each material of minimum crystallization temperature, crystallization velocity between m.p. and room temperature, nucleation rates?

E. Melt sample of nylon or Dacron on hot stage. Observe during heating and note:

1. dimensional changes, color changes, evidence of decomposition;

2. changes in birefringence;

3. final m.p.

On basis of observations, what would you predict as maximum safe ironing temperature, softening temperature, temperature of significant loss of strength?

F. p-Nitrophenol-fiber melting points

1. Prepare a slide prep with a few fibers of nylon and a few scattered p-nitrophenol crystals under a coverslip and heat on the hot stage to determine melting behavior.

FUSION METHODS

The following experiments may be done using an alcohol lamp or hot plate (or matches); no hot stage or hot bar is necessary although either could be used.

Characterize each compound by observations of the prep melted between slide and coverslip over an alcohol lamp, then cooled spontaneously on the microscope stage.

For each compound note the following:

sublimation (stop heating and check any cloudiness on underside of coverslip for crystals (polymorphism?)
decomposition, decrepitation
estimate melting point (50°C, 75°C, 100°C, 150°C etc.)
degree of supercooling (seed with crystals, if necessary)
remelt 75-80% of the crystals ("Meltback")
form of the crystal-front as a function of temperature
crystallization velocity as a function of temperature
crystal angles, extinction angles, refractive indices relative to the melt and estimated birefringence (thickness = 20-25 μm)
"lens crystal" to see anomalous polarization colors
polymorphic transformations (mono- or enantiotropic)
thymol "mixed fusion" for crystal morphology and optics relative to saturated thymol solution

Perform these operations for:

A. General

1. thymol (m.p. 50°C)
2. TNT (m.p. 81°C)
3. phenanthrene (m.p. 99°C)
4. picric acid (m.p. 122°C)
5. TNF (m.p. 176°C)
6. anthracene (m.p. 215°C)

B. Polymorphism

7. p,p'-DDT (m.p. 109°C)
8. mononitronaphthalene (m.p. 60°C)
9. benzotriazole (m.p. 99°C)
10. mercuric iodide (m.p. 354°C)
11. ammonium nitrate (m.p. 170°C)
12. cholesteryl acetate (m.p. 114°C)

C. Composition diagrams — study mixed fusions of the following binary systems:

1. ammonium nitrate (m.p. 176°C)—TNT (m.p. 82°C)
2. phenanthrene (m.p. 101°C)—benzotriazole (m.p. 99°C)
3. TNT (m.p. 82°C)—mononitronaphthalene (m.p. 60°C)
4. anthracene (m.p. 217°C)—phenanthrene (m.p. 101°C)
5. anthracene (m.p. 217°C)—picric acid (m.p. 122°C)
6. trinitrobenzene (m.p. 121°C)—picric acid (m.p. 122°C I; 75°C II) (use a very small quantity of picric acid).

PHYSICAL CHEMISTRY

The following exercises are a potpourri of experiments which can be carried out with a polarizing microscope; they are generally related to physical chemistry, but represent a variety of phenomena. Any industrial process, procedure or reaction can be carried out on the microscope stage. It is obviously cheaper, easier, quicker and often the best way to study such processes.

1. Effect of concentration on particle size of precipitate

Prepare a relatively concentrated drop of lead nitrate on a microscope slide, and add a small crystal of potassium iodide. Note the very fine yellow crystals of lead iodide and, sometimes, the longer white needles of a $PbI_2 \cdot KI \cdot 2H_2O$ double salt. Note the transformation of PbI_2 to the double salt in some areas. Repeat this experiment, using drops more dilute, by a factor of 20 or more, and observe the size of the resulting lead iodide crystals.

2. Crystal habit change of sodium chloride on addition of urea

Recrystallize sodium chloride from a drop of distilled water and nearby recrystallize the same solute from an aqueous solution containing 1% urea. Note the crystal form of the sodium chloride in the two cases and explain the difference.

3. Isomorphous solid solutions

Prepare two saturated drops of ammonium perchlorate and potassium permanganate near each other on a microscope slide. Help them run together so that a concentration gradient is formed between the two drops. Let the preparation go to dryness, and observe the crystals obtained throughout the mixed drop. Explain your observations.

4. Stability of suspensions

On opposite ends of a full slide, make up a one drop slurry of kaolin clay, using water for one drop and dimethylsulfoxide (DMSO) for the other drop. Cover each of the drops with a coverslip and disperse the particles, using a pencil eraser on top of the coverslip with a circular shearing motion. Then remove the coverslip from each of the preps and observe what happens in each drop over the next minute or so. Explain what you see, in terms of the phenomenon itself and its significance in terms of suspensions of any solid particles in a liquid vehicle.

5. Boundary migration

Melt DDT between coverslip and slide and set it to one side for 15 minutes to crystallize as spherulites. After crystallization is complete, place the slide on a hot bar or hot plate at about 95°C. Explain what you observe after 30 minutes of annealing. If you have trouble, see the Fusion Methods book (p. 192).

6. Liquid crystals

Add an excess of solid INTAL® to a drop of water and cover with a coverslip. Note the behavior of the liquid crystalline hydrate of INTAL at the H_2O:INTAL interface. This is the room temperature stable nematic phase. Be sure that plenty of excess INTAL is present so that you can heat it nearly to 100° and observe the cooling process. You should see solid needles of the INTAL:water solid solution at temperatures above 70°C. The solid needles should change to a "middle" (M.)

liquid crystalline hydrate at about 70°C. On further cooling, this mesophase changes to the nematic mesophase at about 42°C. These changes are reversible.

7. Orientation of polymers by stretching

Baggies® are polyethylene polymer in an unoriented state. This means they were not stretched during manufacture. Observe what happens macroscopically as you try to push your finger through the side of the Baggie. Cut a 1 inch square from another portion of the Baggie and try stretching it in the two major directions. When partially elongated, observe the retardation between crossed polars and determine the birefringence of the stretched and unstretched portions. Finally, observe the interference figures on the stretched and unstretched portions. You should be able to determine the high refractive index direction relative to the direction you stretched the sample. (Note: you can assume the interference figure is a Bx_o figure.)

8. Polymorphism (solution phase transformation)

Melt a few micrograms of sulfur to just fill about one-quarter of the area beneath a coverslip. Without waiting for cooling and crystallization, add a drop of aniline to fill the remaining area. Reheat until the sulfur melts completely, then cool on a cool surface at a temperature $\leqslant$20°C (fridge?).

The cooled prep can be transferred in one minute to the polarizing microscope. Observe growth of the rhombic dendrites to and through the supercooled sulfur droplets. Some will be seeded and some will dissolve in the path of the dendrites. Between crossed polars the scene often looks like lights going on in a darkened city as each droplet suddenly crystallizes. In some area of the prep flat blades of monoclinic sulfur will be slowly transforming through the solution phase to the stable orthorhombic form. The latter transformation may take several hours before completion.

9. Plastic crystals

A number of compounds having nearly spherical molecules, e.g., carbon tetrabromide, hexachlorethane, adamantane and camphor. Most such compounds have very low entropies of fusion, sublime very readily and even smell similarly (like camphor). Furthermore, many show polymorphism with the high temperature form either cubic or closely related hexagonal. Hexachloroethane melts at 186.6°C with two enantiotropic transitions at 71°C (I $\rightleftharpoons$ II) and 43.6°C (II $\rightleftharpoons$ III). Form I (high temperature form) is cubic, II is triclinic and III is orthorhombic. Confirm as much of this as you can qualitatively.

KEY TO REAGENT BLOCK

1. acetate rayon
2. ammonium chloride
3. ammonium dihydrogen phosphate
4. ammonium nitrate
5. anthracene
6. benzotriazole
7. calcite
8. chloroplatinic acid
9. cholesteryl acetate
10. cupric sulfate pentahydrate
11. Dacron
12. p,p'-DDT
13. p-dibromobenzene
14. p-dichlorobenzene
15. 1,8-dinitronaphthalene
16. gypsum
17. hexamethylenetetramine
18. β-HMX
19. HND
20. INTAL
21. isotropic unknown
22. isotropic unknown mixture
23. iodoform
24. kaolin
25. lead nitrate
26. magnesium powder
27. mercuric iodide
28. mica
29. mononitronaphthalene
30. mylar
31. nicotinamide
32. p-nitrophenol
33. nylon
34. nylon, undrawn
35. phenanthrene
36. picric acid
37. potassium dichromate
38. potassium iodide
39. potassium mercuric thiocyanate
40. quartz
41. silver nitrate
42. sodium chloride
43. sodium nitrate
44. sodium sulfite
45. sulfur
46. tetracyclone
47. thymol
48. TNB
49. TNF
50. TNT
51. uniaxial unknown
52. uranyl acetate
53. vanillin
54. zinc acetate
55.
56.
57.
58.
59. hexachloroethane
60. ammonium perchlorate

REFERENCE LIST OF SUPPLIERS

AO Instrument Company[7,12,15*]
Instrument Division
Eggert and Sugar Roads
Buffalo, NY 14215 716/895-4000

Bausch & Lomb, Inc.[2,12,15]
Scientific Instrument Division
635 St. Paul Street
Rochester, NY 14602 716/232-6000

Brinkmann Instruments, Inc.[9,12,14]
Cantiague Road
Westbury, NY 11590 516/334-7500

R. P. Cargille Laboratories, Inc.[1]
55 Commerce Road
Cedar Grove, NJ 07009 201/239-6633

Ehrenreich Photo-Optical Industries, Inc.[4,12]
623 Stewart Avenue
Garden City, NY 11530 516/248-5200

Ernest F. Fullam, Inc.[5]
P. O. Box 144
Schenectady, NY 12301 518/785-5533

Gelman Instrument Company[12]
600 South Wagner Road
Ann Arbor, MI 48106 313/665-0651

Glyco Chemicals Division[10]
Charles L. Huisking and Company, Inc.
Williamsport, PA 17701 717/322-4681

William J. Hacker and Company[7,8,12]
P. O. Box 646
Fairfield, NJ 07007 201/226-8450

International Micro Optics ("Zeiss" Jena)[14]
5 Daniel Road
Fairfield, NJ 07006 201/227-6767

A. M. de La Rue[18]
3406 Lancer Drive
Hyattsville, MD 20782 301/927-3018

E. Leitz, Inc.[3,12,14]
Rockleigh, NJ 07647 201/767-1100

Walter C. McCrone Associates, Inc.[1,6,8,11,12,17]
2820 South Michigan Avenue
Chicago, IL 60616 312/842-7100

McCrone Research Associates Ltd.[1,12,19]
2 McCrone Mews, Belsize Lane
London NW3 5BG, England 011-441 435-2282

McCrone Research Institute, Inc.[19]
2508 South Michigan Avenue
Chicago, IL 60616 312/842-7105

Mettler Instrument Corporation[8]
20 Nassau Street
Princeton, NJ 08540 609/448-3000

Mico Instrument Corporation[18]
80 Trowbridge Street
Cambridge, MA 02138 617/547-8660

Microscope Publications Ltd.[17]
2820 South Michigan Avenue
Chicago, IL 60616 312/842-7100
AND
28 Southway, Carshalton Beeches, SM5 4HW
Surrey, England

Millipore Corporation[12]
Bedford, MA 01730 617/275-9200

Olympus Corporation of America[4,12]
2 Nevada Drive
New Hyde Park, NY 11040 516/488-3880

Science and Mechanics[9]
Instruments Division
229 Park Avenue South
New York, NY 10003 212/673-1300

Arthur H. Thomas Company[8,12,14]
Vine Street at Third
P. O. Box 779
Philadelphia, PA 19105 215/627-5600

Unitron Instrument Company[4,12]
66 Needham Street
Newton Highlands, MA 02161 617/969-8600

Vickers Instruments, Inc.[2,12]
15 Waite Court, P.O. Box K
Malden, MA 02148 617/935-3113

Wild Heerbrugg Instruments, Inc.[12,16]
465 Smith Street
Farmingdale, NY 11735 516/293-7400

Carl Zeiss, Inc.[12,13,14]
444 Fifth Avenue
New York, NY 10018 212/736-6070

* For code, please see next page.

Code for Suppliers

1. Refractive index liquids, prepared slides etc.
2. British microscope
3. Represents Leitz
4. Japanese microscopes
5. Electron microscope supplies
6. Aroclors
7. Represents Reichert
8. Hot stages, hot bars, accessories
9. Exposure meter
10. Aquaresin
11. Coverslips (round, 1-1/2)
12. Accessories for microscopists
13. Represents Zeiss (Oberkochen)
14. German microscopes
15. American microscopes
16. Swiss microscopes
17. Microscopy books, journal
18. Hardy microtomes
19. Courses

35-mm COLOR SLIDES

1. A set of 35-mm color slides on interference figures can be purchased from Ward's Natural Science Establishment, Inc., P. O. Box 24, Beechwood Station, Rochester, New York 14609. Notes by Dr. E. E. Jelley are provided with the slides.
2. Three sets of 35-mm color slides on general microscopy for teaching purposes available from Wild Heerbrugg Instruments, Inc., 465 Smith Street, Farmingdale, New York 11735.
3. A set of 48 35-mm color slides covering the applications of polarized light microscopy with notes by W. C. McCrone.

PARTICULATE SUBSTANCES IN REFERENCE SET NO. 1
(available from MAC, 2820 S. Michigan Ave., Chicago, IL 60616

Apatite	81
Arnel	64
Asbestos	18
Ashes, cigarette	73
Asphalt	80
Bakelite	54
Bituminous coal	27
Calcite	20
Carborundum	38
Catalyst dust	40
Cement	28
Cereal grain dust	68
Cigarette ashes	73
Cleanser, abrasive	58
Coal, bituminous	27
Coal boiler, pulverized	49,50
Coal, domestic furnace	47
Cotton	10
Cottonwood pollen	100
Cork dust	31
Corn starch	29
Corundum	56
Cryolite	82
Dacron	63
Dandruff	90
Diatoms	1
Domestic furnace, coal	47
Dust, catalyst	40
Dust, cereal grain	68
Dust, cork	31
Dust, enamel	39
Dust, engraving	51
Dust, paint spray	78
Dust, pencil sharpener	72
Dust, tea	34
Dust, tobacco	77
Dynel	65
Enamel dust	39
Engraving dust	51
Eraser, ink	75
Eraser, pencil	74
Face powder	76
Feldspar, orthoclase	83
Fern spores	93
Fertilizer (inorganic)	35
Fertilizer (organic)	36
Fiber, glass	16
Fibers, wood, chemical coniferous	8

INDEX

NOW IN SIX UNIQUE VOLUMES . . . THE DEFINITIVE SYSTEM OF PARTICLE IDENTIFICATION AND ANALYSIS

Culminating More Than Twenty Years of Microscopy Research and Pioneering Technology

- Six 9″ x 12″ Individually Slipcased Volumes
- 4,500 Micrographs (Full-Color/Black-and-White)
- 1,021 Common and Uncommon Particles
- 1,700 Pages
- Index for Each Volume and a Master Index
- Over 4,000 References
- Clearly Defined Techniques

Users of THE PARTICLE ATLAS have learned this direct microscopic approach yields the quickest results with reliable accuracy. Discover for yourself this pragmatic and stimulating science of ultramicro analysis.

Become an Expert in Your Own Laboratory

The new six-volume PARTICLE ATLAS contains everything you need to master particle identification in your own laboratory, at your own microscope. Step-by-step techniques guide you in what to do. Meticulously exact micrographs (full color as well as black-and-white) show you every detail of all common particles and hundreds of others as well.

Accompanying the photomicrographs are elaborate and thorough descriptions of every particle. Study the photomicrographs and the descriptions . . . apply the techniques. You'll be amazed with the ATLAS and impressed with yourself as **you learn to distinguish successfully among all the particles challenging** and **often plaguing us today.**

THE PARTICLE ATLAS Guides You To a New Technical Skill

Effective particle identification and analysis are the natural result of complementing your microscope with THE PARTICLE ATLAS. Keep THE PARTICLE ATLAS available for instant reference, use it regularly; and you'll become your own in-house expert, with time and cost savings for your organization, and the satisfaction of a new aptitude for yourself.

How to Use the Six-Volume Particle Atlas for Conclusive Particle Identification

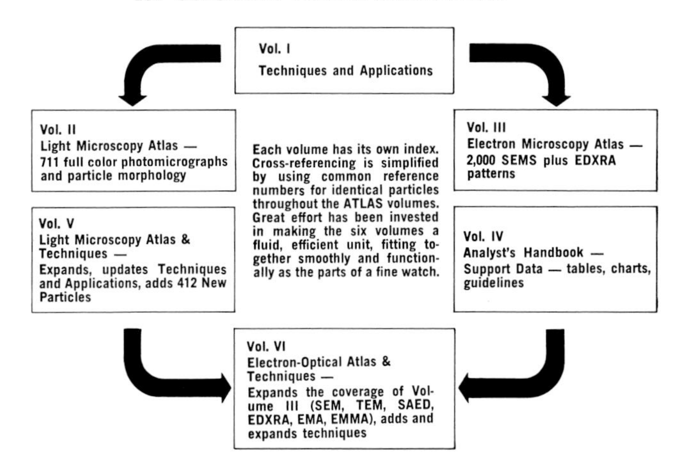

Vol. I
Techniques and Applications
Vol. II
Light Microscopy Atlas — 711 full color photomicrographs and particle morphology
Vol. III
Electron Microscopy Atlas — 2,000 SEMS plus EDXRA patterns
Each volume has its own index. Cross-referencing is simplified by using common reference numbers for identical particles throughout the ATLAS volumes. Great effort has been invested in making the six volumes a fluid, efficient unit, fitting together smoothly and functionally as the parts of a fine watch.
Vol. V
Light Microscopy Atlas & Techniques — Expands, updates Techniques and Applications, adds 412 New Particles
Vol. IV
Analyst's Handbook — Support Data — tables, charts, guidelines
Vol. VI
Electron-Optical Atlas & Techniques — Expands the coverage of Volume III (SEM, TEM, SAED, EDXRA, EMA, EMMA), adds and expands techniques

PARTICLE ATLAS CONTENTS

VOLUME V — LIGHT MICROSCOPY ATLAS & TECHNIQUES

LASER RAMAN MICROPROBE A. History & Background B. General Theory C. Instrumentation D. Sampling & Sample Preparation E. Data Presentation F. Interpretation of the Data G. Applications DISPERSION STAINING AND COLOR A. Effect of Light Source Color Temperature B. The C.I.E. Color Chart C. Application to Particle Identification MICROCHEMICAL REACTIONS AND PARTICLE IDENTIFICATION A. Importance of Microchemical Reactions in Modern Light Microscopy B. Analytical Approach and Chemical Thinking C. Preparation D. Selected Techniques and Reactions (inorganic compounds • organic compounds • metals, minerals & refractories • complex substances) COLOR ATLAS—PHOTOMICROGRAPHS AND DESCRIPTIONS—409 PARTICLES DETERMINATION OF GEOGRAPHICAL ORIGIN OF A DUST SAMPLE A. Introduction B. Techniques C. Analysis & Interpretation (minerals • botanical components • zoological • combustion products • industrial markers • man-made fibers • chemicals • miscellaneous components) MICROSCOPICAL ANALYSIS OF AIRBORNE DUST A. Scope & Applications B. The Program C. Qualitative Analyses D. Quantitative Methods E. Precautions F. Methods of Reporting IDENTIFICATION OF ASBESTOS A. Survey of Methods B. Dispersion Staining C. Asbestos Types and Associated Minerals D. Analytical Data APPLICATION OF PARTICLE STUDY IN FORENSIC MICROSCOPY A. Types of Trace Evidence B. Techniques for Characterization of Trace Evidence C. Microscopical Procedures D. Recommended Methods APPLICATION OF PARTICLE STUDY IN ART CONSERVATION A. Areas of Application B. Sampling Procedures C. Techniques D. Data Summaries LITERATURE SURVEY INDEX (Cumulative for 5 volumes)

VOLUME VI — ELECTRON OPTICAL ATLAS & TECHNIQUES

THE ELECTRON MICROSCOPY MICROPROBE ANALYZER A. Introduction B. Instrumentation C. Techniques D. Applications IDENTIFICATION OF SUBMICROMETER PARTICLES A. Introduction B. Special Methods of Sample Preparation for Analysis of Submicrometer Particles ELECTRON OPTICAL ATLAS — ELECTRON MICROGRAPHS AND DESCRIPTIONS—409 Particles ASBESTOS IDENTIFICATION A. Chrysotile Asbestos B. Amphibole Asbestos C. Sample Preparation D. Data Output & Interpretation AUTOMATION OF PARTICLE ANALYSIS INSTRUMENTATION A. On-line vs Time-sharing Computers B. Interfacing Problems C. Automation of the Electron Microprobe Analyzer D. Automation of the Ion Microprobe Analyzer E. Automation of the Mass Spectrometer F. Data Reduction G. Future Automation Goals LITERATURE SURVEY INDEX (Cumulative for 6 volumes)

THE 6-VOLUME SET INCLUDES A TOTAL OF 4,000 REFERENCES